LEARNING RESOURCES CTR/NEW ENGLAND TECH.
GEN TK7871.15.S54 P76 1995
Properties of strained and
3 0147 0002 2066 8

EMIS Datareviews Series No. 12

Series Advisor: Dr. B. L. Weiss

PROPERTIES OF STRAINED AND RELAXED Silicon Germanium

ELECTRONIC MATERIALS INFORMATION SERVICE

Other books in the EMIS Datareviews Series from INSPEC:

No. 1 Properties of Amorphous Silicon (2nd Edition, 1989)
No. 2 Properties of Gallium Arsenide (2nd Edition, 1990)
No. 3 Properties of Mercury Cadmium Telluride (1987; out of print)
No. 4 Properties of Silicon (1988)
No. 5 Properties of Lithium Niobate (1989; out of print)
No. 6 Properties of Indium Phosphide (1991)
No. 7 Properties of Aluminium Gallium Arsenide (1993)
No. 8 Properties of Lattice-Matched and Strained Indium Gallium Arsenide (1993)
No. 9 Properties and Growth of Diamond (1994)
No. 10 Properties of Narrow Gap Cadmium-based Compounds (1994)
No. 11 Properties of Group III Nitrides (1994)

PROPERTIES OF STRAINED AND RELAXED

Silicon Germanium

Edited by

ERICH KASPER
University of Stuttgart, Germany

Published by: INSPEC, the Institution of Electrical Engineers,
London, United Kingdom

Institution of Electrical Engineers
Michael Faraday House,
Six Hills Way, Stevenage,
Herts. SG1 2AY, United Kingdom

British Library Cataloguing in Publication Data

A CIP catalogue record for this book
is available from the British Library

ISBN 0 85296 826 4

Printed in England by Short Run Press Ltd., Exeter

Contents

Foreword

Fundamental limits imposed by the nanoscale dimensions of matter are acknowledged in silicon integrated circuit technology in which doubling of chip capacity every few years is no longer possible. Simultaneously, the new functionality in established silicon-based device structures of optoelectronics has stimulated integration and progress in expansion of the processing capacity of ICs. First steps in convergence of the technological gap - because of dissimilar materials - between silicon-based devices and optoelectronics were made in the study of lattice-mismatched epitaxy for proposed possible new electronic and optical properties of strained-layer structures, absent in their unstrained constituents. Thus was introduced the realisable concept of strain as a predictable and controllable modification of the semiconductor band structures, the most important strained layer superlattice systems of the group-IV elements being Si and Ge. Appreciation of the operation and structure of SiGe/Si devices necessitates rigorous understanding of misfit-dislocation nucleation and motion and the established fabrication techniques for device quality Si and Ge. Progress in strained-layer epitaxy of planar heterostructures a few atomic layers thick is primarily due to advances in materials characterisation techniques.

The basic properties of the SiGe/Si system have had significant impact on device applications, and the basic science of semiconductor materials, which includes the original concept of direct optical transitions in an indirect semiconductor with superlattice-induced band structure, provides optoelectronic capabilities for integration with standard VLSI. Further superlattice-induced effects in the SiGe/Si system are clearly discernible in the photoluminescence and Raman spectra of optical and folded acoustic modes. The various device structures treated here result from strained layer band structure modifications and the resulting enhanced optical and transport properties. Of particular significance are the surface properties of reconstruction, bonding, segregation and step growth where the real space techniques of Scanning Probe Microscopies will prove indispensable.

This volume represents the most significant contribution to knowledge of the SiGe/Si devices system so far and is a necessary requirement for scientists and engineers in this field.

Hamish G. Maguire
Department of Electrical and Electronic Engineering
The Nottingham Trent University
Burton Street
Nottingham, NG1 4BU
UK

October 1994

Introduction

A few years after the invention of the bipolar transistor the basic electronic semiconductor material changed from germanium to silicon. During that switch around 1960 considerable interest was focused on bulk, unstrained SiGe alloys. Advanced epitaxy methods like molecular beam epitaxy have enabled the growth of high quality, thin, strained SiGe layers on Si substrates since around 1985. The availability of strained SiGe/Si structures stimulated heavily the research on silicon-based heterostructure devices resulting within a few years in the fastest silicon-based transistors and other very attractive options.

The book is organised to meet three different demands of a reader. In Chapter 1 some general properties of strained layer systems which need caution are summarised. SiGe/Si heterostructures can be considered as a model system for stress driven phenomena because of the chemical similarity of the involved material partners which minimises additional chemical effects.

The specific material data for strained and relaxed SiGe alloys are given in Chapters 2 to 6. Basically the different properties are given as functions of the parameters Ge content x and film strain ε. To a first approximation some properties, e.g. the elastic stiffness constants, can be considered as linear functions of the chemical composition (Vegard's law). Some properties, e.g. the lattice constants, vary monotonically, but non-linearly with composition. Some other properties, e.g. the thermal conductivity, depend even non-monotonically on the chemical composition. Strain dependence can often be approximated by a linear law at least for a given sign of the strain (compressive or tensile). For a few known cases, but not in general, the temperature dependence of the properties is explicitly given. Usually, the doping dependence of the alloy properties has not yet been explored. For the parent materials see either Landolt-Börnstein, New Series, Group III, Volume 17a (Springer-Verlag, Berlin, 1982) or 'Properties of Silicon', vol.4 in the EMIS Datareviews Series (IEE, INSPEC, London, 1988).

In Chapter 7 some device relevant structures are selected out from a much larger variety. Doping effects, adjustment of strain in multiple layer structures, formation of quantum wells and superlattices should be demonstrated for a certain set of parameters.

Finally, I would like to express my thanks to the authors and to the following for their critical comments and advice:

M. Cardona, Max Planck Institut für Festkörperforschung Stuttgart

M. Ershov, The University of Aizu

M. Jaros, University of Newcastle

H. Lüth, KfA Kernforschungszentrum Jülich

H. Maguire, Nottingham Trent University

H.J. Queisser, Max Planck Institut für Festkörperforschung Stuttgart

G. Theodorou, Aristotle University of Thessaloniki

M. Thorpe, Michigan State University, East Lansing

K. Wang, University of California, Los Angeles

Also, I acknowledge assistance from M. Rinner and continuous support from J. Sears. Suggestions and remarks from readers are very welcome.

Erich Kasper
University of Stuttgart
Germany

November 1994

Contributing Authors

G. Abstreiter	Walter Schottky Institut, Am Coulombwall, D-85748 Garching, Germany	4.4
E. Arzt	Universitat Stuttgart, Institut fur Metallkunde, Seestrasse 71, 70174 Stuttgart, Germany	3.1
S.P. Baker	Universitat Stuttgart, Institut fur Metallkunde, Seestrasse 71, 70174 Stuttgart, Germany	3.1
G. Bauer	Johannes Kepler Universitat Linz, Institut fur Halbleiterphysik, Altenbergerstrasse 69, Linz, A-4040, Austria	4.1, 5.3
C.M. Engelhardt	Walter Schottky Institut, Am Coulombwall, D-85748 Garching, Germany	4.4
T. Fromherz	Johannes Kepler Universitat Linz, Institut fur Halbleiterphysik, Altenbergerstrasse 69, Linz, A-4040, Austria	4.1
M.A. Grinfeld	Rutgers University, Busch Campus, New Brunswick, NJ 08903, USA	1.1
H.-J. Herzog	Daimler-Benz AG, Ulm Research Center, Wilhelm-Runge-Strasse 11, Ulm D-7900, Germany	2.1
R. Hull	University of Virginia, Department of Materials Science, Charlottesville, VA 22903, USA	1.2, 1.3
J. Humlicek	Masaryk University of Brno, Department of Solid State Physics, Kotlarska 2, 61137 Brno, Czech Republic	4.6, 4.7
W. Jager	Forschungszentrum Julich, Institut fur Festkorperforschung, Postfach 1913, D-52425 Julich, Germany	2.2
H. Jorke	Daimler-Benz AG, Ulm Research Center, Wilhelm-Runge-Strasse 11, Ulm D-89081, Germany	5.2, 6.3
E. Karra	Aristotle University of Thessaloniki, Department of Physics, 54006 Thessaloniki, Greece	6.2

P.C. Kelires	Aristotle University of Thessaloniki, Department of Physics, 54006 Thessaloniki, Greece	2.3, 6.1
D.K. Nayak	Advanced Micro Devices, M/S 79, One AMD Place, PO Box 3453, Sunnyvale, CA 94088-3453, USA	7.2, 7.3
J.F. Nutzel	Walter Schottky Institut, Am Coulombwall, D-85748 Garching, Germany	4.4
T.P. Pearsall	University of Washington, Department of Materials Science and Engineering, 302 Roberts Hall, FB-10, Seattle, WA 98195, USA	7.4
F. Schaffler	Daimler-Benz AG, Ulm Research Center, Wilhelm-Runge-Strasse 11, Ulm D-7900, Germany	5.1
R. Schorer	Walter Schottky Institut, Am Coulombwall, D-85748 Garching, Germany	3.3
Y. Shiraki	RCAST, University of Tokyo, 4-6-1 Komaba, Meguro-ku, Tokyo 153, Japan	7.2
D.J. Srolovitz	Rutgers University, Busch Campus, New Brunswick, NJ 08903, USA	1.1
G. Stoger	Johannes Kepler Universitat Linz, Institut fur Halbleiterphysik, Altenbergerstrasse 69, Linz, A-4040, Austria	5.3
J.C. Sturm	Princeton University, Department of Electrical Engineering/Applied Science, Engineering Quadrangle, Princeton, NJ 08544-5263, USA	7.1
G. Theodorou	Aristotle University of Thessaloniki, Department of Physics, 54006 Thessaloniki, Greece	2.3, 6.2
C.G. Van de Walle	Xerox Corporation, Palo Alto Research Center, 3333 Coyote Hill Road, Palo Alto, CA 94304, USA	4.2, 4.3, 4.5
K. Wang	University of California, Los Angeles, Electrical Engineering Department, Engineering Building IV, Room 66-147, Los Angeles, CA 90024-1594, USA	3.2

X. Zheng | University of California, Los Angeles,
Electrical Engineering Department,
Engineering Building IV, Room 66-147,
Los Angeles, CA 90024-1594, USA | 3.2

Abbreviations

The following abbreviations are used in this book.

AC	alternating current
AES	Auger electron spectroscopy
AFM	atomic force microscopy
BC	base-collector
BE	base-emitter
BJT	bipolar junction transistor
CMOS	complementary metal oxide semiconductor
CPA	coherent potential approximation
CR	cyclotron resonance
CVD	chemical vapour deposition
D	double-layer
2D	two-dimensional
3D	three-dimensional
DBRT	double barrier resonant tunnelling
DC	direct current
2DCG	two-dimensional carrier gas
2DEG	two-dimensional electron gas
DF	dark field TEM imaging
2DHG	two-dimensional hole gas
DLTS	deep level transient spectroscopy
DOS	density of states
ECL	emitter-coupled logic
EXAFS	extended X-ray-absorption fine structure spectroscopy
fcc	face centred cubic
FQHE	fractional quantum Hall effect
HBT	heterojunction bipolar transistor
hh	heavy hole
HRTEM	high-resolution transmission electron microscopy
IC	integrated circuit
IR	infrared
LA	longitudinal acoustic
LED	light emitting diode
LEED	low energy electron diffraction
lh	light hole
LO	longitudinal optical
LPE	liquid phase epitaxy
LRO	long-range order

MB	Matthews Blakeslee (model)
MBE	molecular beam epitaxy
MC	Monte Carlo
MCPA	molecular coherent potential approximation
MODFET	modulation doped field effect transistor
MODQW	modulation doped quantum well
MOS	metal oxide semiconductor
MOSFET	metal oxide semiconductor field effect transistor
NMOS	n-channel metal oxide semiconductor
NTL	non-threshold logic
PL	photoluminescence
PMOS	p-channel metal oxide semiconductor
QHE	quantum Hall effect
RBS	Rutherford backscattering spectroscopy
RHEED	reflection high energy electron diffraction
ROM	rule of mixtures
S	single-layer
SAD	selected area electron diffraction
SdH	Shubnikov-de Haas
SIMOX	separation by implanted oxygen
SIMS	secondary ion mass spectrometry
SL	strained layer or superlattice
SLS	strained layer superlattice
STEM	scanning transmission electron microscopy
STM	scanning tunnelling microscopy
TA	transverse acoustic
TEM	transmission electron microscopy
TO	transverse optical
UHV-CVD	ultrahigh-vacuum chemical vapour deposition
UV	ultraviolet
VLSI	very large scale integration
WKB	Wentzel-Kramers-Brillouin
XPS	X-ray photoemission spectroscopy
XRD	X-ray diffraction

CHAPTER 1

INTRODUCTION

1.1 Stress driven morphological instabilities and islanding of epitaxial films

1.2 Equilibrium theories of misfit dislocation networks in the SiGe/Si system

1.3 Metastable strained layer configurations in the SiGe/Si system

1.1 Stress driven morphological instabilities and islanding of epitaxial films

M.A. Grinfeld and D.J. Srolovitz

September 1993

A INTRODUCTION

For many decades, morphological instabilities associated with interfaces have been intensively studied. Traditionally, morphological instabilities have been of primary interest in the fields of crystal growth, metallurgy, fracture, geology, petrology, etc. Within the last few years, however, interest in morphological instabilities has extended into several new areas, including epitaxy, electronic packaging, tribology, biology, etc. (see reviews [1-4] and references therein).

The recent efforts of several researchers have resulted in rapid progress and the development of a much deeper understanding of stress driven instabilities which occur via matter transport at interfaces and free surfaces of solids. Because of its universal nature, this new understanding of stress driven morphological instability provides new insights into phenomena in different branches of materials and other sciences. It was recently suggested that Stranski-Krastanov epitaxial growth may be related to this stress driven morphological instability [5]. Several authors [5-9] have demonstrated, using different approaches, that regardless of specific symmetry, geometry and elastic moduli, accumulated elastic (either linear or non-linear) energy in a solid can always be reduced by means of appropriate mass rearrangement in the vicinity of the free surface. Neglecting surface energy, all non-hydrostatically stressed solids bounded by a smooth, traction-free boundary are unstable against appropriate mass rearrangement of their constituent particles. Surface energy, on the other hand, can stabilise the system against some types (short wavelength) of instability. The specific form of the stress driven instabilities depends on geometry, stress state, body forces (e.g. gravity), mass transport mechanisms, etc. Recently several elegant, quantitative experiments [10] have demonstrated that these instabilities do happen, and have validated the earlier theoretical predictions.

While this stress driven morphological instability has found widespread applications in different fields (e.g. metallurgy [11], gels [12], geology [13] and fracture [14]), it has also received serious attention in problems related to the growth of epitaxial films (see [15]). The reason for this interest can be traced to some of the recent observations of dislocation-free Stranski-Krastanov growth [16]. These experiments have clearly demonstrated that the Stranski-Krastanov growth morphology may not be triggered by nucleation of misfit dislocations, as many had previously thought (see e.g. [17]). The key roles played by misfit stresses and diffusion in this phenomenon were previously discussed by several authors exploiting different approaches [18]. The most interesting parameter in the problems of epitaxy is the critical thickness - the critical thickness for stress relief either by dislocation generation or by morphological instabilities. These two types of critical thickness are quite distinct in stress relief mechanism and in quantitative value.

In this Datareview, we treat the phenomena relating to the observations on the dislocation-free Stranski-Krastanov growth from the viewpoint of stress driven morphological instabilities. We pay special attention to the 2D in-plane morphologies and their changes at the critical film thicknesses and/or critical misfit stress.

B INTUITIVE ANALYSIS OF STRESS DRIVEN MORPHOLOGICAL INSTABILITY

Let us consider the processes of crystallisation or sublimation at the free surface of a uniaxially prestressed, elastic crystal (see FIGURE 1). We explicitly consider two physical effects: namely, elasticity and surface energy. The stresses within the solid can be generated by an applied stress or be internal stresses, such as those associated with heteroepitaxy. For the sake of simplicity, we consider that the solid is only two-dimensional and assume that deposition takes place in the form of elementary square cells of material, as in FIGURE 1. We view each cell as a continuum so as not to have to justify the application of elasticity and surface energy at an atomic scale. Assume that the material being deposited has a different lattice parameter from the substrate due to the presence of the uniaxial, lateral stress. When cell A attaches to the uniform ad-layer under it, its bottom stretches to match the lattice parameter of the strained ad-layer. Its top on the other hand remains at its initial unstrained width and the initially rectangular cell distorts into a trapezoidal shape.

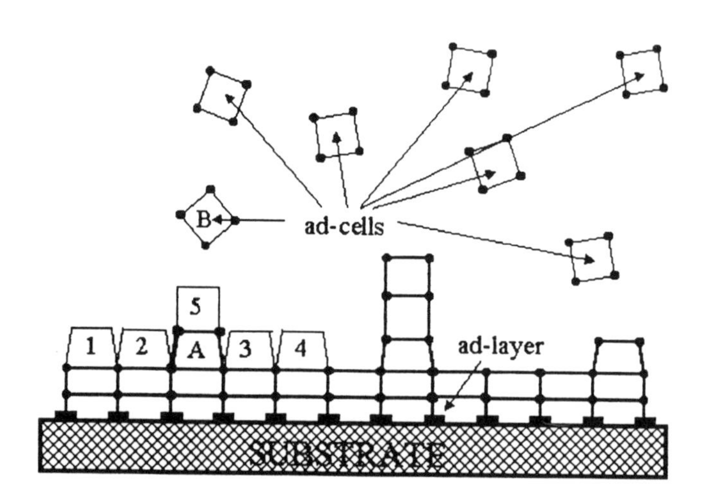

FIGURE 1 The mechanism of the stress driven rearrangement instability.

Consider now the possible locations for cell B to attach to the film in the vicinity of cell A. Particle B may attach itself to the ad-layer in, for example, positions 1, 2, 3 or 4. Since surface energy favours as large a number of nearest neighbours as possible, sites 2 and 4 are preferable to 1 and 3 due to the proximity of cell A. This is why surface energy favours the growth of as smooth a surface as possible. If cell B attaches to site 1 or 4 it will take on the strained, trapezoidal shape of cell A. If, on the other hand, it attaches to site 2 or 3, the wall B shares with A becomes vertical and therefore both cells A and B become more strained than if cell B was at either site 1 or 4. Therefore, strain energy works against the surface smoothing tendencies of surface energy. Now, consider cell B becoming attached to site 5, on top of A. Since site 5 has the same number of nearest neighbours as sites 1 and 4, the surface energy associated with B is the same for attachment to sites 1, 4 or 5. However, while the bottom of cell B would be stretched at sites 1 or 4, its bottom is unstretched at site 5 because the top of cell A is unstretched. Therefore, consideration of strain energy favours site 5 over sites 1 to 4. Depending on the ratio of the surface energy to the strain energy, site 5 may (small ratio) or may not (large ratio) be favoured over sites 2 or 3. If the misfit strain and/or the elastic content are large, the roughness of the surface will increase with continued film growth. This is the stress driven morphological instability.

C THE 'MISFIT' DEFORMATION AND STRESS

Let us consider an elastic solid with elastic energy density $e(\varepsilon_{ij})$ per unit volume in its originally unstressed state: $\varepsilon_{ij} = (D_j u_i + D_i u_j)/2$ is the strain (i.e. the dependence of the linear deformation tensor on the displacement ui gradient), D_i is the symbol for differentiation with respect to the Lagrangian (material) coordinates x^i (the spatial indices i, j, k, l take on the values 1, 2, 3, whereas the indices a, b, c, d are either 1 or 2), and the standard Einstein convention (summation over repeated indexes, e.g. $\kappa^i_i = \kappa^1_1 + \kappa^2_2 + \kappa^3_3$, $\kappa^c_c = \kappa^1_1 + \kappa^2_2$) is implied.

Consider an elastic, crystalline solid in the form of a relatively thin film which is attached coherently to a solid, crystalline substrate which is modelled as an elastic half-space (see FIGURE 2). The interface where film and substrate meet is initially flat. We first examine the deformation for the case of a 2D film (i.e. the film's thickness is negligible). Because of the mismatch in the lattice parameters between the unstressed crystalline film and the substrate, the 2D film is subjected to the in-plane 'misfit' deformation M^2. Choosing the coordinate system with the (x^1,x^2)-plane coinciding with the interface, one can express the affine misfit displacements in the film particles as:

$$U_a(x^b) = \kappa_{ab}\, x^b \quad (\text{or } U_1(x^1,x^2) = \kappa_{11}x^1 + \kappa_{12}x^2,\ U_2(x^1,x^2) = \kappa_{21}x^1 + \kappa_{22}x^2) \tag{1}$$

where κ_{ab} is the 2D-tensor defined by the in-plane lattices (as we use the Cartesian coordinates only, there are no distinctions in the co- and contra-variant components). Thus, the in-plane components of the 3D-displacements in the film $u^i(x^j)$ should obey the following boundary conditions at the interface Γ $(x^3 = 0)$:

$$u_1(x^1,x^2,0) = \kappa_{11}x^1 + \kappa_{12}x^2,\quad u_2(x^1,x^2,0) = \kappa_{21}x^1 + \kappa_{22}x^2,$$
$$u_3(x^1,x^2,0) = 0 \ (\text{or } u_a(x^b,0) = U_a(x^b) = \kappa_{ab}\, x^b) \tag{2}$$

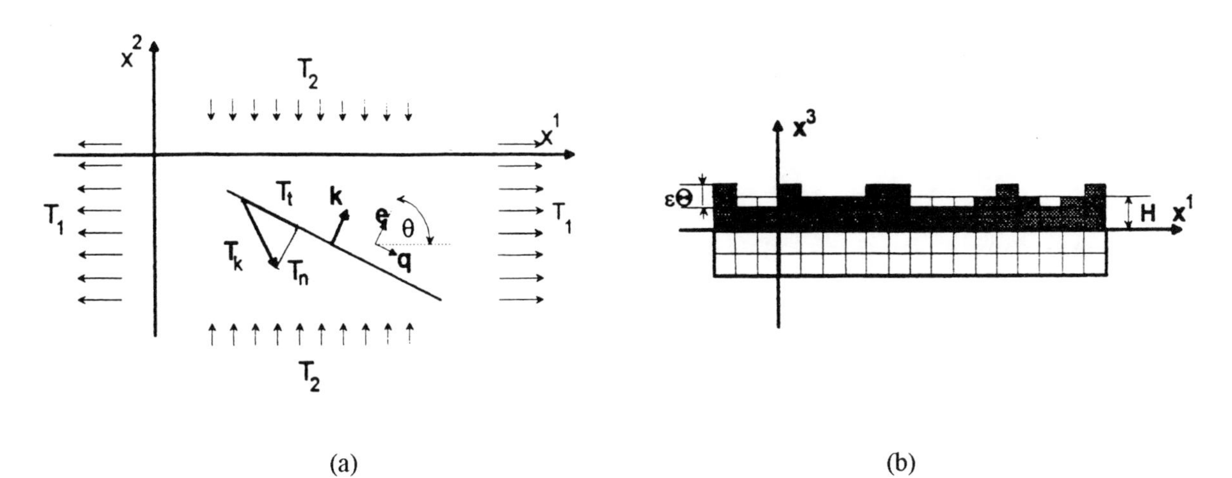

FIGURE 2 The geometry of the corrugated film: (a) top view; (b) side view.

The deformation described considered only a 2D film (or, equivalently, a 3D film with no variation in the deformation κ_{ab} in the x^2 direction). However, it is convenient to extend our original notion of the 2D-misfit deformations M^2 to 3D-misfit deformation, with the components kij defined in the following way. Let us consider the equilibrium displacements of the unbounded elastic layer with a traction-free and perfectly flat surface and flat interface where the film is coherently attached to the substrate. The lattice misfit M^2 generates 3D uniform strains and stresses within the film. We identify the components of the 3D misfit strain κ_{ij} and stress p^{ijo}. The indices of κ_{ij} running over 1 and 2 are obviously equal to the components of the 2D misfit strain tensor κ_{ab}. Since all material particles of any arbitrary horizontal plane in the film experience the same vertical displacement in the 3D uniform deformation, the components κ_{31} and κ_{32} vanish identically in crystals of any symmetry. The three remaining components κ_{i3} should guarantee the absence of tractions at the free surface (and, because of uniformity, at the interfaces as well): $p^{3jo} = p^{i3o} = 0$. These algebraic equations can be rewritten in terms of the elastic constant tensor as $c^{i3kl}\kappa_{kl} = 0$. The misfit strains and stresses can be easily computed for the case of an isotropic elastic solid. In this case, the elastic constant tensor is simply $c^{ijkl} = \lambda\delta^{ij}\delta^{kl} + \mu(\delta^{ik}\delta^{jl} + \delta^{il}\delta^{jk})$ where λ and μ are the Lame constants and δ^{ij} is the Kroneker delta function. Simple manipulations lead to the following components of the misfit strain and stress:

$$\kappa_{a3} = 0, \qquad \kappa_{33} = -\frac{\nu}{1-\nu}\,\kappa^{a}_{\;a},$$

$$p^{abo} = 2\mu\left(\frac{\nu}{1-\nu}\,\kappa^{c}_{\;c}\,\delta^{ab} + \kappa^{(ab)}\right), \qquad p^{j3o} = p^{3jo} = 0 \tag{3}$$

where ν is the Poisson ratio, and $\kappa^{(ab)} = (\kappa^{ab} + \kappa^{ba})/2$ is the symbol for symmetrisation.

In the following sections, we use the notation T_1, T_2 for the principle in-plane misfit stress (i.e. the eigenvalues of stresses p^{abo}) and choose the principal in-plane directions (i.e. the eigenvectors) as the x^1 and x^2 axes. We refer to 'shear-like' misfit stresses when T_1 is approximately $-T_2$, and to 'dilatation-like' misfit stresses when T_1 is approximately T_2.

D THE ELASTIC ENERGY ASSOCIATED WITH CORRUGATIONS OF THE FILM SURFACE

Consider the 'uncorrugated' (i.e. flat) elastic film with a free surface at $x^3 = H$ which is coherently attached to an infinitely thick elastic substrate at $x^3 = 0$. The infinitely thick substrate is stress-free and has zero strain-energy, while the film is subjected to the misfit strain κ_{ij} and therefore has a non-zero elastic energy E_{reg} proportional to its volume and the elastic energy density

$$e_{reg} = \frac{1}{2}\,c^{ijkl}\kappa_{ij}\kappa_{kl}.$$

Imagine the free surface of the film γ becomes 'corrugated' via rearrangement of the material particles. The amplitude of the corrugation is assumed sufficiently small in the sense that the corrugated surface γ is close to the original surface γ^o:

$$\text{(a) } \gamma\text{: } x^3 = H + Z(x^a), \ (Z/H << 1); \quad \text{(b) } \int_\gamma dx^1 dx^2 Z(x^a) = 0 \tag{4}$$

Eqn (4b) reflects conservation of film mass. We denote the disturbance of the equilibrium displacement field and the accumulated elastic energy of the system with the 'corrugated', traction-free surface γ as $v_i(x^k)$ and E^e_{irreg}, respectively. The change in the elastic energy of the system associated with the corrugation is [5,6]:

$$E^e_{irreg}(\varepsilon) - E^e_{reg} \simeq -\frac{1}{2}\int_{\omega_o} d\omega\, c^{ijkl}\, D_{(j}v_{i)}\, D_{(l}v_{k)} = -\frac{1}{2}\, p^{abo} \int_{\gamma^o} d\gamma\, v_b\, D_a Z \tag{5}$$

Eqn (5) shows that any corrugations of the initially flat, traction-free surface diminish the total strain energy of the system.

This computation can be made more explicit by considering an isotropic elastic film and substrate. Omitting the mathematical details, we find [19]:

$$E^e_{irreg} - E^e_{reg} = \int_{-\infty}^{\infty}\int_{-\infty}^{\infty} \mathbf{K}_e\, \Delta(\mathbf{k})\Delta^*(-\mathbf{k})\, dk_1 dk_2 \tag{6}$$

where

$$\mathbf{K}_e = -\frac{|\mathbf{k}|}{2\mu_f}\left\{ \frac{1-\nu_f}{D^*}\left[h - \frac{1}{2}\sinh 2h\left(1 + \frac{4(1-\nu_f)[\chi(1-2\nu_s)+1]}{(\chi-1)[\chi(3-4\nu_s)+1]}\right) - \right.\right.$$

$$\left.\left. \cosh 2h\, \frac{4\chi(1-\nu_f)(1-\nu_s)}{(\chi-1)[\chi(3-4\nu_s)+1]}\right] T_n^2 + \frac{\sinh h + \chi\cosh h}{\cosh h + \chi\sinh h}\, T_t^2 \right\};$$

$$D^* = h^2 - \frac{1}{(\chi-1)[\chi(3-4\nu_s)+1]}\,[\chi\sinh h + \cosh h +$$

$$(1-2\nu_f)(\sinh h + \cosh h)] \times$$

$$[\chi\sinh h + \cosh h + (1-2\nu_f)(-\sinh h + \cosh h) + 2\chi(1-2\nu_s)\sinh h]$$

and where the following notation has been employed. $\Delta(\mathbf{k})$ are the Fourier components of the corrugations with the in-plane wave-vector $\mathbf{k}$; μ_f, ν_f, μ_s, ν_s are the shear moduli and Poisson ratios of the film and substrate, respectively ($\chi = \mu_f/\mu_s$), θ is the angle between $\mathbf{k}$ and the principal direction of the in-plane stress T_1; $\mathbf{e}$, $\mathbf{q}$ are the unit in-plane vectors parallel and orthogonal to $\mathbf{k}$, respectively, T_n ($= T_1 e_1^2 + T_2 e_2^2 = T_1\cos^2\theta + T_2\sin^2\theta$) and T_t ($= T^1 e_1 q_1 + T^2 e_2 q_2 = (T_1 - T_2)\sin\theta\cos\theta$) are the normal and tangential components of the stress $\mathbf{T}_k$ acting at the cross-section orthogonal to the wave-vector $\mathbf{k}$, and $h = |\mathbf{k}|H$. This change in elastic energy (Eqn (6)) has a particularly simple form for the case of a uniaxial stress (see [5]).

E THE COMPETITION BETWEEN ELASTIC AND SURFACE ENERGY EFFECTS

In addition to the elastic energy, the surface E^s also makes a significant contribution to the total energy of the film/substrate systems. Other energetic contributions, such as gravity and van der Waals interactions, are omitted from consideration since they are small compared with the surface and elastic energy contributions for systems such as Ge on Si. The surface energy is assumed proportional to the product of the surface energy density σ and the total area of the free surface in the stress free, reference configuration. We explicitly neglect the contribution of the surface stress to the elastic energy [20]. The change in total energy associated with forming surface corrugation on an initially flat surface is:

$$E^t_{irreg} - E^t_{reg} = \varepsilon^2 \int_{-\infty}^{+\infty} dk_1 \int_{-\infty}^{+\infty} dk_2\, K(\mathbf{k},H)\Delta(\mathbf{k})\Delta^*(-\mathbf{k}) \qquad (7)$$

which can be computed explicitly for the isotropic case. The function $K(\mathbf{k},H)$ includes (i) the bulk (elastic) energy K_b and (ii) the surface energy K_s, which to within an insignificant positive multiplier is:

$$K(\mathbf{k},H) = K_b + K_s = -|\mathbf{k}|\, J_b(\mathbf{e},h) + \sigma|\mathbf{k}|^2 \qquad (8)$$

In contrast to the surface energy, the bulk energy term $J_b(\mathbf{e},h)$ depends on the mechanical properties of the film and substrate, and on their thicknesses.

We can label all types of corrugations as either stable, neutral or destabilising depending on their wave-vector, the nature of the stress field and the physical properties of the materials. We find the following conditions for stability, neutrality or instability:

$$K(\mathbf{k},H) > 0, \quad K(\mathbf{k}_{ne},H) = 0, \quad K(\mathbf{k},H) < 0 \qquad (9)$$

respectively. Corrugations which are unstable grow in amplitude, corrugations which are stable decay in amplitude and neutral corrugations neither grow nor shrink. A film thickness for which all corrugations are stable is referred to as stable. Situations in which at least one destabilising corrugation exists are unstable. The critical thickness of the uniform film corresponds to the situation in which there are no destabilising corrugations and for which at least one mode of surface corrugation is neutral. The corrugations with wave-vector **k** that minimises K are maximally unstable. In other words, this is the corrugation wave-vector which decreases the energy of the system with increasing amplitude faster than corrugation of any other wave-vector. This is the corrugation wave vector that will most likely dominate the surface morphology. However, different material transport mechanisms may shift the fastest growing corrugation to nearby wave-vectors. The function $G \equiv -K$ is proportional (in a certain range of **k**-vectors) to the (quasi-static) rate of growth of the Fourier components of the corrugations.

F MORPHOLOGICAL PATTERNS FROM SURFACE INSTABILITIES

We first consider the case of a film bonded to a rigid substrate ($\nu_s = 1/2$, $\chi = 0$). In this case, Eqns (6) to (8) yield:

$$G(\mathbf{k},H) = -\sigma\,|\mathbf{k}|^2 + \frac{|\mathbf{k}|}{\mu_f}\Big[\frac{(1-\nu_f)[h + (3-4\nu_f)\sinh h \cosh h]}{4(1-\nu_f)^2 + h^2 + (3-4\nu_f)\sinh^2 h}\,T_n^2 \qquad (10)$$

$$+ \frac{\sinh h}{\cosh h}\,T_t^2\Big]$$

Introducing the angle θ between **e** and the principle in-plane stress T_1, Eqn (10) can be rewritten as ($\tau_a = T_a/\mu_f$ are the dimensionless principal in-plane stresses):

$$\frac{4G}{\mu_f|\mathbf{k}|(\tau_1 - \tau_2)^2} = -A + B\sin^2 2\theta + C(s + \cos 2\theta)^2 \qquad (11)$$

where

$$A = \frac{4\sigma|\mathbf{k}|}{\mu_f(\tau_1 - \tau_2)^2},\; B = \frac{\sinh h}{\cosh h},\; C = \frac{(1-\nu_f)[h + (3-4\nu_f)\sinh h\cosh h]}{4(1-\nu_f)^2 + h^2 + (3-4\nu_f)\sinh^2 h},$$

$$s = \frac{\tau_1 + \tau_2}{\tau_1 - \tau_2}$$

The dimensionless parameter s characterises the in-plane misfit stresses and plays a crucial role in determining the dominant surface morphology; s equals ± 1 in the case of an applied uniaxial stress, vanishes for pure in-plane shear, and is infinite for pure in-plane dilatation. The surface energy (first) term in Eqn (10) prevails as $|\mathbf{k}|$ approaches infinity and, thus, corrugations with sufficiently small wavelength always raise the total energy of the system. In order for long wavelength corrugations ($|\mathbf{k}| \ll 1$) to be stable the function K (G) must be positive (negative). This condition leads to the following formula for the first critical thickness (Grinfeld [7,19], Spencer et al [15])

$$H_{crit} = \frac{\sigma}{\mu_f \tau_1^2} = \frac{\sigma \mu_f}{T_1^2} \qquad (12)$$

above which the free surface becomes rough.

We now examine the extrema of the RHS of Eqn (10), which we denote as $\varphi(\theta)$. Setting the first derivative of $\varphi(\theta)$ to zero, we find that there are two types of extrema, satisfying:

$$\text{(a) } \sin 2\theta = 0 \quad \text{and} \quad \text{(b) } \cos 2\theta = \frac{C}{B - C}\,s \qquad (13)$$

The 'a'-solutions always exist and correspond to the k-vectors which are parallel to the directions of the lateral principal stresses (i.e. corrugations with valley perpendicular to the directions of the lateral principal stresses). The 'b'-solutions may or may not exist, depending on whether or not the inequality $-1 \leq Cs/(B - C) \leq 1$ is satisfied. The 'a'-solution gives the maximum values of $\varphi(\theta)$, provided that the 'b'-solutions do not exist. If the 'b'-solutions do exist and B > C, then the 'b'-solutions correspond to maxima in $\varphi(\theta)$, whereas the 'a'-solutions deliver $\varphi(\theta)$ minima (if B < C, then the situation is opposite: the 'b'-solutions yield maximum $\varphi(\theta)$). It is obvious that at fixed $|\mathbf{k}|$ and h the solutions maximising $\varphi(\theta)$ determine the most probable surface morphology. The 'b'-solutions exist for 'shear-like' misfit stresses, but do not exist for 'dilatation-like' misfit stresses. In order to make these results more concrete, we now consider several specific applications of these results.

F1 Pure In-Plane Dilatation

In this case, the principal in-plane stresses are equal, i.e. $\tau_1 = \tau_2 = \tau$. As a result, the dimensionless parameter s is infinite and Eqn (11) should be rewritten as:

$$\frac{GH}{\mu_f \tau^2} = -\frac{1}{\Phi\alpha^H_{dil}} h^2 + C h \quad \text{where} \quad \Phi\alpha^H_{dil} \equiv \frac{\mu_f \tau^2 H}{\sigma} \tag{14}$$

Thus, given h (= H|**k**|) and |**k**|, all the Fourier components of the corrugation have the same energy dependence on amplitude, independent of the angle θ. In other words, the corrugations are equally likely to appear in any direction. While the present results are for isotropic elasticity and surface energy, anisotropy would break the symmetry and make certain corrugation directions more favourable than others. Several plots of the RHS of Eqn (14) are shown in FIGURE 3. Eqn (14) shows that the wave-numbers $|\mathbf{k}|_{ne}$ and $|\mathbf{k}|_p$ corresponding to the 'neutral' and most probable (at which the G-function vanishes and reaches its maximum, respectively) are:

$$|\mathbf{k}|_{ne} = (1 - \nu_f)\frac{\mu_f \tau^2}{\sigma}; \; |\mathbf{k}|_p = \frac{1}{2}(1 - \nu_f)\frac{\mu_f \tau^2}{\sigma} \tag{15}$$

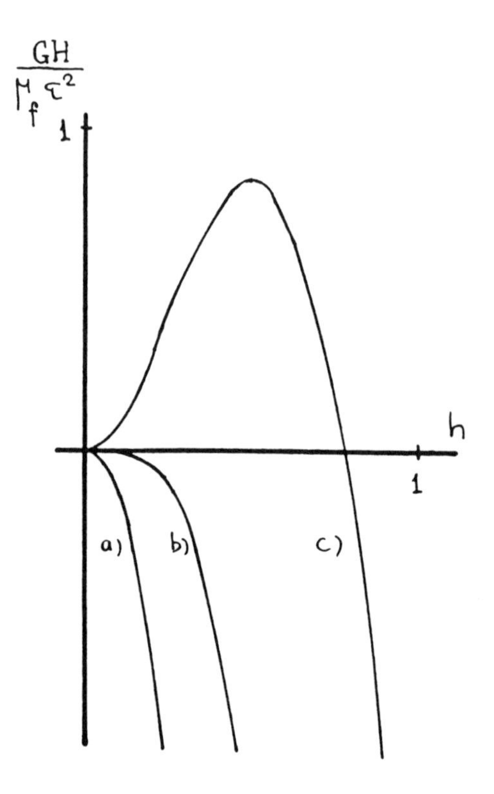

FIGURE 3 The dependence of the contributed energy on h = kH.

If we view the corrugations as cracks (see below), these corrugations correspond to Mode I cracks, i.e. the stress tends to separate the crack faces normal to the crack plane.

F2 Pure In-Plane Shear

In this case, the principal in-plane stresses are of the same absolute value and of opposite sign: $\tau_1 = -\tau_2 = \tau$. In pure in-plane shear the dimensionless parameter s is zero, such that Eqns (11) and (13b) can be rewritten as:

$$\text{(a)}\ \frac{4G}{\mu_f|\mathbf{k}|\ \tau^2} = -\frac{\sigma|\mathbf{k}|}{\mu_f\tau^2} + B\sin^2 2\theta + C\cos^2 2\theta, \qquad \text{(b)}\ \cos 2\theta = 0 \tag{16}$$

Eqn (16b) shows that the 'b'-solutions always exist in pure shear and have optimal k-vectors bisecting the principal in-plane directions. If we view the corrugations as cracks, these corrugations correspond to Mode II cracks, i.e. the stress tends to slide the crack faces in opposite directions and parallel to the crack plane.

F3 The Infinitely Thick Film (Arbitrary Stress)

For infinitely thick films, the parameters B and C are equal to 1 and $1 - \nu_f$, respectively. Inserting these values into Eqn (11) and considering the 'a'-modes of the corrugations, we find:

$$\frac{G^a_{extr}}{\mu_f(\tau_1 - \tau_2)^2} = \left[-\frac{4}{\Phi\alpha^k} + (1 - \nu_f)\,(s \pm 1)^2\right]|\mathbf{k}| \quad \text{where}$$

$$\Phi\alpha^k \equiv \frac{\mu_f(\tau_1 - \tau_2)^2}{\sigma\,|\mathbf{k}|} \tag{17}$$

which leads to the following analogies of Eqn (15):

$$|\mathbf{k}|_{ne} = \frac{1}{4}\,(1 - \nu_f)\,(s \pm 1)^2\,\frac{\mu_f(\tau_1 - \tau_2)^2}{\sigma} \quad \text{or} \quad \Phi\alpha^k_{ne} = \frac{4}{(1 - \nu_f)\,(s \pm 1)^2}$$

$$|\mathbf{k}|_p = \frac{1}{8}\,(1 - \nu_f)\,\frac{\mu_f(\tau_1 - \tau_2)^2}{\sigma} \tag{18}$$

Eqn (18) gives the most probable corrugations when the 'b'-modes do not exist. In the case at hand the equations for the orientation of the **k**-vectors, the existence inequality and equation for G^b_{extr} can be reduced to the following forms, respectively:

$$\cos 2\theta = \frac{\nu_f}{1 - \nu_f}\,s; \quad \left|\frac{\nu_f}{1 - \nu_f}\,s\right| < 1;$$

$$\frac{4G^b_{extr}}{\mu_f(\tau_1 - \tau_2)^2} = -\frac{4\sigma|\mathbf{k}|^2}{\mu_f(\tau_1 - \tau_2)^2} + \left(1 + \frac{1 - \nu_f}{\nu_f}\,s^2\right)|\mathbf{k}| \tag{19}$$

These special cases clearly demonstrate that different patterns of islanding can occur in even the simplest of isotropic incompressible elastic films.

G LATE STAGE MORPHOLOGICAL EVOLUTION

The general analysis of the growth of corrugations on the surface of stressed films was based upon the assumption that the amplitude of the corrugations is small compared with the wavelength of the corrugations, i.e. small slope surface profiles. This is clearly valid for early times, when the initial surface is flat. However, within the framework of the theory, unstable corrugations tend to grow exponentially with time [5-9]. Therefore, the unstable corrugations rapidly grow to the extent that the small amplitude assumption in the theory is violated. The small slope assumption in the theory was made in order to linearise the equations. In order to treat large slope surface morphologies and late times, we must treat the non-linear surface evolution equations. Spencer et al [15] and Nozieres [20] recently performed a weakly non-linear analysis of a misfitting film on a substrate. Their results showed that there was some tendency for the corrugations to sharpen as they grew into the film.

Yang and Srolovitz [21] performed a numerical study of the growth of corrugations on the surface of a uniaxially stressed elastic solid, based upon the full, non-linear partial differential equations, using a boundary integral formulation of the elasticity and surface diffusion kinetics. That study showed that the initial growth of the corrugations (initial growth rates, critical wavelengths) accurately follows the linear analysis. However, as the amplitude of the corrugations grew, they rapidly sharpened and accelerated. Starting at very early times, the effects of the non-linearities become apparent in the surface morphology when the corrugations are unstable. The initial small amplitude corrugations develop sharp cusps at the lowest point in the profile (i.e. in the valleys). As time proceeds, these cusps sharpen as they become deeper while the remainder of the surface profile shows very little change. After a relatively short time, the morphology begins to resemble that of an array of cracks. The velocity of the cusps continues to accelerate and appears to diverge at a finite cusp depth (crack length). Analysis of these critical crack lengths as a function of applied stress, surface energy and elastic constants shows that the critical crack length is equal to that predicted by the classical Griffith theory of linear elastic fracture mechanics.

In a self-stressed thin film geometry, such as commonly found in heteroepitaxy, the corrugations cannot grow indefinitely. In fact, the depth will be effectively limited by the film thickness, since growth beyond this will change the sign of the stress field on the crack tip. It is unclear whether these cracks or cusps actually reach the substrate. As a result, the 'islands' formed may appear either as Volmer-Weber or Stranski-Krastanov islands. The type of surface morphology that should be observed consists of relatively smooth corrugations at early times with sharp cusps developing at long times. Since the cusps will progress down to near the substrate at times when the amplitude of the remainder of the corrugations is still small, this will lead to the formation of 'islands' with relatively flat tops. At still longer times, the flat tops will become rounded due to surface energy effects. Whether any particular experiment is within the short or long time regime is largely dependent on matter transport kinetics. Since diffusivities scale exponentially with temperature, a small change in temperature could shift the film morphology from one with small amplitude corrugations to an array of flat top or rounded top islands.

H HETEROEPITAXIAL Si-Ge FILMS

The above analysis of corrugations that develop on the surfaces of stressed films and islanding is very general and, hence, can be applied to a wide variety of heteroepitaxial/stressed films. We now turn to the specific case of heteroepitaxial Si-Ge films and analyse several experimentally observed thin film morphologies in the light of the above theoretical discussion. Heteroepitaxial growth of a Si-Ge alloy film on Si creates bi-axial compression within the film. Since the misfit stress is pure (or nearly pure) in-plane compression (Section F1) pertains. According to Eqn (14), the stress state is such that corrugations are independent of in-plane orientation; however crystalline anisotropy may bias the corrugation orientation. The expected corrugations consist of two orthogonal sets with the wavelength given by $2\pi/|k|_p$ of Eqn (15). Therefore, a two-dimensional, square array of corrugations or islands, depending on the kinetics and duration of the experiment, should be observed.

Kuan and Iyer [22] grew $Si_{0.5}Ge_{0.5}$ on Si (100), (111) and (110) by MBE. The films were only partially strain relieved by a low density array of misfit dislocations. For growth at low temperatures (400°C), they achieved flat, uniform (100) films. However, when the growth temperature was raised to 580°C, the surface 'undulated with a characteristic wavelength of about 100 nm and an amplitude of about 20 nm'. When Si was grown over the undulating SiGe alloy film, the corrugations disappeared. These results suggest (a) the presence of the small amplitude corrugations of stressed surfaces and (b) that the formation of corrugations is controlled by kinetics [5,15]. The small oscillations are consistent with the analysis that led to Eqns (14) and (15) in Section F. The fact that these oscillations were not observed for growth at the lower temperature and were observed at a slightly higher temperature is consistent with the time constant for the growth of corrugations scaling inversely with the surface diffusivity and with temperature as exp(Q/kT), where Q is the activation energy for surface diffusion. However, since these corrugations most likely occurred during film growth, we cannot rule out the possibility that these corrugations are, in some way, associated with the growth process. Further, since these films were relatively thick and a small density of dislocations was observed at the interface (the stress was only partially relaxed), the possibility that the corrugations are associated with dislocation slip steps cannot be ruled out.

Robbins et al [23] used in-situ laser light scattering, ex-situ TEM and AFM to analyse $Si_{1-x}Ge_x$ ($0.19 < x < 0.26$) vapour phase epitaxially grown on Si (100) at 610°C < T < 750°C. They found a 'rippled surface morphology ..... preceding misfit dislocation generation'. This indicates that the corrugations can and do develop in the absence of dislocation slip steps on the surface. The corrugations had amplitudes between 0.3 and 4.5 nm and a periodicity of between 100 and 200 nm, with the surface roughness developed during growth 'much greater at 750°C than at 610°C' for the same concentration. Their analysis of the existing experimental data through 1991 indicates that the wavelength of the corrugations scales as the inverse square of the strain, which is consistent with Eqn (15) above. They also found a 'clear tendency for the growth temperature needed to observe undulations to increase with increasing surface wavelength', which is also consistent with our analysis (see e.g. equation (16) of [5]).

Pike et al [24] grew 200 nm thick $Si_{0.7}Ge_{0.3}$ on (100) and (110) Si which was partially strain relieved by interfacial misfit dislocations. They observed surface corrugations with a wavelength approximately ten times larger than the interfacial dislocation spacing. Their analysis also showed that in order for the observed surface oscillation amplitude to be associated with slip steps, over one hundred dislocations would have to propagate along each corrugation. This is inconsistent with the dislocation and corrugation spacings. High temperature, post film growth annealing was found to increase the surface roughness and eventually led to partial islanding along the pre-existing surface undulations. Since this growth of the corrugation continued after film growth was complete, the growth of surface corrugations is not a film 'growth' phenomenon, although it can occur simultaneously with film growth. The observation of island formation with post growth annealing at a significantly higher temperature is consistent with the numerical results on late stage morphological evolution.

The recent work of Jesson et al [25] on $Si_{0.5}Ge_{0.5}$ grown on Si (100) also observed surface corrugations on the unrelaxed film. They found that the alloy film remained strained up to a 'factor of twenty above the critical thickness anticipated from equilibrium concentrations'. However, 'after strain relaxation in the film by the introduction of misfit dislocations, the surface smooths'. This indicates that the misfit stress is responsible for the formation of the corrugations. When the stress is relieved, surface energy flattens the corrugations, as originally proved by Mullins [26].

Because of the large misfits and early island formation associated with unalloyed Ge growth on Si, the resultant morphologies are much more difficult to directly compare with our analysis. Ge on Si is often referred to as the classical example of Stranski-Krastanov growth (three monolayers, followed by islanding [27,28]). Mo et al [29] used an STM to demonstrate that the transition from 2D layer growth to 3D growth occurs by the formation of small, crystallographically faceted, rectangular islands which are a precursor to the formation of large, 'equilibrium' islands. Tersoff and Tromp [30] analysed these small clusters and showed that their rectangular shapes are size dependent and are a compromise between strain energy and surface energy. This further demonstrates that surface morphologies of stressed bodies are, given sufficient transport, dictated by a balance between strain energy and surface energy - as indicated above in our analysis of surface corrugations.

I CONCLUSION

In the absence of surface tension a flat boundary of non-hydrostatically stressed elastic solids is always unstable with respect to 'mass rearrangement'. Combining the mechanism of the stress driven, mass rearrangement instability and stabilising effects of surface tension, one obtains a very general picture of the morphology of stressed films that can lead to dislocation free, Stranski-Krastanov growth of heteroepitaxial solid films on substrates. This instability is purely thermodynamic in nature and does not depend on the specific mechanisms of mass transport. There may exist two critical thicknesses associated with this instability in prestressed elastic films attached to a solid substrate. The first one corresponds to the destabilisation of flat films in favour of long, parallel, periodic corrugations. For some types of stress, a second critical thickness exists, corresponding to the formation of another non-parallel set of corrugations. A non-linear analysis of these instabilities shows that these corrugations develop increasingly sharp bottoms as they grow, leading to crack-like defects which can propagate down to very close to the substrate, thereby creating islands and an apparent Stranski-Krastanov growth morphology. The present analysis was shown to be consistent with a wide range of observations of the morphology of SiGe alloys grown on Si.

ACKNOWLEDGEMENTS

The authors would like to thank Professors L.B. Freund and P.W. Voorhees for stimulating discussions on the morphological instabilities in epitaxial films. D.J. Srolovitz gratefully acknowledges the hospitality of The Weizmann Institute of Science and the support of the Michael Visiting Professorship.

REFERENCES

[1] D.P. Woodruff [*The Solid-Liquid Interface* (Cambridge University Press, Cambridge, 1973)]

[2] J.W. Christian [*The Theory of Transformations in Metals and Alloys* (Pergamon Press, Oxford, 1975)]

[3] J.W. Martin, R.D. Doherty [*Stability of Microstructure in Metallic Systems* (Cambridge University Press, Cambridge, 1976)]

[4] C. Godriche (Ed.) [*Solids Far From Equilibrium* (Cambridge University Press, Cambridge, 1991)]

[5] D.J. Srolovitz [*Acta Metall. (USA)* vol.37 (1989) p.621]

[6] R.J. Asaro, W.A. Tiller [*Metall. Trans. (USA)* vol.3 (1972) p.1789]

[7] M.A. Grinfeld [*Sov. Phys.-Dokl. (USA)* vol.31 (1986) p.831]; M.A. Grinfeld [*Fluid Dyn. (USA)* vol.22 (1987) p.169]; M.A. Grinfeld [*Thermodynamic Methods in the Theory of Heterogeneous Systems* (Longman, Sussex, 1991)]; M.A. Grinfeld [*J. Nonlinear Sci. (USA)* vol.3 (1993) p.35]

[8] P. Nozieres [unpublished lectures (1988)]; P. Nozieres [in *Solids Far From Equilibrium* Ed. C. Godriche (Cambridge University Press, Cambridge, 1991)]

[9] L.B. Freund, F. Jonsdottir [*J. Mech. Phys. Solids (UK)* (in press)]

[10] M. Thiel, A. Willibald, P. Evers, A. Levchenko, P. Leiderer, S. Balibar [*Europhys. Lett. (Switzerland)* vol.20 (1992) p.707]; R.H. Torii, S. Balibar [*J. Low Temp. Phys. (USA)* vol.89 (1992) p.391]; J. Berrehar, C. Caroli, C. Lapersonne-Meyer, M. Schott [*Phys. Rev. B (USA)* vol.46 (1992) p.13487]

[11] B. Caroli, C. Caroli, B. Roulet, P.W. Voorhees [*Acta Metall. (USA)* vol.37 (1989) p.257]; P.H. Leo, R.F. Sekerka [*Acta Metall. (USA)* vol.37 (1989) p.3119]; J. Grilhe [*Acta Metall. Mater. (USA)* vol.41 (1993) p.909]

[12] A. Onuki [*Phys. Rev. A (USA)* vol.39 (1989) p.5932]

[13] W.K. Heidug [*J. Geophys. Res. (USA)* vol.21 (1991) p.909]; W.K. Heidug, Y.M. Leroy [*J. Geophys. Res. (USA)* (in press)]

[14] H. Gao [*J. Mech. Phys. Solids (UK)* vol.39 (1991) p.443]; H. Gao [*Int. J. Solids Struct. (UK)* vol.28 (1991) p.701]

[15] C.W. Snyder, B.G. Orr, D. Kessler, L.M. Sander [*Phys. Rev. Lett. (USA)* vol.66 (1991) p.3032]; B.J. Spencer, P.W. Voorhees, S.H. Davis [*Phys. Rev. Lett. (USA)* vol.67 (1991) p.3696]; B.J. Spencer, S.H. Davis, P.W. Voorhees [*Phys. Rev. B (USA)* vol.47 (1993) p.9760]; M.A. Grinfeld [*J. Intell. Mater. Syst. Struct. (USA)* vol.4 (1993) p.76]

[16] D.J. Eaglesham, M. Cerullo [*Phys. Rev. Lett. (USA)* vol.64 (1990) p.1943]; S. Guha, A. Madhukar, K.C. Rajkumar [*Appl. Phys. Lett. (USA)* vol.57 (1990) p.2110]; F.K. LeGoues, M. Copel, R.M. Tromp [*Phys. Rev. B (USA)* vol.42 (1990) p.11690]

[17] J.H. van der Merwe [*J. Appl. Phys. (USA)* vol.34 (1963) p.117]; J.W. Matthews, A.E. Blakeslee [*J. Cryst. Growth (Netherlands)* vol.27 (1974) p.118]; G.H. Gilmer, M.H. Grabow [*J. Met. (USA)* vol.39 (1987) p.19-23]; R. Bruinsma, A. Zangwill [*Europhys. Lett. (Switzerland)* vol.4 (1987) p.1729]

[18] L.C. Stoop [*Thin Solid Films (Switzerland)* vol.24 (1974) p.229 and 243]; D. Vanderbilt, O.L. Alerhand, R.D. Meade, J.D. Joannopoulos [*J. Vac. Sci. Technol. B (USA)* vol.7 (1989)

p.1013]; E. Kasper, H. Jorke [*J. Vac. Sci. Technol. (USA)* vol.10 no.4 (1992) p.1927]; J. Tersoff [*Phys. Rev. B (USA)* vol.43 (1991) p.9377]
[19] M.A. Grinfeld [submitted papers]
[20] P. Nozieres [unpublished lectures on the Stranski-Krastanov epitaxial growth (1992)]; P Nozieres [*J. Phys. (France)* vol.3 (1993) p.681]
[21] W. Yang, D.J. Srolovitz [*Phys. Rev. Lett. (USA)* vol.71 (1993) p.1593]
[22] T.S. Kuan, S.S. Iyer [*Appl. Phys. Lett. (USA)* vol.59 (1991) p.2242]
[23] D.J. Robbins, A.G. Cullis, A.J. Pidduck [*J. Vac. Sci. Technol. B (USA)* vol.9 (1991) p.2048]; A.J. Pidduck, D.J. Robbins, A.G. Cullis, W.Y. Leong, A.M. Pitt [*Thin Solid Films (Switzerland)* vol.222 (1992) p.78]
[24] W.T. Pike, R.W. Fathauer, M.S. Anderson [*J. Vac. Sci. Technol. B (USA)* vol.10 (1992) p.1990]
[25] D.E. Jesson, S.J. Pennycock, J.Z. Tischler, J.D. Budai, J.M. Baribeau, D.C. Houghton [*Phys. Rev. Lett. (USA)* vol.70 (1993) p.2293]
[26] W.W. Mullins [*J. Appl. Phys. (USA)* vol.30 (1959) p.77]
[27] B.Y. Tsaur, M.W. Geis, J.C.C. Fan, R.P. Gale [*Appl. Phys. Lett. (USA)* vol.38 (1981) p.799]
[28] F.K. LeGoues, M. Copel, R.M. Tromp [*Phys. Rev. B (USA)* vol.42 (1990) p.11690]
[29] Y.W. Mo, D.E. Savage, B.S. Swarzentruber, M.G. Lagally [*Phys. Rev. Lett. (USA)* vol.65 (1990) p.1020]
[30] J. Tersoff, R.M. Tromp [*Phys. Rev. Lett. (USA)* vol.70 (1993) p.278]

1.2 Equilibrium theories of misfit dislocation networks in the SiGe/Si system

R. Hull

May 1994

A INTRODUCTION

The coherency strain which exists in pseudomorphic lattice-mismatched epitaxial heterostructures can drive the formation of an interfacial misfit dislocation array, which acts so as to relieve elastic strain in the strained layers. For planar structures in a given materials system, strain and interface orientation, a strain-dependent critical thickness, h_c, will exist. For epilayer thicknesses above h_c, the misfit dislocation array is energetically favoured, and for thicknesses below h_c, the dislocation array is not favoured. In this Datareview, we will summarise the parameters determining the microstructure of the misfit dislocation array and models for predicting the critical thickness for planar structures, with particular reference to the Ge_xSi_{1-x}/Si system. The primary other mechanisms for lattice-mismatch strain relief, islanding of the epitaxial layer and interfacial interdiffusion, will be discussed elsewhere in this volume.

B MISFIT DISLOCATION GEOMETRY

In the diamond cubic lattice structures of Ge, Si and Ge_xSi_{1-x}, dislocations propagate primarily by glide on {111} planes inclined to the interface (for a review of fundamental dislocation properties see [1]). The geometry of the interfacial dislocation array is therefore defined by the intersection of these glide planes with the interface. This produces orthogonal, uniaxial, and hexagonal misfit dislocation arrays, with line directions along in-plane $<01\bar{1}>$ directions, for (100), (011) and (111) interfaces respectively. This has been confirmed experimentally [2] and is illustrated in FIGURE 1. At very high lattice-mismatches (corresponding to Ge concentrations >80% in Ge_xSi_{1-x}/Si), secondary slip systems may operate, with dislocation glide occurring on other inclined planes [3]. For example, interfacial dislocation arrays with line directions along <010> have been observed at high strain Ge_xSi_{1-x}/Si(100) interfaces, corresponding to slip on {011} planes [3].

It is a geometrical property of a dislocation that it cannot terminate within the bulk of a crystal, but must rather terminate upon itself (forming a closed loop), at a node with another defect, or most generally in the case of strained layer epitaxy, at the nearest free surface. Finite dislocation propagation velocities generally prohibit formation of sufficiently long misfit dislocation segments to traverse an entire substrate diameter (a reported exception is in the growth by chemical vapour deposition (CVD) of Ge_xSi_{1-x}/Si ($x \leq 0.05$) structures by high temperature, ~1120°C, where misfit dislocation velocities will be very high [4]). Misfit dislocation segments therefore generally terminate by connection to a threading dislocation segment, which traverses the strained layer. This is illustrated schematically for Ge_xSi_{1-x}/Si and Si/Ge_xSi_{1-x}/Si structures in FIGURES 2(a) and 2(b) respectively. For an uncapped Ge_xSi_{1-x}/Si

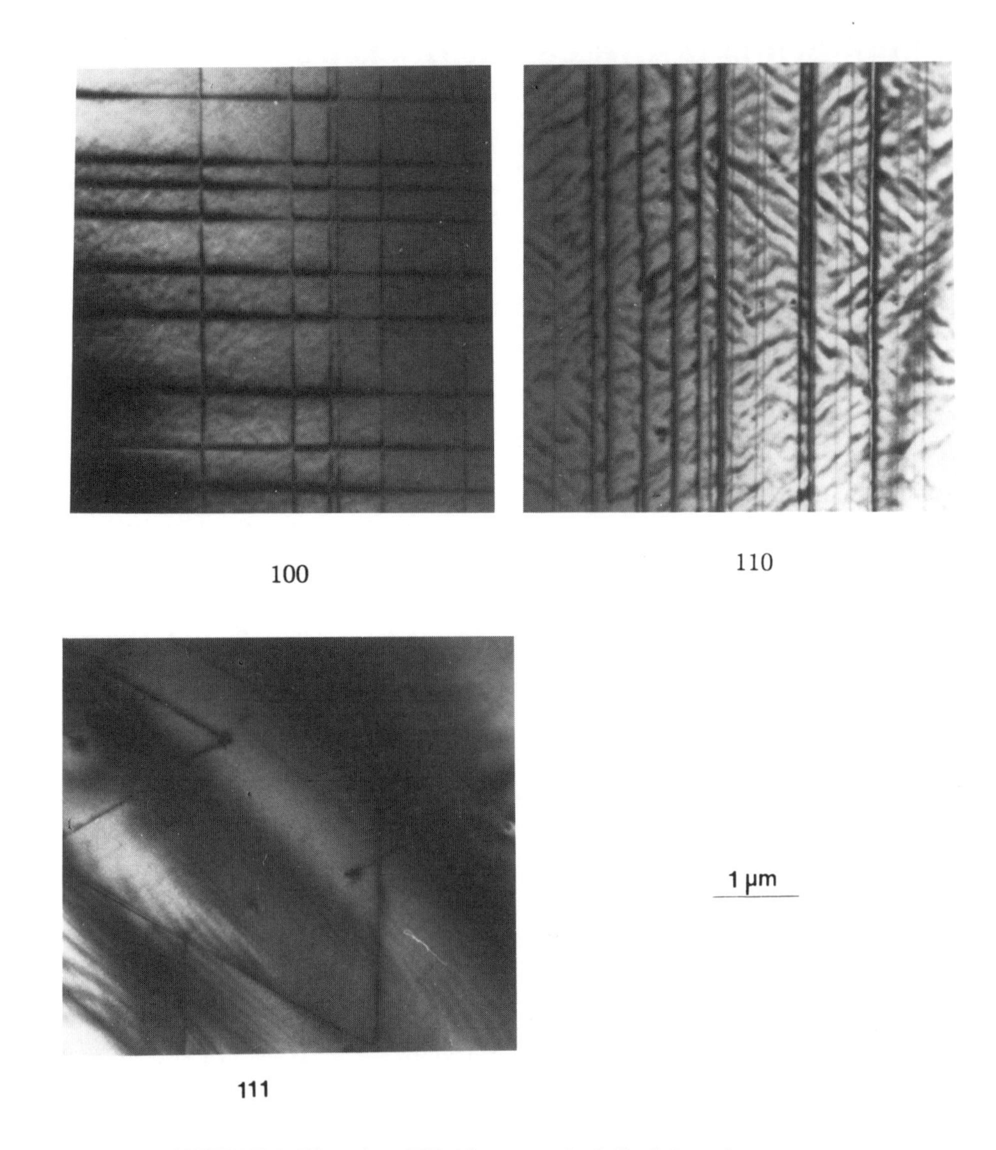

FIGURE 1 Plan view TEM images of misfit dislocations at Ge_xSi_{1-x}/Si (100), (110) and (111) interfaces.

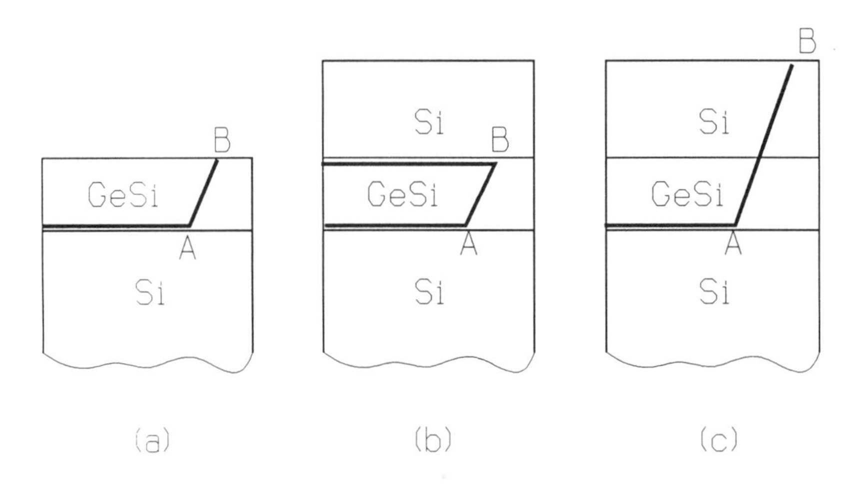

FIGURE 2 Schematic illustration of misfit/threading arm dislocation geometries at (a) uncapped Ge_xSi_{1-x}/Si structures and (b) and (c) capped Si/Ge_xSi_{1-x}/Si structures. The threading dislocations are the segments A .. B in each image. The misfit segments are those parallel to the interface(s).

structure, the threading segment AB connects the end of the interfacial segment to the free surface. For capped structures, the misfit dislocation generally propagates by simultaneous formation of interfacial misfit segments at both top and bottom Ge_xSi_{1-x}/Si interfaces, these interfacial segments being connected by a threading segment traversing the intervening Ge_xSi_{1-x} layer. (Note that the interfacial segment at the top Ge_xSi_{1-x}/Si interface must eventually be connected to another threading segment which reaches to the free surface, or must form a complete loop within the buried layer.) In principle, for very thin capping Si layers, only one interfacial misfit segment may exist at the lower Ge_xSi_{1-x}/Si interface, with the threading segment joining this interface to the free surface as illustrated in FIGURE 2(c). The regimes of epilayer and capping thicknesses, h_{ep} and h_{cp} respectively, where models (b) and (c) dominate have been modelled by Twigg [5], who demonstrated that (b) is the general configuration and (c) only applies to a relatively narrow range of structures with low h_{cp}. A more detailed analysis by Gosling et al [6] suggests that the two variants (b) and (c) may co-exist in the same structure.

C MISFIT DISLOCATION MICROSTRUCTURE

The general structure of a perfect dislocation (i.e. whose Burgers vector, b, is a lattice translation vector, and thus leaves no fault in the crystal lattice through which it has propagated) in Si, Ge or Ge_xSi_{1-x} is to have a Burgers vector of a/2<101>, the minimum lattice translation vector in the diamond cubic structure. It has long been recognised, however, that such a defect is energetically unstable with respect to dissociation into Shockley partial dislocations [1], according to Burgers vector reactions of the type:

$$a/2\langle 101\rangle = a/6\langle 211\rangle + a/6\langle 1\bar{1}2\rangle \qquad (1)$$

The Shockley partial dislocations, with Burgers vector $a/6\langle 211\rangle$ and $a/6\langle 1\bar{1}2\rangle$ in Eqn (1), are mutually repelled from each other by the interaction of their crystal stress fields and glide apart from each other on their common {111} glide plane. This produces a ribbon of stacking fault between them on the {111} glide plane. The equilibrium partial separation, s_o, is determined by the balance of their repulsive interaction energy with the energy of the intervening stacking fault. In the absence of an applied stress, s_o is of the order of a few nm in bulk Ge or Si. For (100), (011) and (111) interfaces in the Ge_xSi_{1-x}/Si system, motion by glide on {111} planes results in angles, θ, between the misfit dislocation Burgers vector, b, and line direction, u, of 60° for the parent a/2<101> dislocation and 30° and 90° respectively for the two Shockley partial dislocations in Eqn (1). (Note all angles here and in subsequent discussion are referred to with respect to the interfacial misfit dislocation line direction.) The lattice mismatch stress, arising from the lattice parameter difference between Ge_xSi_{1-x} and Si, is resolved differently onto these two partials, and thus they experience different resolved shear stresses (higher on the 90° partial than on the 30°). This difference in applied stress on the two partials can change the magnitude of the partial separation, s_o. Depending upon the interface orientation and upon the magnitude and sign (compressive or tensile) of the applied stress, s_o can vary from zero to infinity.

The two Shockley partials of Eqn (1) must move in a defined order, for formation of a low energy stacking fault, with only second nearest-neighbour stacking violation of type ABC/BCA, in the diamond cubic lattice (if the partials move in the incorrect order, a very high energy fault involving nearest neighbour stacking violation of type ABC/CAB is obtained).

The correct order for a given interface orientation and sign of stress is given by the Thompson tetrahedron construction [7]. For a (100) interface with the epitaxial layer under compressive stress, such as for the Ge_xSi_{1-x}/Si(100) interface, the 30° partial leads and the 90° partial trails. The trailing 90° partial then experiences a greater resolved applied stress than the leading 30° partial. Thus, the partial separation is reduced (and may even become zero) and the misfit/threading defect combination propagates as a narrowly dissociated or undissociated 60° a/2<101> dislocation. For tensile stress at a (100) interface, as in the Ge_xSi_{1-x}/Ge(100) system, the 90° partial now leads, and the 30° partial trails. The leading partial now experiences a greater resolved applied stress than the trailing partial, and s_o is now increased with respect to the zero-stress case, and may even become infinite. In this case, the relevant misfit dislocation microstructure consists of isolated 90° a/6<211> Shockley partial dislocations, which leave stacking faults behind in the lattice as they propagate. (Note that it may also be possible in this case for the misfit dislocations to nucleate as 90° Shockley partials [8].) Such defects have indeed been observed for Ge_xSi_{1-x}/Ge(100) epitaxy [8].

Changes in interface orientation can also affect the order of the Shockley partial motion, and thus the observed misfit dislocation microstructure. For compressively strained Ge_xSi_{1-x} layers grown on Si (110) and (111) substrates, the 90° partial is again predicted to lead. In both cases, the expected 90° a/6<211> partial misfit dislocations have been observed [9,10]. An illustration for the Ge_xSi_{1-x}/Si(110) interface is shown in FIGURE 3. Reversing the sign of stress for these interface orientations should again reverse the order of partial motion, and thus for Ge_xSi_{1-x}/Ge(110) and Ge_xSi_{1-x}/Ge(111) structures the misfit dislocation microstructure should again revert to a Burgers vector type of 60° a/2<101>, although to our knowledge this has not been experimentally confirmed. General rules for misfit dislocation dissociation and microstructure have been summarised by Kvam and Hull [11].

FIGURE 3 Cross-sectional TEM image ([220] bright field near the $(1\bar{1}0)$ pole) of a 410 Å $Ge_{0.33}Si_{0.67}$/Si(110) structure, showing stacking faults produced by passage of a/6<211> partial misfit dislocations.

Even for configurations where the 90° partial is expected to lead, formation of isolated 90° a/6<211> partial misfit dislocations requires that the difference in resolved applied stress between the 90° and 30° partials be greater than the back stress exerted by the stacking fault energy formed between the partials. These conditions have been calculated by Hull et al [12] and results for a typical areal stacking fault energy, γ, of 65 mJ m^{-2} are shown in FIGURE 4 (note that these calculations are very sensitive to the exact magnitude of γ). FIGURE 4 shows that the 90° a/6<211> partial misfit dislocations are favoured at higher strains and lower epitaxial layer thicknesses. There is a minimum strain below which the isolated partial misfit dislocations will not be expected to appear. These trends will be qualitatively followed for all configurations where the 90° partial leads the 30° partial.

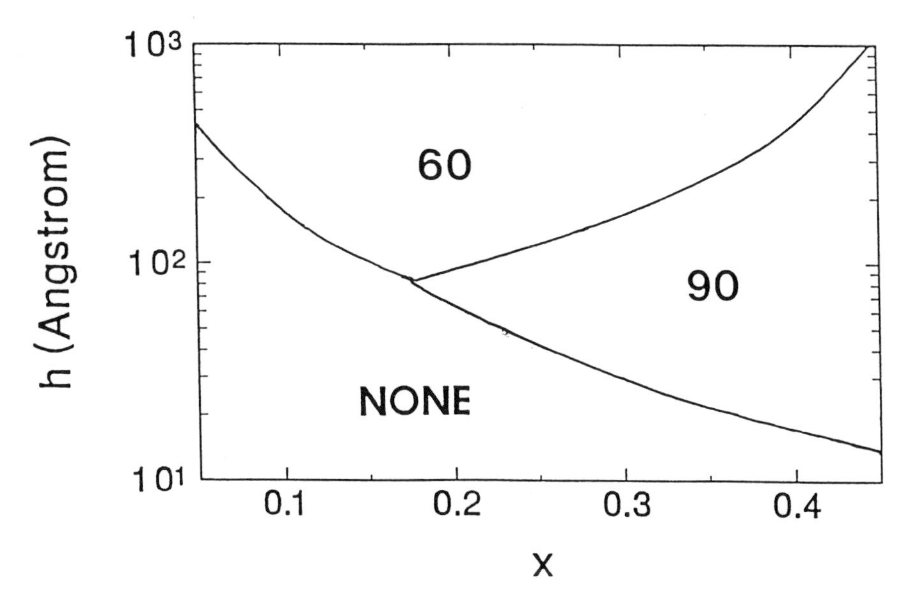

FIGURE 4 Predicted dislocation microstructure in the Ge_xSi_{1-x}/Si(011) system as a function of epilayer thickness, h, and Ge concentration, x, for a stacking fault energy of $\gamma = 65$ mJ m^{-2}. Regions labelled 'NONE', '60' and '90' respectively refer to regions where no dislocations are expected, regions where 60° a/2<101> dislocations are most favoured, and regions where 90° a/6<211> defects are most favoured.

D CALCULATIONS AND MEASUREMENT OF THE CRITICAL THICKNESS

Models of the calculation of the critical thickness, h_c, (the minimum layer thickness at which introduction of misfit dislocations into a lattice-mismatched interface is energetically favoured) have been developed for over four decades. Virtually all models interpret some form of energy balance (or, equivalently, a balance of forces or stresses), comparing the strain energy relaxed by introduction of the misfit dislocation with the energy cost associated with the self energy of the dislocation. When the relaxed strain energy is greater than the defect self energy, misfit dislocations are energetically favoured; the critical thickness corresponds to equality of these two energy terms. Early models by Frank and Van der Merwe [13] used a one-dimensional Fourier series to represent interactions between atoms in the epilayer and substrate, and demonstrated the concept of critical thickness. Subsequent development by Van der Merwe and co-workers [14-16] developed the theme of calculation of the energy of a misfit dislocation array, including dislocation interaction energies, and comparison to strain energy relaxed within the epitaxial layer. Many of these models were mathematically complex, and had analytical solutions only in the limits of very thin and very thick epitaxial layers. Nevertheless, the energy balance approach derived in these papers set the foundations for subsequent development in this field.

The most conceptually simple model for predicting the critical thickness, and thus not coincidentally the most frequently quoted, is that due to Matthews and Blakeslee (MB) [17]. They adopted a force balance approach on a propagating misfit/threading dislocation, which is essentially equivalent to the energy balance concept. By reference to FIGURE 2(a) for an uncapped epitaxial layer, the lattice mismatch between epilayer and substrate exerts a force, F_σ, on the threading dislocation arm AB. This will drive threading arm motion, so as to increase the length of the interfacial misfit dislocation and relax strain energy in the epitaxial layer. Isotropic linear elasticity theory gives:

$$F_\sigma = 2\,Gbh\varepsilon\cos\lambda\,\frac{(1+\nu)}{(1-\nu)} \tag{2}$$

The self energy of the extra interfacial misfit dislocation created produces a restoring line tension force, which is given by dislocation theory as:

$$F_T = Gb^2\,\frac{(1-\nu\cos^2\theta)}{4\pi(1-\nu)}\,\ln\frac{\alpha h}{b} \tag{3}$$

In these equations h is the epilayer thickness, G is the epilayer shear modulus, ν is the epilayer Poisson ratio, ε is the lattice mismatch strain between epitaxial layer and substrate, λ is the angle between the misfit dislocation Burgers vector and a line in the interface drawn perpendicular to the dislocation line direction, and α is a factor which describes the energy of the dislocation core, where linear elasticity theory does not apply.

The critical thickness for perfect dislocations for a given lattice mismatch strain and interface configuration is found by solving $F_\sigma = F_T$ using Eqns (2) and (3). In FIGURE 5, we plot the prediction of the Matthews-Blakeslee theory for 60° a/2<101> misfit dislocations in the Ge_xSi_{1-x}/Si(100) system, using G = 64 GPa, $\nu = 0.28$, $\cos\lambda = 0.5$, $\cos\theta = 0.5$, b = 3.9 Å, $\varepsilon = 0.041x$. A range of values for α = 1, 2 and 4 are shown: these represent the significant uncertainty in the core energy of Si and Ge. A value of $\alpha = 4$ is quoted by Hirth and Lothe [1]; this value dates from an earlier reference to atomistic calculations for cubic diamond by Maradudin [18]. Given the lack of available accurate potentials for covalent crystals at the time of Maradudin's work, some uncertainty attaches to this value for α although it is often

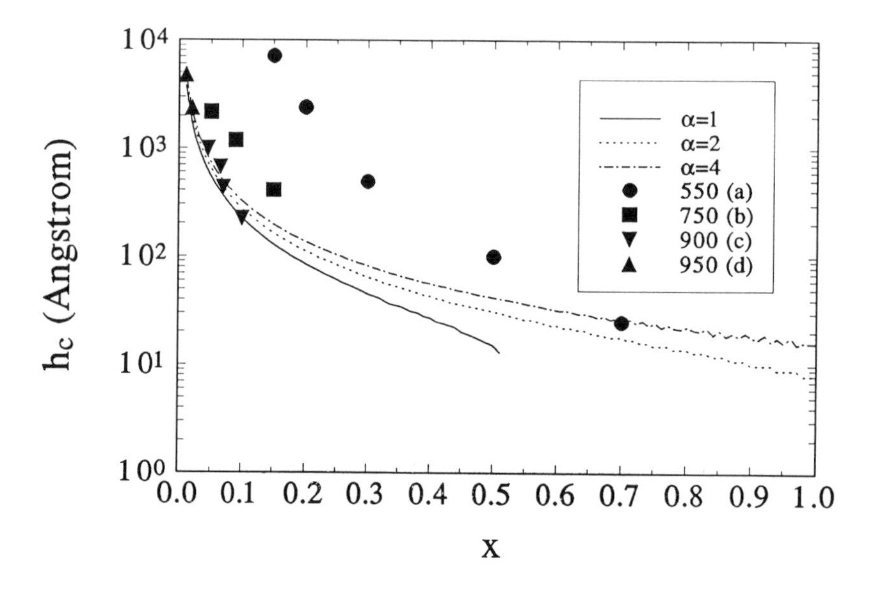

FIGURE 5 Predictions of the MB theory for the critical thickness, h_c, in Ge_xSi_{1-x}/Si(100) structures for different values of the core energy parameter, α. Also shown are experimental measurements of h_c for different growth/annealing temperatures from the work of (a) Bean et al [23], (b) Kasper et al [26], (c) Green et al [28], (d) Houghton et al [27].

quoted in the literature. More recent and sophisticated atomistic calculations of dislocation core structure in Si are likely to provide more reliable calculations of this quantity, and typically yield values of the order $\alpha = 1$, e.g. [19-21]. For a more extensive discussion of the dislocation core energy parameter see Perovic and Houghton [22]. The uncertainty in α has the greatest effect at high lattice strains, where h_c decreases sufficiently to become not much greater than b, and the dependence of the logarithmic term in Eqn (3) upon α becomes more significant.

Note that the Matthews-Blakeslee theory also predicts the equilibrium interfacial misfit dislocation density as a function of epilayer thickness, $h > h_c$. By solving the equation $F_\sigma = F_T$ for any $h > h_c$, an equilibrium residual strain ε_o can be determined. This can then be translated into an equilibrium average interfacial dislocation spacing, p_o, using the relation

$$\varepsilon_i - \varepsilon_o = \frac{b\cos\lambda}{p_o}$$

where ε_i is the initial lattice mismatch strain before formation of any dislocations.

Also shown in FIGURE 5 are experimental values of h_c for Ge_xSi_{1-x}/Si(100) structures. Early, low growth temperature (550°C by Molecular Beam Epitaxy, MBE) measurements of h_c by Bean et al [23] showed substantially greater values of h_c than predicted by the Matthews-Blakeslee theory, particularly at lower Ge concentrations. It is now known that this apparent discrepancy is resolved by taking into account finite experimental sensitivity to dislocation densities [24] and sluggish dislocation nucleation and relaxation kinetics (see Datareview 1.3 in this volume [25]). The techniques used to determine the onset of dislocation generation in the measurements of Bean et al were X-Ray Diffraction (XRD), Rutherford Backscattering Spectroscopy (RBS) and Transmission Electron Microscopy (TEM) with an average sensitivity to strain relaxation of perhaps one part in 10^3. Earlier, higher temperature, MBE growth of low Ge concentration Ge_xSi_{1-x}/Si by Kasper et al [26] had yielded experimental measurements of h_c much closer to the MB prediction, as illustrated in FIGURE 5. A cost to exact correspondence between h_c theory and experiment has been demonstrated by Houghton et al [27], who studied low Ge concentration Ge_xSi_{1-x}/Si(100) structures which had been annealed to very high temperatures (950°C). This very high temperature anneal allowed dislocation activation barriers to be overcome, and as illustrated in FIGURE 5, this produces very close agreement between experiment and theory. Similar correspondence between experiment and theory has been observed for high temperature (900°C) CVD growth by Green et al [28].

The critical thickness for Ge_xSi_{1-x}/Si will vary as a function of interface orientation through (i) the magnitude of $\cos\lambda$ in Eqn (2), and (ii) the possibility of a different Burgers vector magnitude, a/6<211>, for e.g. (011) and (111) interfaces, as discussed in the previous section. In FIGURE 6, we present data for h_c at a Ge_xSi_{1-x}/Si(011) interface, for an MBE growth temperature of 550°C. Experimental measurements are made using careful plan-view TEM experiments. The measured h_c values for the (011) interface are significantly lower than comparable values for the Ge_xSi_{1-x}/Si(100) interface from the work of Bean et al [23]. This is consistent with the lower MB predictions for h_c at the (011) vs. (100) interface, due to a higher value of $\cos\lambda$ at the (011) interface, and the difference in Burgers vector type between the two interfaces. Note that, in FIGURE 6, a range of calculations for the (011) interface are shown, corresponding to different values of the core energy parameter, α, and the areal stacking fault energy, γ (arising from passage of a/6<211> dislocations at this interface). We are not aware of experimental data for h_c at other Ge_xSi_{1-x}/Si interface orientations. Existing data for the measured onset of misfit dislocation introduction into Ge_xSi_{1-x}/Ge(100) interfaces [29] suggests lower values of h_c than those measured for comparable temperatures at the Ge_xSi_{1-x}/Si(100) interface. As discussed in Datareview 1.3 in this volume [25], this is consistent with lower kinetic barriers for dislocation motion in Ge-rich compared to Si-rich Ge_xSi_{1-x} alloys.

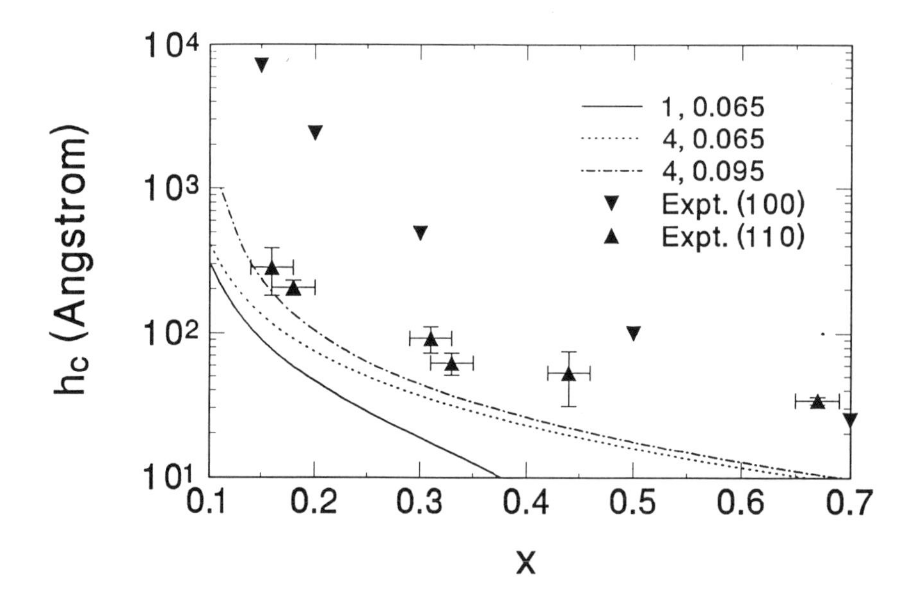

FIGURE 6 Predictions of the MB theory for h_c at the Ge_xSi_{1-x}/Si(011) interface. Calculations are for a/6<211> dislocations with a range of different values of α, γ (in J m^{-2}). Isotropic elasticity theory is used. Also shown are experimental measurements from plan view TEM of h_c for the (011) interface, compared with earlier measurements by Bean et al [23] for the (100) interface.

We now summarise the salient features of some other critical thickness models which have been developed as refinements of, or alternatives to, the MB framework.

(1) The model of People and Bean [30] which attempted to reconcile theory with the early low temperature measurements of h_c in the Ge_xSi_{1-x}/Si system [23]. This model assumed the self-energy of the dislocation to be localised within a certain region (of dimension 5b in the original People-Bean reference) centred on the dislocation core. This is in contrast to the usual elasticity treatment of dislocation self energy, and to the subsequent verification that the MB theory describes the Ge_xSi_{1-x}/Si system well in the limits of infinitely sensitive experimental techniques or very high temperatures.

(2) The model of Cammarata and Sieradzki [31], who incorporated the concept of surface stress into the MB framework. For the Ge_xSi_{1-x}/Si system they argued that surface stress will be inward along the surface normal, and thus reduce the tetragonal distortion of the Ge_xSi_{1-x} epilayer and increase h_c. These effects are significant only in the range of high Ge concentration where h_c becomes relatively small.

(3) The work of Willis et al [32], who derived a more precise expression for the energy of an array of misfit dislocations than the original Van der Merwe formulations.

Subsequent development of this work using both energy minimisation and force balance analyses [33] has enabled refinement of both the Matthews-Blakeslee and Van der Merwe models.

(4) The work of Chidambarrao et al [34] considered the effect of the orientation of the threading arm within the glide plane, and also analysed a quasi-static Peierls force on the misfit dislocation. Certain orientations (e.g. screw) of the threading arm were found to significantly enhance the predicted critical thickness.

For capped $Si/Ge_xSi_{1-x}/Si$ structures, as discussed in Section B of this Datareview, the general dislocation configuration corresponds to that illustrated by FIGURE 2(b). In this case, it is necessary to modify the F_T term in Eqn (3) to include the energies of both interfacial misfit dislocations, and the interaction energy between them. For a capping layer thickness h_{cp} and a strained layer thickness h_{ep}, this yields:

$$F_{Tb} = Gb^2 \frac{(1 - \nu\cos^2\theta)}{4\pi(1 - \nu)} [\ln(\frac{\alpha h_{cp}}{b}) + \ln(\frac{\alpha(h_{cp} + h_{ep})}{b}) - 2\ln(\frac{\alpha(h_{cp} + h_{ep})}{b}) + 2\ln(\frac{\alpha h_{ep}}{b})] \quad (4)$$

The last pair of logarithmic terms in this equation represents the interaction force between the two interfacial dislocations [1]. For the limit $h_{cp} >> h_{ep}$, the first three terms cancel and the entire logarithmic expression simplifies to

$$\sim 2\ln\left(\frac{\alpha h_{ep}}{b}\right)$$

The line tension term for the buried layer, F_{Tb}, is then a factor of 2x higher than for the single layer, F_T, and the critical thickness of $h_c >> b$ is also approximately a factor of two higher for the uncapped vs. capped structure. In fact this approximation is reasonably accurate for all $h_{cp} > h_{ep}$ if $h_{ep} >> b$. For a detailed description of critical thickness calculations for buried strained layers see [6].

Extension of critical thickness concepts to multilayer structures is highly complex due to the greater degrees of freedom involved (individual layer strains and dimensions, total number of layers, total multilayer thickness etc.). Experimentally, it is generally observed that providing each of the individual layers are thinner than the critical thickness for that particular layer grown directly onto the substrate, then the great majority of the strain relaxation occurs via a misfit dislocation network at the interface between the substrate and the first strained multilayer constituent. (If this condition is not met, then individual interfaces within the superlattice will relax.) Thus, in general the relaxation may be regarded as occurring primarily between the substrate and the multilayer structure as a whole. A simple energetic model has been developed [35], based upon reduction of the multilayer to an equivalent single strained layer. In this model, a multilaycr structure consisting of n periods of bilayers A and B, of thickness d_A and d_B, compound elastic constants k_A and k_B, and lattice mismatch strains with respect to the substrate of ε_A and ε_B respectively is considered. The amount of strain energy which may be relaxed by a misfit dislocation array at the substrate/superlattice interface is found to be:

$$E_{rel} = \frac{n(k_A d_A \varepsilon_A + k_B d_B \varepsilon_B)^2}{d_A k_A + d_B k_B} \quad (5)$$

The form of this relaxable energy is equivalent to the average strain energy in the superlattice, weighted over the appropriate elastic constants and layer thicknesses. For the Ge_xSi_{1-x}/Si(100) system, if we assume $k_A \sim k_B$ and that Vegard's Law applies (i.e. that the strain of Ge_xSi_{1-x} with respect to Si varies linearly with x), then the relaxable strain energy simplifies to that in an equivalent uniform layer of the same total thickness as the superlattice = $n(d_{GS} + d_{Si})$, and the average superlattice composition,

$$x_{av} = \frac{x d_{GS}}{d_{GS} + d_{Si}}$$

E CONCLUSION

In this Datareview we have summarised the geometry and microstructure of misfit dislocations in Ge_xSi_{1-x}/Si-based heterostructures, and reviewed theoretical and experimental understanding of the critical thickness for misfit dislocation introduction. Classic rules for dislocation glide and dissociation generally can be applied to predict the operative geometry and type of misfit dislocations. Although classic a/2<101> misfit dislocations dominate at the most widely studied Ge_xSi_{1-x}/Si(100) interface, a/6<211> misfit dislocations may exist when either the sign of the strain is reversed (i.e. Ge_xSi_{1-x}/Ge(100) structures) or the interface orientation is changed (e.g. (011) or (111)). Adequate theoretical frameworks exist for predicting the equilibrium critical thickness (at least in the limit $h_c \gg b$) for dislocation introduction, although there are continuing refinements of the original Matthews-Blakeslee and Van der Merwe models. Experimental measurements of h_c are strongly influenced by finite dislocation nucleation and propagation kinetics, and by finite experimental sensitivity to low dislocation densities. Nevertheless, in the limit where either high growth temperatures and/or highly sensitive experimental techniques are applied, a reasonable reconciliation between experiment and theory is obtained.

REFERENCES

[1] J.P. Hirth, J. Lothe [*Theory of Dislocations* (McGraw-Hill, New York, 1968)]

[2] R. Hull, J.C. Bean, L.J. Peticolas, Y.H. Xie, Y.F. Hsieh [*Mater. Res. Soc. Symp. Proc. (USA)* vol.220 (1991) p.153]

[3] M. Albrecht, H.P. Strunk, R. Hull, J.M. Bonar [*Appl. Phys. Lett. (USA)* vol.62 (1993) p.2206]

[4] G.A. Rozgonyi et al [*J. Cryst. Growth (Netherlands)* vol.85 (1987) p.300]

[5] M.E. Twigg [*J. Appl. Phys. (USA)* vol.68 (1990) p.5109]

[6] T.J. Gosling, R. Bullough, S.C. Jain, J.R. Willis [*J. Appl. Phys. (USA)* vol.73 (1993) p.8267]

[7] N. Thompson [*Proc. Phys. Soc. London B (UK)* vol.66 (1953) p.481]

[8] W. Wegscheider, K. Eberl, U. Menczigar, G. Abstreiter [*Appl. Phys. Lett. (USA)* vol.57 (1990) p.875]

[9] R. Hull, J.C. Bean, J.M. Bonar, L.J. Peticolas [in *Microscopy of Semiconducting Materials* vol.117 (Institute of Physics, Bristol, England, 1991) p.497]

[10] R. Hull, J.C. Bean, L.J. Peticolas, D. Bahnck [*Appl. Phys. Lett. (USA)* vol.59 (1991) p.964]

[11] E.P. Kvam, R. Hull [*J. Appl. Phys. (USA)* vol.73 (1993) p.7407]

[12] R. Hull, J.C. Bean, L.J. Peticolas, D. Bahnck, B.E. Weir, L.C. Feldman [*Appl. Phys. Lett. (USA)* vol.61 (1993) p.2802]

[13] F.C. Frank, J.H. Van der Merwe [*Proc. R. Soc. Lond. A (UK)* vol.198 (1949) p.205 and p.216; vol.200 (1949) p.125]

[14] J.H. Van der Merwe [*J. Appl. Phys. (USA)* vol.34 (1963) p.117]

[15] J.H. Van der Merwe, C.A.B. Ball [in *Epitaxial Growth Part B* Ed. J.W. Matthews (Academic, New York, 1975) p.493-528]

[16] W.A. Jesser, J.H. Van der Merwe [*J. Appl. Phys. (USA)* vol.63 (1988) p.1928]

[17] J.W. Matthews, A.E. Blakeslee [*J. Cryst. Growth (Netherlands)* vol.27 (1974) p.118; vol.29 (1975) p.273; vol.32 (1976) p.265]; J.W. Matthews [*J. Vac. Sci. Technol. (USA)* vol.12 (1975) p.126 and references contained therein]. Note that some of these papers contain inconsistencies in the magnitude and form of angular factors in Eqns (2) and (3) of this Datareview. The formulae and values we have quoted in this Datareview are the generally accepted values.

[18] A. Maradudin [*J. Phys. Chem. Solids (UK)* vol.9 (1959) p.1]

[19] M. Heggie, R. Jones [*Inst. Phys. Conf. Ser. (UK)* vol.87 (1987) p.367]

[20] A.S. Nandedkar, J. Narayan [*Philos. Mag. A (UK)* vol.61 (1990) p.873]

[21] R. Jones, A. Umerski, P. Sitch, M.I. Heggie, S. Oberg [*Phys. Status Solidi A (Germany)* vol.138 (1993) p.369]

[22] D.D. Perovic, D.C. Houghton [*Mater. Res. Soc. Symp. Proc. (USA)* vol.263 (1992) p.391]

[23] J.C. Bean, L.C. Feldman, A.T. Fiory, S. Nakahara, I.K. Robinson [*J. Vac. Sci. Technol. A (USA)* vol.2 (1984) p.436]

[24] I.J. Fritz [*Appl. Phys. Lett. (USA)* vol.51 (1987) p.1080]

[25] R. Hull [Datareview in this book: 1.3 Metastable strained layer configurations in the GeSi/Si system]

[26] E. Kasper, H.-J. Herzog, H. Kibbel [*Appl. Phys. (Germany)* vol.8 (1975) p.199]

[27] D.C. Houghton, C.J. Gibbings, C.G. Tuppen, M.H. Lyons, M.A.G. Halliwell [*Appl. Phys. Lett. (USA)* vol.56 (1990) p.460]

[28] M.L. Green et al [*J. Appl. Phys. (USA)* vol.69 (1991) p.745]

[29] J.Y. Tsao, B.W. Dodson, S.T. Picreaux, D.M. Cornelison [*Phys. Rev. Lett. (USA)* vol.59 (1987) p.2455]

[30] R. People, J.C. Bean [*Appl. Phys. Lett. (USA)* vol.47 (1985) p.322; vol.49 (1986) p.229]

[31] R.C. Cammarata, K. Sieradzki [*Appl. Phys. Lett. (USA)* vol.55 (1989) p.1197]

[32] J.R. Willis, S.C. Jain, R. Bullough [*Philos. Mag. A (UK)* vol.62 (1990) p.115]

[33] T.J. Gosling, S.C. Jain, J.R. Willis, A. Atkinson, R. Bullough [*Philos. Mag. A (UK)* vol.66 (1992) p.119]

[34] D. Chidambarrao, G.R. Srinivasan, B. Cunningham, C.S. Murthy [*Appl. Phys. Lett. (USA)* vol.57 (1990) p.1001]

[35] R. Hull, J.C. Bean, F. Cerdeira, A.T. Fiory, J.M. Gibson [*Appl. Phys. Lett. (USA)* vol.48 (1986) p.56]

1.3 Metastable strained layer configurations in the SiGe/Si system

R. Hull

May 1994

A INTRODUCTION

In this Datareview, we will summarise the kinetic factors which define metastable strained layer configurations in the Ge_xSi_{1-x}/Si system. We will consider the separate processes of misfit dislocation nucleation, propagation and interaction, and discuss how these combine to determine the overall strain relaxation kinetics by misfit dislocation introduction.

B THE CONCEPT OF METASTABILITY

The concept of activation barriers to the formation of an interfacial misfit dislocation array has long been recognised [1,2]. Specifically, finite energy is generally required, typically >> kT, for both nucleation and propagation of misfit dislocations. In the later stages of strain relief, where dislocation densities are relatively high, dislocation interactions also become increasingly important.

In FIGURE 1, we show a summary of the prediction of equilibrium modelling and experimental measurement of critical thickness, h_c, in the Ge_xSi_{1-x}/Si(100) system. As discussed in Datareview 1.2 in this volume [3], the apparent discrepancy between theory and

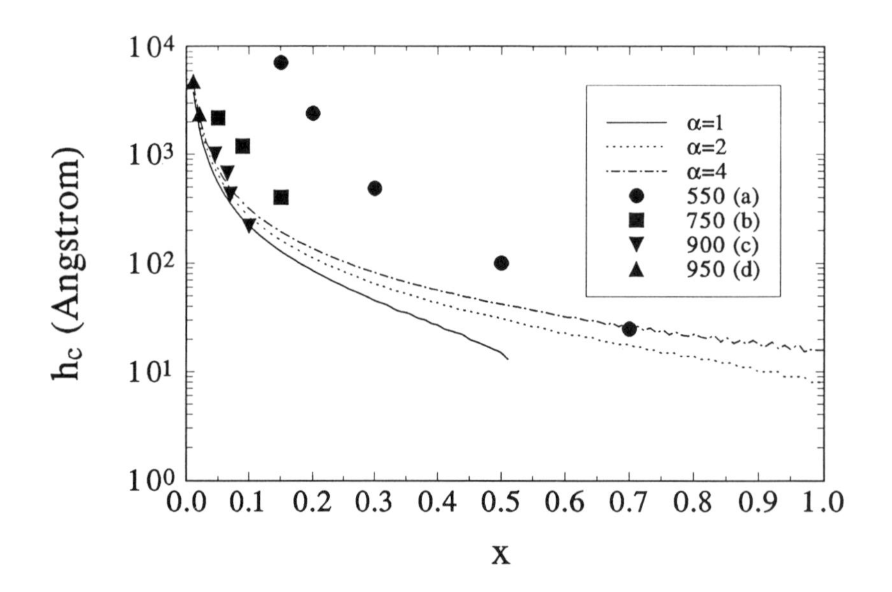

FIGURE 1 Predictions of the MB theory for the critical thickness, h_c, in Ge_xSi_{1-x}/Si(100) structures for different values of the core energy parameter, α. Also shown are experimental measurements of h_c for different growth/annealing temperatures from the work of (a) Bean et al [42], (b) Kasper et al [43], (c) Green et al [49], (d) Houghton et al [50].

experimental measurements from low temperature growth arises from the sluggish relaxation kinetics, providing an initial 'incubation time' during which a detectable misfit dislocation density is built up. The regime between equilibrium theory and experimental measurement of h_c is often termed the 'metastable' regime. The extent of this metastable regime will depend on several factors and will generally increase with:

(i) Decreasing growth temperature.

(ii) Increasing growth rate (as this will ensure less time at temperature).

(iii) Decreasing strain: misfit dislocation nucleation and propagation rates generally increase with increasing strain, as will be discussed in later sections of this Datareview.

(iv) Interface orientation, because of the angular factors resolving the lattice mismatch stress onto the misfit dislocation Burgers vector (see Section C). For isotropic elasticity, this factor will increase the metastable regime for different surfaces in the order (011) to (100) to (111) (highest).

(v) Increasing Peierls barrier [4] (essentially the inter-atomic barrier height to dislocation motion). Of common semiconductors, Si has one of the highest Peierls barriers. Thus Ge_xSi_{1-x} alloys with low Ge concentration, x, will have high resistance to dislocation motion. The Peierls barrier is significantly lower for Ge than Si. Thus high x Ge_xSi_{1-x} alloys will relax more rapidly than lower x Ge_xSi_{1-x} alloys, even for equivalent lattice mismatches and epilayer thicknesses (e.g. $Ge_{0.8}Si_{0.2}$/Ge structures will have a smaller metastable window than equivalent thicknesses of $Ge_{0.2}Si_{0.8}$/Si structures).

(vi) The sign of the stress. As discussed in Datareview 1.2 in this volume [3], changing the sign of the stress raises the possibility of changing the dislocation microstructure, e.g. for Ge_xSi_{1-x} alloys grown in the (100) orientation, a/2<101> Burgers vector misfit dislocations will dominate for growth on Si, whilst a/6<211> Burgers vector dislocations may dominate for growth on Ge. In general the a/6<211> misfit dislocations experience a greater net stress than a/2<101> dislocations, so structures in which misfit dislocations are of the former type will have a smaller metastable window.

C CONCEPT OF EXCESS STRESS

A crucial concept for understanding and modelling strain relaxation kinetics, that of 'excess stress', σ_{ex}, was introduced by Dodson and Tsao [5]. The excess stress acting on a perfect dislocation is defined (by analogy to the Matthews-Blakeslee force balance model [6] for the critical thickness, discussed in Datareview 1.2 in this volume [3]), as:

$$\sigma_{ex} = \sigma_a - \sigma_T \tag{1a}$$

$$= 2\,G\varepsilon\cos\lambda\cos\phi\,\frac{(1+\nu)}{(1-\nu)} - Gb\cos\phi\,\frac{(1-\nu\cos^2\theta)}{4\pi h(1-\nu)}\ln\frac{\alpha h}{b} \tag{1b}$$

The component stresses, σ_a and σ_T, are derived directly from the Matthews-Blakeslee forces F_σ and F_T [3] by dividing these forces by hbcosecϕ. The stress σ_a arises from the lattice mismatch strain driving misfit dislocation formation and the stress σ_T is a restoring stress arising from the self energy of the misfit dislocation created. In these equations h is the epilayer thickness, G is the epilayer shear modulus, ν is the epilayer Poisson ratio, ε is the lattice mismatch strain between epitaxial layer and substrate, b is the magnitude of the misfit dislocation Burgers vector, λ is the angle between the misfit dislocation Burgers vector and a line in the interface drawn perpendicular to the dislocation line direction, θ is the angle between the misfit dislocation line direction and its Burgers vector, ϕ is the angle between the interface normal and the glide plane and α is a factor which describes the energy of the dislocation core, where linear elasticity theory does not apply.

For epilayer thicknesses less than the critical thickness, $h < h_c$, the excess stress is negative, and misfit dislocation formation is not energetically favoured. At $h = h_c$, $\sigma_{ex} = 0$. For increasing $h > h_c$, the excess stress becomes increasingly positive and misfit dislocation formation becomes increasingly energetically favoured. The beauty of the concept of the excess stress is that it defines a net stress acting on the dislocation, and it serves as a quantitative measure for how much a strained layer wants to dislocate. The excess stress will also be expected to appear quantitatively in expressions for misfit dislocation nucleation, propagation and interactions (see later sections of this Datareview).

D NUCLEATION OF MISFIT DISLOCATIONS

The mechanism for nucleation of misfit dislocations in the Ge_xSi_{1-x}/Si system remains uncertain and controversial. Existing experimental data, e.g. [7-9], suggests that very high densities (of the order 10^6 - 10^8 cm^{-2}) of sources of misfit dislocations are required to describe the observed relaxation behaviour. There are three generic candidates for misfit dislocation nucleation:

(i) 'Homogeneous' nucleation, in which the intrinsic strain in the epitaxial layer is sufficiently high to allow a finite rate of dislocation loop nucleation, or half-loop nucleation at a free surface. The intrinsic problem with this mechanism is that high epilayer strains are required to produce a significant nucleation rate.

The calculation of activation energies for homogeneous dislocation loop nucleation goes back several decades (see e.g. [10]). For the case of a strained epitaxial heterostructure, the activation barrier essentially arises from the sum of the self energy of the dislocation loop, E_T, and the strain energy relaxed by it, E_σ. Thus, for nucleation of a dislocation half-loop at a free surface, the simplest expression for the total loop energy, E(R), as a function of loop radius, R, is given by:

$$E(R) = E_T - E_\sigma \tag{2a}$$

where

$$E_\sigma = \frac{\pi R^2 Gb\cos\phi\cos\lambda(1+\nu)\varepsilon}{(1-\nu)} \tag{2b}$$

An expression for the dislocation self energy has been developed by Bacon and Crocker [11]:

$$E_T = \frac{Gb^2R}{4(1-\nu)}\left(1-\frac{\nu}{2}\right)\left[\ln\left(\frac{2\pi R\alpha}{b}\right) - 1.758\right] - \frac{Gb^2R(1-2\nu)}{32(1-\nu)^2} \qquad (2c)$$

The form of the total energy E(R) as a function of R is to initially increase from zero, then pass through a maximum energy, ΔE, at a critical radius, R_c, and next to rapidly decrease (eventually becoming negative, at which point the dislocation is energetically favoured in the structure; see the discussion on critical thickness in [3]). The quantity ΔE may be regarded as an activation barrier for homogeneous nucleation.

There are many refinements which can be incorporated into Eqn (2a), including surface step energies created or removed by the nucleation event, entropic contributions, alloy effects and stacking fault energies if the dislocation is nucleated either in a dissociated form or as a separate partial dislocation. Another significant uncertainty in these calculations is the value of the dislocation core energy parameter, α. For detailed discussions of these parameters in Ge_xSi_{1-x}-based heterostructures see [12-15].

An example of a homogeneous surface half-loop calculation, using materials parameters appropriate to the Ge_xSi_{1-x}/Si(100) system, a dislocation core parameter of $\alpha = 1$, and assuming the nucleation event removes a surface step (following [2], the surface step energy is approximated by $Rb^2\sin\beta G/4$, where β is the angle between the Burgers vector and the free surface), is shown in FIGURE 2.

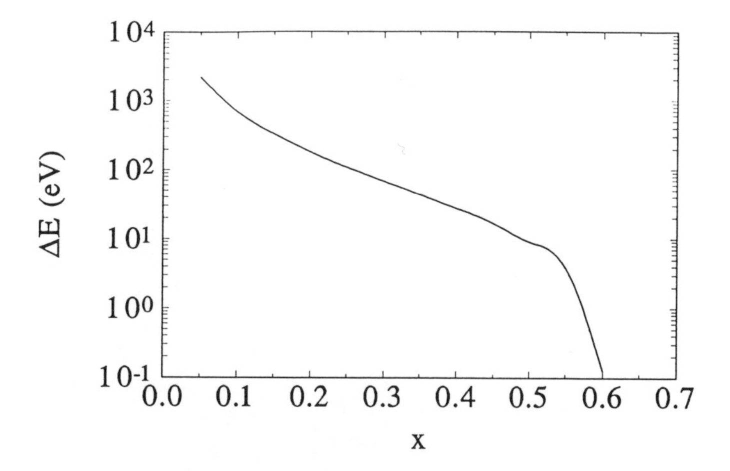

FIGURE 2 Calculation of the activation barrier for surface half-loop nucleation of a dislocation in coherently strained Ge_xSi_{1-x}/Si(100) heterostructures, assuming that the dislocation annihilates a surface step during the nucleation process. Materials parameters are: G = 64 GPa, $\nu = 0.28$, b = 3.9 Å, $\varepsilon = 0.041x$, $\alpha = 1$.

The uncertainties in the homogeneous nucleation calculation result in quantitatively different calculation results from different groups. However, qualitatively all calculations agree that there is a minimum strain above which homogeneous nucleation looks physically plausible and below which it is not plausible. This minimum strain is calculated on the basis of a prefactor and activation barrier which produce a reasonable nucleation rate. For example, if we assume the prefactor is given by the product of an attempt frequency (which we approximate by the Debye frequency, $\sim 10^{13}$ s^{-1} in Si) and the surface density of atoms ($\sim 10^{15}$ cm^{-2}), we obtain a prefactor of 10^{28} cm^{-2} s^{-1}. A total nucleation rate of 10^2 cm^{-2} s^{-1} then translates into an activation barrier of ~60 kT, or about 5 eV at 600°C. The strain that this activation barrier corresponds to has been calculated by different groups [12-14] to be in the range 0.01 - 0.06 (cf. the value of ~0.02 at x ~ 0.5 in FIGURE 2).

(ii) 'Heterogeneous' nucleation: if the epilayer strain is too small to allow a significant homogeneous nucleation rate, then heterogeneous nucleation mechanisms must be considered. In this context, 'heterogeneous' refers to a local site in the crystal where the local strain is significantly higher than the matrix, and dislocation nucleation probability is correspondingly higher. Such heterogeneous sites generally correspond to crystal growth defects such as transition metal precipitation, surface or interface contamination, incomplete substrate cleaning, particulate inclusion in the film, etc. Examples of such heterogeneous sites have been described, for example, by Eaglesham et al [12], and an illustrative example (a polycrystalline Si inclusion arising from flaking from the walls of the MBE chamber) is shown in FIGURE 3. The nature of the predominant heterogeneous source may vary between different materials systems, different growth conditions and even different growth techniques and chambers. This makes quantitative comparison between different groups difficult. Nevertheless, a common theme will be that for high quality crystal growth the density of available heterogeneous sites will be relatively sparse (say in the range 10^1 - 10^3 cm^{-2}). Depending on the local stress around the defect, however, they may be able to operate at arbitrarily low epilayer strains and thus may be the dominant nucleation mechanism in low strain structures.

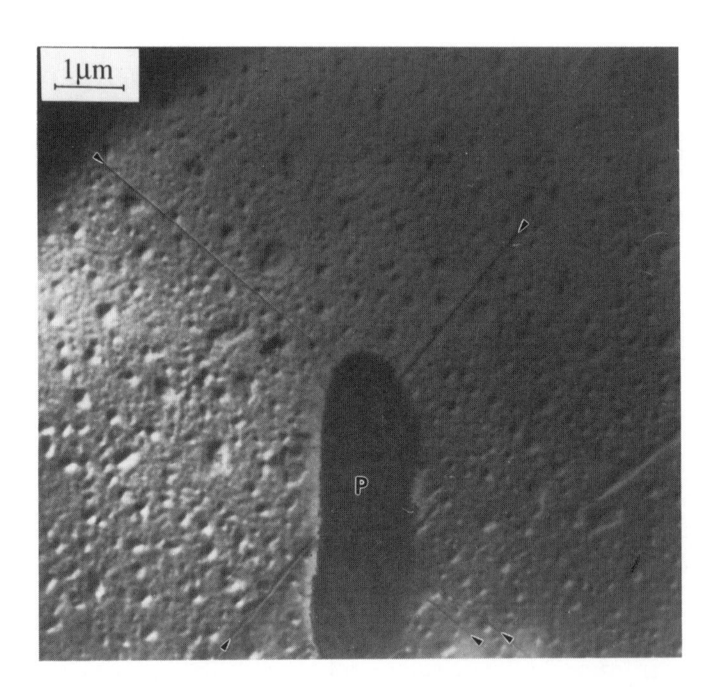

FIGURE 3 Example of a heterogeneous dislocation source in a 2000 Å $Ge_{0.15}Si_{0.85}$/Si(100) heterostructure. The feature labelled P is an inclusion of polycrystalline Si which arises from flaking from the walls of the MBE growth chamber. The typical density of these inclusions in our material is ~10^3 cm^{-2}. Arrows show misfit dislocations emanating from the inclusion.

The boundary between 'homogeneous' and 'heterogeneous' source mechanisms can become indistinct, and to some extent no nucleation event will be truly homogeneous: loop nucleation will always be favoured at some non-periodic event, such as a surface step. Other events which may enhance apparent homogeneous nucleation are formation of surface cusps [16-18] and statistical variations in alloy concentration [14].

(iii) Multiplication events: there is growing evidence that the dominant dislocation nucleation mechanism at relatively low epilayer strain and high epilayer thickness is dislocation multiplication. The concept of dislocation nucleation goes back to the Frank-Read source (see chapter 20 in [10] for a review), and the first documented source in strained layer epitaxy of which we are aware is the Hagen-Strunk [19] source in the Ge/GaAs system (this source has also been invoked in Ge_xSi_{1-x}/Si structures [20]).

In recent years there have been several reports of dislocation multiplication in Ge_xSi_{1-x}/Si structures (e.g. [21-23]), as well as in other strained layer systems (e.g. the extensive work of Lefebvre et al [24] in the InGaAs/GaAs system). All these mechanisms involve interaction between dislocations with orthogonal line directions in the (100) interface. The resultant dislocation intersections can evolve into an unstable configuration which can form a regenerative multiplication source. Typically such sources require relatively thick epilayers, as the intermediate unstable loop configurations involved require relatively large radii of curvature. The requirement for high epitaxial layer thicknesses in which relatively little strain has been relaxed by other mechanisms therefore translates into relatively low epilayer strains.

Although the evidence for dislocation multiplication in Ge_xSi_{1-x} heteroepitaxy is becoming increasingly compelling, it should be emphasised that several groups (including ours) have not observed systematic evidence of such sources, probably because of the epilayer thickness/strain regime in which they are working. Also, dislocation intersections themselves require precursor dislocations for their formation; these precursor dislocations presumably must initially form from heterogeneous sources.

In summary of the mechanistics of dislocation nucleation in Ge_xSi_{1-x}/Si epitaxy, homogeneous or quasi-homogeneous (surface steps or cusps, alloy clustering etc.) mechanisms are expected to dominate at higher strains (say greater than ~0.02). At lower strains, dislocation multiplication sources may become efficient at high epitaxial layer thicknesses (of order several thousand Å or greater). In other regimes (predominantly the low thickness, low strain regime), only heterogeneous sources may be available and strain relaxation by misfit dislocations is then strongly nucleation limited.

Experimentally, there remains a paucity of nucleation data in the Ge_xSi_{1-x}/Si (or any other strained layer) system. The most complete data is that of Houghton [8], as illustrated in FIGURE 4. This data was obtained from careful etching and optical microscopy experiments. Houghton interpreted this data on the basis of a thermally activated process:

$$N = BN_o \left(\frac{\sigma_{ex}}{G}\right)^p e^{-E_n/kT} \qquad (3)$$

In this equation, B_o is a constant of the order 10^{18} cm^{-2} s^{-1} and p was found to be 2.5. The activation energy, E_n, was found to be of the order 2.5 eV for $0.0 < x < 0.3$ in

(Si)/Ge_xSi_{1-x}/Si(100). The parameter N_o was described as the density of pre-existing heterogeneous nucleation sites. In Houghton's data, the quoted values of N_o also generally increased with Ge concentration x (and hence strain and stress). Limited data from in-situ TEM observations of misfit dislocation nucleation in the work of Hull et al [7,25] are reasonably numerically consistent with those of Houghton in the relatively narrow temperature and Ge concentration range where the data overlap, although lower activation energies in the range 0.3 - 1.0 eV were inferred. The differences in activation energy between the two sets of data may arise from different sample thickness or strain relaxation regimes. Also, Perovic and Houghton [15] have recently re-interpreted the original Houghton data in terms of a 'barrierless' activation process and interpreted the value of $E_n \sim 2.5$ eV as essentially due to the glide activation energy for dislocation motion (~2.2 eV in Si, as described in Section E). The implication is that E_n then corresponds to the activation barrier associated with dislocation growth, which defines the incubation times until the defect becomes optically visible.

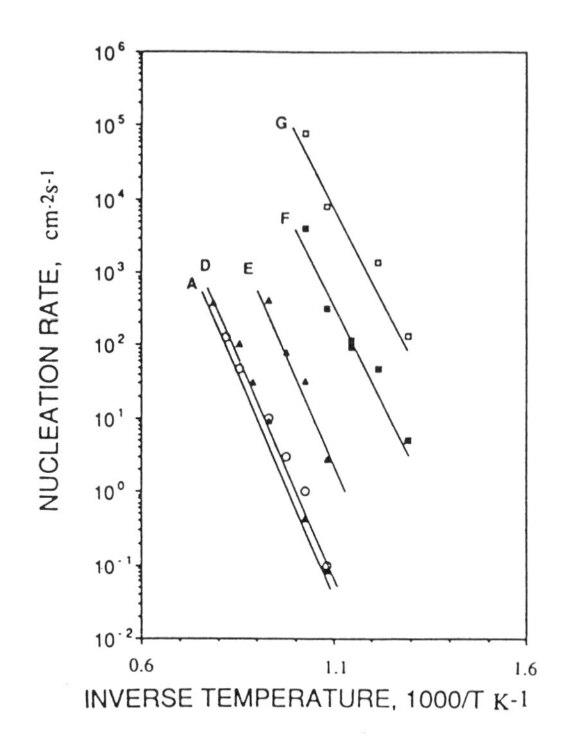

FIGURE 4 Measured misfit dislocation nucleation rates during annealing of Ge_xSi_{1-x}/Si(100) heterostructures from the work of Houghton [8]. Key to labels: A = 20 period superlattice of 32 nm Si/10 nm $Ge_{0.10}Si_{0.90}$, D = 500 nm $Ge_{0.035}Si_{0.965}$, E = 3000 nm $Ge_{0.035}Si_{0.965}$, F = 190 nm $Ge_{0.17}Si_{0.83}$, G = 100 nm $Ge_{0.23}Si_{0.77}$. Reproduced with permission from D.C. Houghton [Appl. Phys. Lett. (USA) vol.57 (1990) p.2125].

The only other quantification of misfit dislocation nucleation in the Ge_xSi_{1-x}/Si system of which we are aware is the recent work of LeGoues et al [9] in which the Frank-Read-like multiplication mechanism proposed by this group was studied. They determined an activation energy of ~5 eV and a total dislocation nucleation rate of 6×10^5 cm^{-3} s^{-1}.

The availability of more quantitative data on misfit dislocation nucleation mechanisms would certainly help improve our understanding of the regimes where the generic mechanisms of homogeneous, heterogeneous and multiplication nucleation occur. Such studies should be an experimental priority.

E MISFIT DISLOCATION PROPAGATION

The mechanisms of misfit dislocation propagation in Ge_xSi_{1-x}/Si structures are relatively well understood, both experimentally and mechanistically. In addition, there is a copious literature on dislocation glide from deformation experiments on bulk Si and Ge (but unfortunately not Ge_xSi_{1-x}) crystals, e.g. [26-30]. These bulk experiments suggest that dislocation glide can be described by the following relation:

$$v = v_o \, \sigma^m e^{-E_v(\sigma)/kT} \qquad (4)$$

In the low stress (10 - 100 MPa) regime for intrinsic, pure crystals, $E_v \sim 1.6$ eV in Ge and 2.2 eV in Si. The prefactors are very similar for the two materials ($v_o \sim 3 \times 10^{-3}$ m^2 kg^{-1} s) and thus dislocations glide much faster in Ge than in Si (by a factor ~5000 at 550°C). Thus we should expect that for Ge_xSi_{1-x} alloys the glide activation energy will decrease, and the glide velocity will increase, with increasing x. The stress dependence of the dislocation velocity is still somewhat uncertain in bulk Si and Ge crystals; undoubtedly part of this uncertainty depends upon how the stress dependence is apportioned between the prefactor and the exponential in Eqn (4). An extensive set of measurements by Imai and Sumino [27] has yielded a linear stress dependence, m = 1.0 in Eqn (4), at stresses of the order tens of MPa, and this is also the dependence predicted by the generally accepted microscopic model of dislocation motion, the Hirth-Lothe diffusive double kink model [10]. There is also evidence from bulk measurements that the activation energy E_v is stress dependent for σ of the order a few hundred MPa or more [29,31]. This is again consistent with the diffusive double kink model which predicts $E_v \sim \sqrt{\sigma}$ [10,32].

In FIGURE 5, we plot measured dislocation propagation velocities vs. excess stress, σ_{ex}, in (Si)/Ge_xSi_{1-x}/Si(100) structures from the data of several groups. The velocities are displayed for a temperature of 550°C, interpolated where necessary from measurements at other temperatures. The experimental techniques for measuring this data are either in-situ TEM observations [33,34] or chemical etching and optical microscopy [8,35,36]. The data between different groups and different techniques agree relatively well. Note that vertical scatter of data in this plot is not necessarily due to experimental error, as there are factors other than just excess stress (primarily Ge concentration in the Ge_xSi_{1-x} alloy) which determine dislocation velocity.

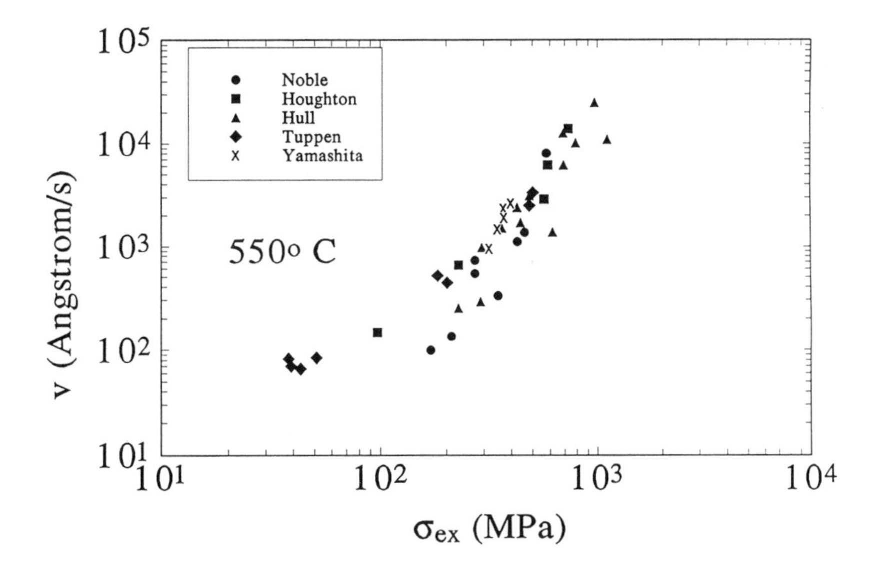

FIGURE 5 Measured dislocation propagation velocities in (Si)/Ge_xSi_{1-x}/Si(100) heterostructures vs. excess stress, σ_{ex}, at 550°C from the work of several groups: Houghton [8]; Hull et al [33]; Tuppen and Gibbings [35]; Nix, Noble and Turlo [34]; and Yamashita et al [36].

In FIGURE 6, we show some of our experimental data [33] for misfit dislocation velocities in a range of Ge_xSi_{1-x}/Si(100) and Si/Ge_xSi_{1-x}/Si(100) heterostructures. The data from different structures can be effectively normalised to each other by plotting the logarithm of the quantity v* vs. inverse temperature, where:

$$v^* = \frac{v_m\ e^{-0.6x/kT}}{\sigma_{ex}} \tag{5}$$

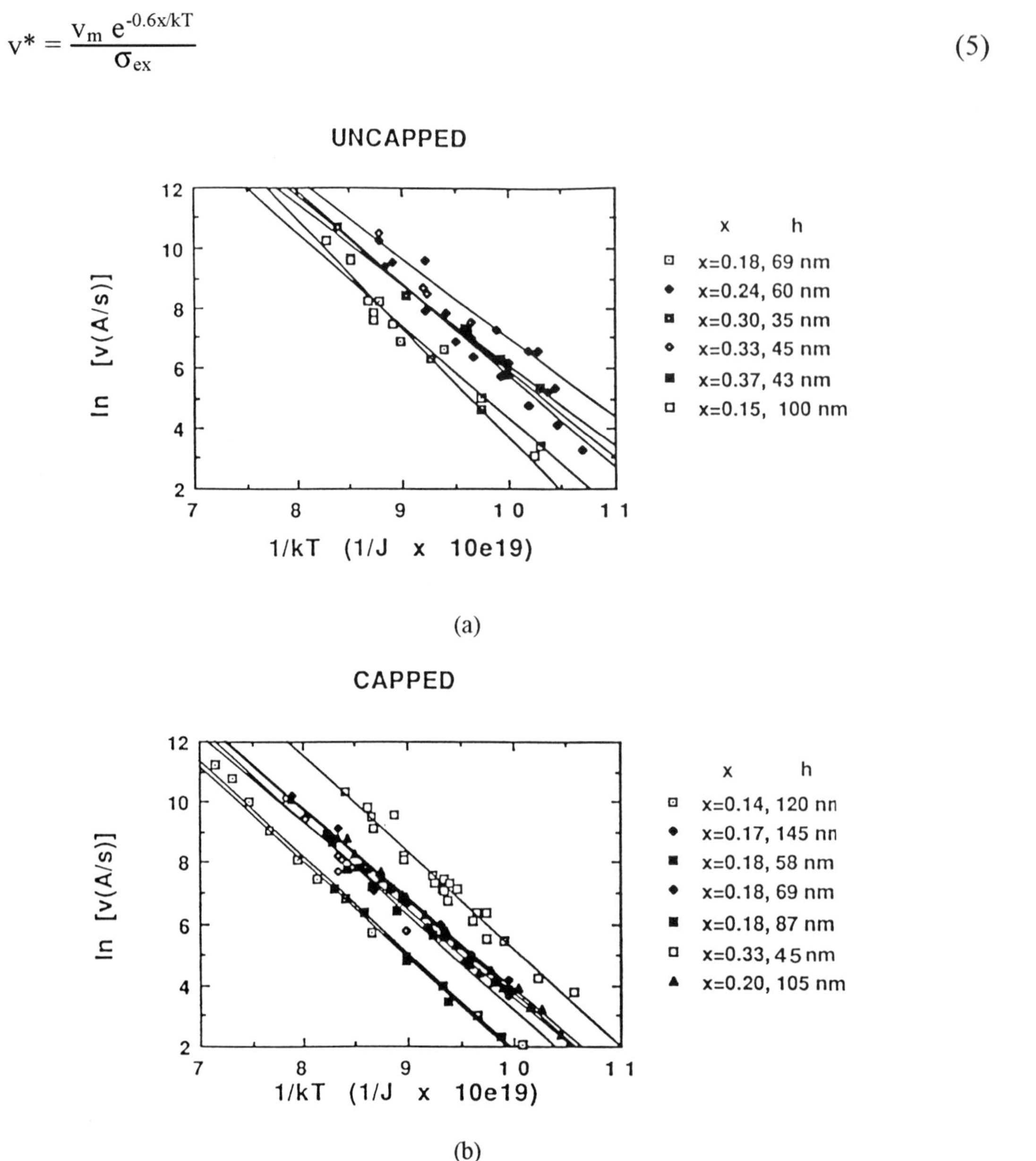

FIGURE 6 Measured dislocation velocities from the work of Hull et al [33] (a) h nm Ge_xSi_{1-x}/Si(100) and (b) 300 nm Si/h nm Ge_xSi_{1-x}/Si(100) structures.

In Eqn (5), v_m is the measured dislocation velocity and the quantity $e^{-0.6x/kT}$ accounts for the 0.6 eV activation energy difference glide between Ge and Si, i.e. we are assuming a linear interpolation of the activation energies for Ge_xSi_{1-x} alloys between the values for Si and Ge ($E_v(x) = 2.2 - 0.6x$ eV). Eqn (5) therefore effectively normalises the measured dislocation velocity to an equivalent velocity at an excess stress of 1 Pa in pure Si. In FIGURE 7, we plot the data of FIGURE 6, normalised according to Eqn (5). Note that there is a systematic difference in normalised velocity for capped vs. uncapped Ge_xSi_{1-x} layers. We have ascribed this difference to different microscopic mechanisms for dislocation motion in these two geometries [33].

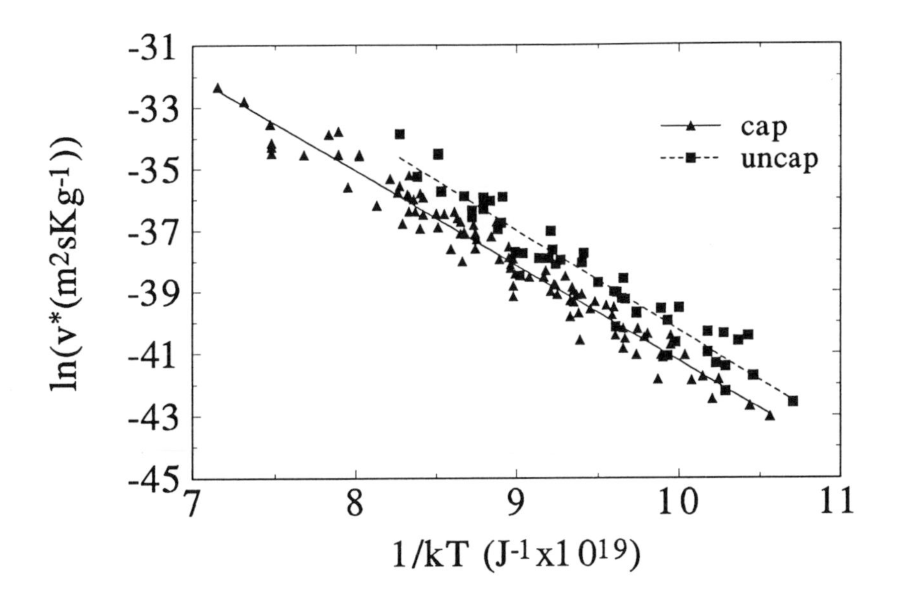

FIGURE 7 Normalised dislocation velocities for uncapped and capped (300 nm Si) Ge_xSi_{1-x}/Si(100) structures. The quantity v* is obtained by multiplying the measured dislocation velocities, v, in FIGURE 6 by $\frac{e^{-(0.6x/kT)}}{\sigma_{ex}}$.

The regression line fits to the data in FIGURE 7 are:

$$v^* = e^{-(7.8\pm1.4)}\, e^{-(2.03\pm0.10\ \mathrm{eV})/kT}\ \mathrm{m^2\ s\ kg^{-1}}\ \text{(uncapped structures)} \tag{6a}$$

$$v^* = e^{-(10.4\pm1.4)}\, e^{-(1.93\pm0.10\ \mathrm{eV})/kT}\ \mathrm{m^2\ s\ kg^{-1}}\ \text{(capped structures)} \tag{6b}$$

The correlation coefficients in these equations are $R^2 = 0.93$ and 0.96 respectively. To calculate the misfit dislocation velocity at a given temperature in a given (Si)/Ge_xSi_{1-x}/Si(100) heterostructure, therefore, one simply multiplies the relevant v* above by $\sigma_{ex}e^{0.6x/kT}$. Note that the difference in activation energies between the regression lines fitted to the capped and uncapped structures is within the limits of experimental error. The absolute magnitude of the dislocation velocity in the capped structures is, however, statistically significantly smaller than for the uncapped structures.

Note that Eqns (6a) and (6b) are still largely empirical, as they ignore further microscopic details of the diffusive double kink model such as the dependence in some regimes of the velocity upon the length of the propagating dislocation line [10,33,35] and single vs. double kink dynamics [10,33]. A stress-independent activation energy is also implied by Eqns (6a) and (6b), although note that the regression calculations give activation energies ~1.9 - 2.0 eV. This is less than the 2.2 eV activation energy which would be expected by the normalisation of our data to equivalent velocities in Si. The lower value of 1.9 - 2.0 eV corresponds effectively to an average 0.2 - 0.3 eV reduction in activation energy over all structures due to the stress dependence. To include these details into a simple analytical expression is not practical, and thus we believe that Eqns (6a) and (6b) offer a reasonably accurate semi-empirical approach to predicting misfit dislocation velocities.

There have been two other attempts to systematically fit measured dislocation propagation velocities to an empirical model. Houghton [8] proposed a description of the velocity which incorporated a power m = 2 in the pre-factor dependence upon the excess stress in Eqn (4), and a constant activation energy of $E_v = 2.3$ eV for $0.0 < x < 0.3$ in (Si)/Ge_xSi_{1-x}/Si(100) heterostructures. The higher pre-exponential factor may be accounted for by the assumption of the constant activation energy: in general, as discussed above, higher Ge concentration structures (which will also in general be structures with higher excess stresses) will have enhanced velocities due to the lower dislocation glide activation energy in Ge than Si. Assuming a constant value for E_v will therefore over-estimate the pre-exponential stress dependence of dislocation velocities in higher Ge concentration films, tending to an artificially high value of m. Tuppen and Gibbings [35] studied misfit dislocation velocities in (Si)/Ge_xSi_{1-x}/Si(100) with lower Ge concentrations (typically $x < 0.15$). They observed a linear pre-exponential dependence of velocity upon excess stress, m = 1 in Eqn (4), and a prefactor consistent with bulk Si and Ge measurements. They measured an activation energy, $E_v(x) = 2.156 - 0.7x$ eV, which is somewhat lower than predicted from bulk values (perhaps due to the theoretically predicted stress dependence of the activation energy) and is linearly dependent upon Ge concentration, x. They also observed in certain structures a dependence of the dislocation velocity upon the threading dislocation length (i.e. Ge_xSi_{1-x} epilayer thickness), consistent with the predictions of the kink model [10]. A similar length dependence has been reported by Yamashita et al [36].

In FIGURE 8, we present results from extension of these studies to growth of Ge_xSi_{1-x}/Ge structures. These structures are of low enough strain ($x > 0.8$) that the dissociation of the a/2<101> dislocations into infinitely extended a/6<211> Burgers vector dislocations, which would be expected at higher strains [3], does not occur. In FIGURE 8(a), we compare comparable Ge_xSi_{1-x} structures with lattice-mismatch strains $\varepsilon \sim 0.008$ grown on Ge(100) and Si(100) structures. The measured dislocation velocities are much higher for the Ge-rich alloy grown on the Ge substrate than for the Si-rich alloy grown on the Si substrate, as expected from the lower activation energy for dislocation glide in Ge than Si. In FIGURE 8(b) we show that the data for uncapped Ge_xSi_{1-x}/Si structures in FIGURE 7 can be accurately extended down to the high x regime in Ge_xSi_{1-x}/Ge structures.

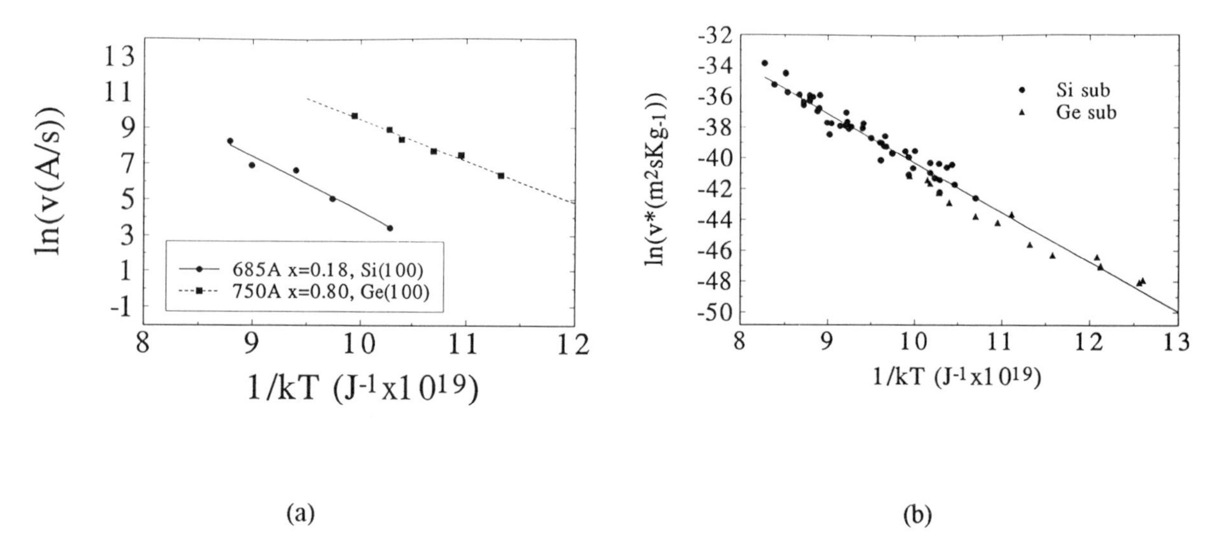

FIGURE 8 (a) Comparison of measured dislocation velocities in Ge_xSi_{1-x} layers with comparable strain and excess stress grown on Si(100) and Ge(100). (b) Normalised dislocation velocities in Ge_xSi_{1-x} layers grown on Si(100) and Ge(100) substrates.

Data have also been reported for velocities of misfit dislocations in Ge_xSi_{1-x}/Si(110) structures [37], including comparison of the velocities of a/6<211> and a/2<101> Burgers vector misfit dislocations in this configuration.

The data on dislocation glide in bulk Si and Ge also reveals a strong influence of dopant and other impurities (e.g. [27,28]). Preliminary studies of such effects in Ge_xSi_{1-x}-based heterostructures have also been made. Gibbings et al [38] studied the effect of both p- and n-type doping in the Ge_xSi_{1-x} layer upon dislocation velocities in Ge_xSi_{1-x}/Si(100) structures. Consistent with studies of bulk Si, it was found that high n-type doping ($>10^{17}$ cm^{-3}) could enhance dislocation motion, whilst p-type doping had relatively little effect. Noble et al [39] and Hull et al [40] have studied the effect of very high ($\sim10^{20}$ cm^{-3}) oxygen concentrations in the Ge_xSi_{1-x} layer, and concluded that misfit dislocation velocities were impeded and critical thicknesses for dislocation introduction increased.

F MISFIT DISLOCATION INTERACTIONS

Misfit dislocation interactions can play a critical role in the metastability of strained layer heterostructures, particularly in the later stages of strain relaxation where dislocation densities are relatively high.

The major interaction between misfit dislocations arises from the interaction between their stress fields in the surrounding crystal. The interaction force between two dislocation segments with Burgers vectors $\underline{b_1}$ and $\underline{b_2}$ varies as:

$$\underline{F_{12}} = K \frac{\underline{b_1}.\underline{b_2}}{R_{12}} \tag{7}$$

where R_{12} is the separation between the two dislocation segments. The constant of proportionality, K, in the above relation depends upon the line directions and Burgers vectors of the two dislocations [10].

The interaction force between two dislocations can pin dislocation propagation. Consider a misfit dislocation at a Ge_xSi_{1-x}/Si(100) interface propagating along the [011] direction, and about to intersect a pre-existing dislocation lying along the $[01\bar{1}]$ direction. If the two dislocations have parallel Burgers vectors, then they will repel each other, and the repulsion stress may be sufficient to overcome the Dodson-Tsao excess stress when the defects approach sufficiently close to each other. The propagating dislocation will then be pinned. Similarly, anti-parallel Burgers vectors will attract each other, and the propagating dislocation will be pulled back towards the pre-existing orthogonal defect after the intersection.

An experimental illustration of this process is shown in FIGURE 9. This mechanism has been modelled in detail by Freund [41], and results of his calculations are shown in FIGURE 10, together with some experimental results. It can be seen that these pinning events are favoured in films with low strain (where the Dodson-Tsao excess stress is relatively low) and low epitaxial layer thickness (where again σ_{ex} is low and also the interaction force is higher along the propagating misfit/threading dislocation as R_{12} is low).

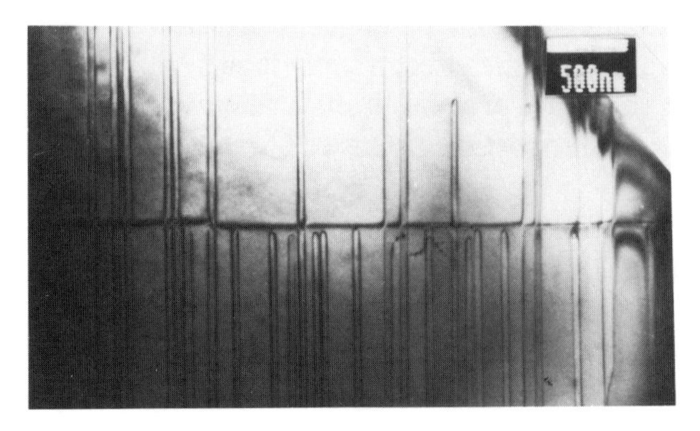

FIGURE 9 Experimental illustration of misfit dislocation blocking events in a 3000 Å Si/585 Å $Ge_{0.18}Si_{0.82}$/Si(100) structure.

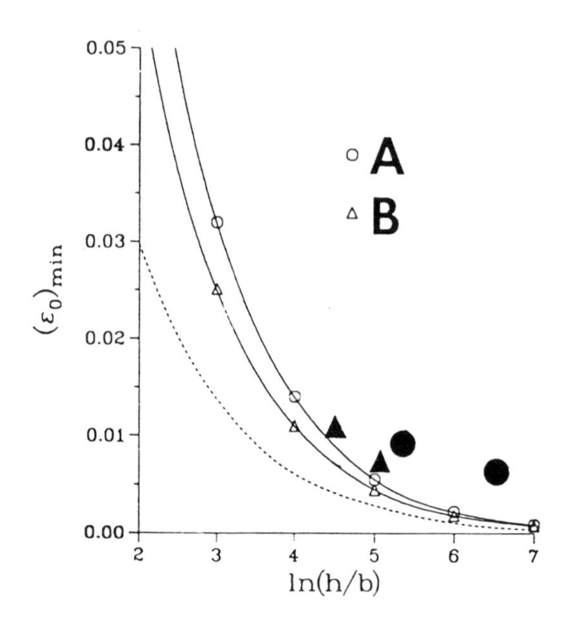

FIGURE 10 Plot of minimum strain required for passage of a misfit dislocation past a pre-existing orthogonal misfit dislocation at a (100) interface vs. epitaxial layer thickness. The solid line marked A corresponds to the case where the misfit dislocations' Burgers vectors are inclined at 60° to each other and the solid line marked B corresponds to the case where the misfit dislocations' Burgers vectors are parallel to each other. The dashed line corresponds to the minimum strain analogue to the critical thickness for misfit dislocation introduction. Reproduced with permission from L.B. Freund [41]. Also shown is experimental data from our own work for blocked (triangle) and non-blocked (circle) configurations in (Si)/Ge_xSi_{1-x}/Si(100) heterostructures. (Note that some experimental data is taken from buried Ge_xSi_{1-x} layers, whilst the calculations in this figure are for uncapped Ge_xSi_{1-x} layers. This makes exact comparison of data and experiment inaccurate, but such comparison does experimentally confirm the essential trends of the calculation.)

Even if the dislocation interaction stress is not sufficient to pin dislocations, it will generally impede motion when dislocations intersect or approach each other. Microscopic interactions of the dislocations, e.g. the formation of kinks or jogs [10] on the propagating dislocation line following intersection events, will also affect the dislocation propagation velocity [33].

G STRAIN RELAXATION MODELS

There have been several attempts to combine the preceding kinetics concepts into predictive models of strain relaxation by misfit dislocations in Ge_xSi_{1-x}-based heterostructures. The original Dodson-Tsao formulation [5] combined the concepts of excess stress, bulk parameters for dislocation propagation, and dislocation multiplication arising from a pre-existing dislocation source density, to produce a predictive equation for strain (and hence dislocation density) for finite time at temperature:

$$\frac{d\varepsilon(t)}{dt} = CG^2 (\varepsilon_o - \varepsilon(t) - \varepsilon_{eq})^2 (\varepsilon(t) + \varepsilon_s) \tag{8}$$

In this equation, $\varepsilon(t)$ is the amount of strain relieved by misfit dislocations, ε_o is the initial lattice mismatch strain, ε_s is an initial 'source' density of dislocations from which multiplication proceeds, ε_{eq} is the equilibrium strain in the structure predicted by Eqns (1a) and (1b) and C is a constant. From fitting to available experimental data, Dodson and Tsao obtained C = 30.1 and $\varepsilon_s = 10^{-4}$. Their model was successful at predicting a relatively wide range of metastable data in Ge_xSi_{1-x}-based heterostructures [5]. An example is shown in FIGURE 11, comparing the predictions of the Dodson-Tsao model with the critical thickness measurements in Ge_xSi_{1-x}/Si(100) of Bean et al at 550°C [42] and Kasper et al at 750°C [43].

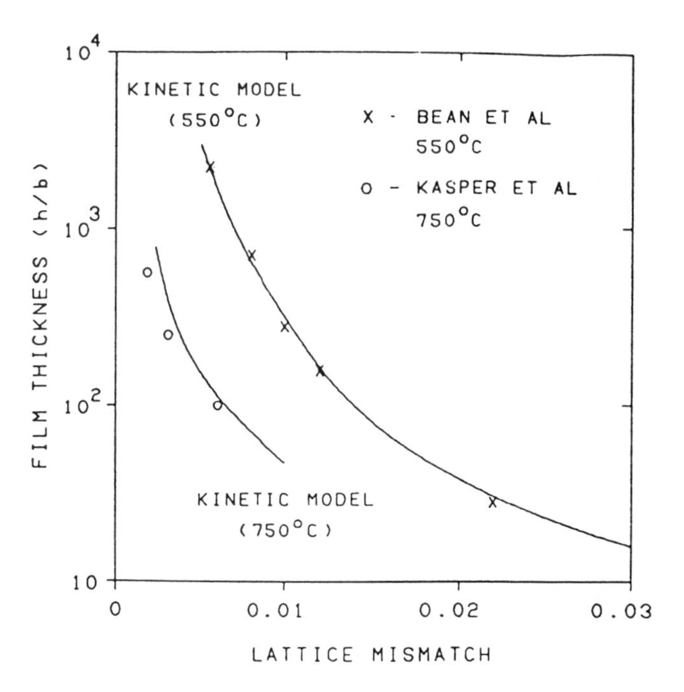

FIGURE 11 Predictions of the Dodson-Tsao model [5] for a strain relaxation of ε_o 10^{-3} (ε_o is the initial coherent lattice mismatch strain) for growth of Ge_xSi_{1-x} at 550°C and 750°C, compared with the equivalent experimental data of Bean et al [42] and Kasper et al [43]. Reproduced with permission from B.W. Dodson and J.Y. Tsao [5].

In subsequent work, Tsao et al [44] studied the kinetics of relaxation of Ge_xSi_{1-x} epilayers grown on Ge(100) substrates, and argued that the strain relaxation rate by misfit dislocation introduction was dominated by a combination of temperature and excess stress. They represented these results and concepts in terms of deformation mechanism maps, as introduced by Ashby and Frost [45]. It was observed empirically, for example, that the onset of detectable strain relaxation (using Rutherford Backscattering) occurred at a temperature-dependent excess stress (e.g. 0.024G and 0.0085G at 494°C and 568°C respectively). Dodson and Tsao subsequently developed empirical relations for strain relaxation in Ge_xSi_{1-x}-based structures, based upon scaling to temperature and excess stress [46]:

$$\frac{\gamma(\sigma_1,T_1)}{\gamma(\sigma_0,T_0)} = \left(\frac{\sigma_1}{\sigma_o}\right)^2 e^{[E(\sigma_o)/kT_o - E(\sigma_1)/kT_1]} \qquad (9)$$

Here, $\gamma(\sigma_1,T_1)$ represents the degree of relaxation for a structure with excess stress, σ_1, at a temperature T_1 during a growth or annealing cycle. The quantity $\gamma(\sigma_0,T_0)$ represents a known degree of relaxation for a structure with excess stress σ_o at a temperature T_o; this may then be compared to the unknown $\gamma(\sigma_1,T_1)$ using Eqn (9). For Ge_xSi_{1-x} alloys, Dodson and Tsao [47] found the effective activation energy, $E(\sigma)$ in Eqn (9), to be stress dependent according to $E(\sigma) = E(0)[1-\sigma/\tau]$ where $\tau = 0.1G$ and $E(0) = 16\ kT_m$, where T_m is an effective alloy melting temperature obtained by a linear weighting of the melting temperatures of the pure components.

Two other groups have attempted to model the strain relaxation process in Ge_xSi_{1-x}/Si(100) heterostructures by direct measurement of the fundamental misfit dislocation parameters. Hull et al [48] modelled the relaxation process by using in-situ TEM observations to quantify the processes describing misfit dislocation nucleation, propagation and interaction, and applying these measurements to the equation:

$$\Delta\varepsilon(t) = \frac{b\cos\lambda}{p(t)} = \frac{\pi b\cos\lambda L(t)}{4D^2} = \frac{\pi b\cos\lambda}{4D^2}\int N(t)\, v(t)\, dt \qquad (10)$$

In this equation D is the diameter of the substrate and $\Delta\varepsilon(t)$ and L(t) are the degree of strain relaxation and total interfacial misfit dislocation length at time t respectively. The quantity p(t) is the average linear separation of misfit dislocations at time t, N(t) is the number of growing dislocations at time t and v(t) is the average dislocation velocity at time t. The limits of the integral are from start to finish of either a growth or annealing cycle for $\sigma_{ex} > 0$. This model incorporated the effects of misfit dislocation interactions by reducing N(t) by the number of dislocations pinned according to the discussion in Section F, and by incorporating an empirical velocity-stress-temperature relation in a regime where dislocation interactions strongly affect propagation rates [33,48]. Examples of results from this analysis are shown in FIGURE 12.

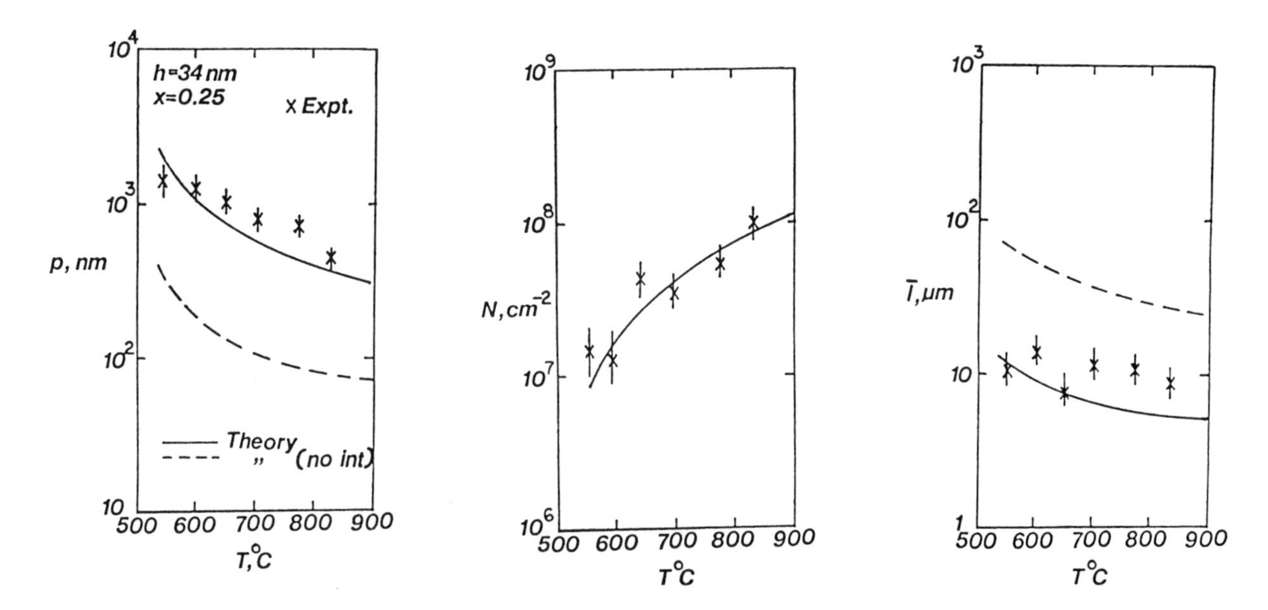

FIGURE 12 Comparison of the predictions of the model of Hull et al [48] with experimental data obtained from annealing (4 minutes at each successive temperature for which an experimental data point is given) of a 34 nm $Ge_{0.25}Si_{0.75}$/Si(100) structure. Experimental data points are given for p, the average distance between misfit dislocations, N, the areal density of misfit dislocations and $\bar{l}$, the average length of misfit dislocations. The solid lines show the model predictions. The dashed lines show the model predictions in the absence of dislocation interactions, establishing the importance of this phenomenon.

Houghton [8] applied Eqn (10) to the initial stages of relaxation, where misfit dislocation densities are low and dislocation interactions relatively unimportant. In this case Eqn (10) simplifies to the product of individual empirical expressions for dislocation nucleation and propagation, which in Houghton's analysis is represented by:

$$\frac{d\varepsilon(t)}{dt} = 1.9 \times 10^4 \, N_o \, \sigma_{ex}^{4.5} \, e^{-4.75 \, eV/kT} \qquad (11)$$

Here N_o is the initial source density of misfit dislocations at time t = 0, as in Eqn (3).

In summary of this section, the different models so far developed for strain relaxation in Ge_xSi_{1-x}-based heterostructures have quite different forms, most probably because they have been developed in comparison with experimental data at different stages of the relaxation process where different kinetic dislocation phenomena dominate. A universal model for determination of metastable dislocation arrays is unlikely to have a simple analytical form due to existence of different mechanistic regimes, and due to the complexity of some (e.g. interactions, multiplication) of the kinetic processes.

H CONCLUSION

In this Datareview we have described the kinetic factors which determine the metastability of misfit dislocation arrays in Ge_xSi_{1-x}-based heterostructures. A central concept in mapping out this metastability is that of excess stress, as defined by Dodson and Tsao [5], which embodies the net stress driving the relaxation process. A second critical parameter is temperature, as many of the kinetic processes (primarily misfit dislocation propagation and nucleation) are thermally activated with activation barriers >> kT.

The individual kinetic processes which define metastability are misfit dislocation nucleation, propagation and interactions. The nucleation process has received much mechanistic treatment, but there remains relatively little quantitative data in the literature. It is unlikely that a single nucleation mechanism dominates over the entire temperature - strain - epilayer thickness regime. Rather, in different regimes homogeneous, heterogeneous and multiplication mechanisms will be expected to dominate as discussed in Section D.

The dislocation propagation process has been relatively well characterised, with a large amount of experimental data from several groups, and a large body of existing literature on dislocation glide in bulk Si and bulk Ge. The analysis of dislocation interactions is complex, but there has been some experimental and theoretical treatment of the relevant processes.

Several models for the overall strain relaxation process by misfit dislocations in Ge_xSi_{1-x}-based heterostructures have been developed. Models embodying the concepts of excess stress and scaling laws appear to provide a useful empirical framework for defining the rate of growth of the misfit dislocation array. Other models have attempted to combine direct experimental measurements of the relevant kinetic dislocation parameters.

REFERENCES

[1] J.W. Matthews, S. Mader, T.B. Light [*J. Appl. Phys. (USA)* vol.41 (1970) p.3800]
[2] J.W. Matthews, A.E. Blakeslee, S. Mader [*Thin Solid Films (Switzerland)* vol.33 (1976) p.253]
[3] R. Hull [Datareview in this book: 1.2 Equilibrium theories of misfit dislocation networks in the GeSi/Si system]
[4] R.E. Peierls [*Proc. Phys. Soc. London (UK)* vol.52 (1940) p.23]
[5] B.W. Dodson, J.Y. Tsao [*Appl. Phys. Lett. (USA)* vol.51 (1987) p.1325]
[6] J.W. Matthews [*J. Vac. Sci. Technol. (USA)* vol.12 (1975) p.126 and references therein]
[7] R. Hull, J.C. Bean, D.J. Werder, R.E. Leibenguth [*Phys. Rev. B (USA)* vol.40 (1989) p.1681]
[8] D.C. Houghton [*J. Appl. Phys. (USA)* vol.70 (1991) p.2136]
[9] F.K. LeGoues, P.M. Mooney, J. Tersoff [*Phys. Rev. Lett. (USA)* vol.71 (1993) p.396]
[10] J.P. Hirth, J. Lothe [*Theory of Dislocations* (McGraw-Hill, New York, 1968)]
[11] D.J. Bacon, A.G. Crocker [*Philos. Mag. (UK)* vol.12 (1965) p.195]
[12] D.J. Eaglesham, E.P. Kvam, D.M. Maher, C.J. Humphreys, J.C. Bean [*Philos. Mag. A (UK)* vol.59 (1989) p.1059]
[13] W. Wegscheider, K. Eberl, U. Menczigar, G. Abstreiter [*Appl. Phys. Lett. (USA)* vol.57 (1990) p.875]
[14] R. Hull, J.C. Bean [*J. Vac. Sci. Technol. A (USA)* vol.7 (1989) p.2580]
[15] D.D. Perovic, D.C. Houghton [*Mater. Res. Soc. Symp. Proc. (USA)* vol.263 (1992) p.391]
[16] D.J. Srolovitz [*Acta Metall. (USA)* vol.37 (1989) p.621]
[17] D.E. Jesson, S.J. Pennycook, J.-M. Baribeau, D.C. Houghton [*Phys. Rev. Lett. (USA)* vol.71 (1993) p.1744]
[18] W.H. Yang, D.J. Srolovitz [*Phys. Rev. Lett. (USA)* vol.71 (1993) p.1593]
[19] W. Hagen, H. Strunk [*Appl. Phys. (Germany)* vol.17 (1978) p.85]
[20] K. Rajan, M. Denhoff [*J. Appl. Phys. (USA)* vol.62 (1987) p.1710]
[21] C.J. Gibbings, C.G. Tuppen, M. Hockley [*Appl. Phys. Lett. (USA)* vol.54 (1989) p.148]
[22] F.K. LeGoues, B.S. Meyerson, J.F. Morar [*Phys. Rev. Lett. (USA)* vol.66 (1991) p.2903]
[23] M.A. Capano [*Phys. Rev. B (USA)* vol.45 (1992) p.11768]
[24] A. Lefebvre, C. Herbeaux, C. Boillet, J. Di Persio [*Philos. Mag. Lett. (UK)* vol.63 (1991) p.23]

[25] R. Hull, J.C. Bean, D.J. Werder, R.E. Leibenguth [*Appl. Phys. Lett. (USA)* vol.52 (1989) p.1605]
[26] H. Alexander, P. Haasen [*Solid State Physics* vol.22 (1968)]
[27] M. Imai, K. Sumino [*Philos. Mag. A (UK)* vol.47 (1983) p.599]
[28] J.R. Patel, A.R. Chaudhuri [*Phys. Rev. (USA)* vol.143 (1966) p.601]
[29] A. George, J. Rabier [*Rev. Phys. Appl. (France)* vol.22 (1987) p.1941]
[30] B.Y. Farber, Y.L. Iunin, V.I. Nititenko [*Phys. Status Solidi (Germany)* vol.97 (1986) p.469]
[31] K.H. Kusters, H. Alexander [*Physica B (Netherlands)* vol.116 (1983) p.594]
[32] A. Seeger, P. Schiller [*Acta Metall. (USA)* vol.10 (1962) p.348]
[33] R. Hull, J.C. Bean, D. Bahnck, L.J. Peticolas, K.T. Short, F.C. Unterwald [*J. Appl. Phys. (USA)* vol.70 (1991) p.2052]
[34] W.A. Nix, D.B. Noble, J.F. Turlo [*Mater. Res. Soc. Symp. Proc. (USA)* vol.188 (1990) p.315]
[35] C.G. Tuppen, C.J. Gibbings [*J. Appl. Phys. (USA)* vol.68 (1990) p.1526]
[36] Y. Yamashita et al [*Philos. Mag. Lett. (UK)* vol.67 (1993) p.165]
[37] R. Hull, J.C. Bean [*Phys. Status Solidi A (Germany)* vol.138 (1993) p.533]
[38] C.J. Gibbings, C.G. Tuppen, V. Higgs [*Mater. Res. Soc. Symp. Proc. (USA)* vol.220 (1992) p.205]
[39] D.B. Noble et al [*Appl. Phys. Lett. (USA)* vol.58 (1991) p.1536]
[40] R. Hull, J.C. Bean, D. Noble, J. Hoyt, J.F. Gibbons [*Appl. Phys. Lett. (USA)* vol.59 (1991) p.1585]
[41] L.B. Freund [*J. Appl. Phys. (USA)* vol.68 (1990) p.2073]
[42] J.C. Bean, L.C. Feldman, A.T. Fiory, S. Nakahara, I.K. Robinson [*J. Vac. Sci. Technol. A (USA)* vol.2 (1984) p.436]
[43] E. Kasper, H.-J. Herzog, H. Kibbel [*Appl. Phys. (Germany)* vol.8 (1975) p.199]
[44] J.Y. Tsao, B.W. Dodson, S.T. Picreaux, D.M. Cornelison [*Phys. Rev. Lett. (USA)* vol.59 (1988) p.2455]
[45] H.J. Frost, M.F. Ashby [*Deformation Mechanism Maps* (Pergamon, Oxford, England, 1982)]
[46] B.W. Dodson, J.Y. Tsao [*Appl. Phys. Lett. (USA)* vol.55 (1989) p.1345]
[47] B.W. Dodson, J.Y. Tsao [*Appl. Phys. Lett. (USA)* vol.53 (1988) p.2498]
[48] R. Hull, J.C. Bean, C. Buescher [*J. Appl. Phys. (USA)* vol.66 (1989) p.5837]
[49] M.L. Green et al [*J. Appl. Phys. (USA)* vol.69 (1991) p.745]
[50] D.C. Houghton, C.J. Gibbings, C.G. Tuppen, M.H. Lyons, M.A.G. Halliwell [*Appl. Phys. Lett. (USA)* vol.56 (1990) p.460]

CHAPTER 2

STRUCTURAL PROPERTIES

2.1 Crystal structure, lattice parameters and liquidus-solidus curve of the SiGe system

H.-J. Herzog

September 1993

A INTRODUCTION

A reliable set of structural data is the essential basis for many investigations of both epilayers and bulk material. In this Datareview we review some crystallographic data (crystal structure, lattice parameters) and a material property (phase diagram) of silicon-germanium alloys.

B CRYSTAL STRUCTURE

Silicon and germanium, both crystallising in the diamond lattice, form a continuous series of $Si_{1-x}Ge_x$ solid solutions with x ranging from 0 to 1. The space lattice of diamond consists of two face-centred cubic (fcc) lattices which are displaced a quarter of the space diagonal. A perspective drawing of the unit cell is depicted in FIGURE 1. The space group of the diamond structure is Fd3m. The cubic unit cell contains eight atoms which occupy the following positions:

$$0\,0\,0 \quad 0\,\tfrac{1}{2}\,\tfrac{1}{2} \quad \tfrac{1}{2}\,0\,\tfrac{1}{2} \quad \tfrac{1}{2}\,\tfrac{1}{2}\,0$$

$$\tfrac{1}{4}\,\tfrac{1}{4}\,\tfrac{1}{4} \quad \tfrac{1}{4}\,\tfrac{3}{4}\,\tfrac{3}{4} \quad \tfrac{3}{4}\,\tfrac{1}{4}\,\tfrac{3}{4} \quad \tfrac{3}{4}\,\tfrac{3}{4}\,\tfrac{1}{4}$$

The fractions denote the height above the base in units of the cube edge. In this structure each atom is bonded to four nearest-neighbours arranged at the corners of a regular tetrahedron

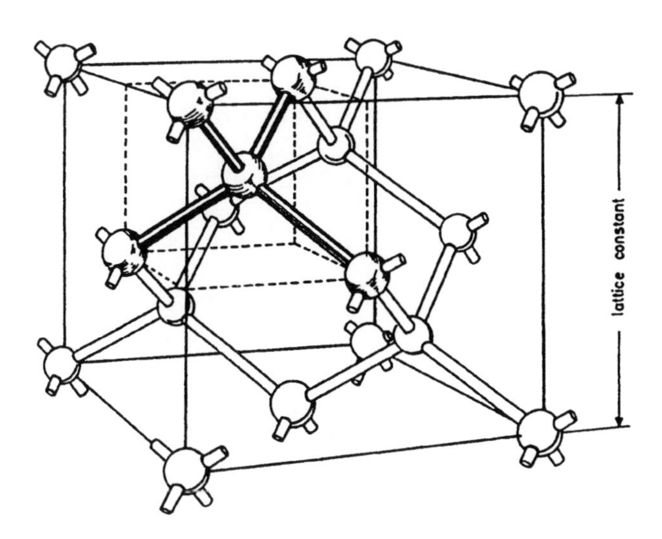

FIGURE 1 Diamond crystal structure. Each atom is tetrahedrally bonded to four nearest neighbours as displayed by the dashed lines.

and twelve next nearest-neighbours. The diamond structure is the result of the covalent bonding between the atoms represented by the rods in FIGURE 1. The diamond lattice is not very compact. Only 34% of the available space is filled with hard spheres.

C LATTICE PARAMETERS

To date, the most precise and comprehensive determination of bulk lattice parameters (and densities) across the whole $Si_{1-x}Ge_x$ system has been carried out by Dismukes et al [1] including the variation of lattice parameters with temperature up to 800°C for a couple of alloys. In TABLE 1 the lattice parameters of $Si_{1-x}Ge_x$ alloys at 25°C are listed for composition intervals of 5 at. % Ge. The data reveal a small deviation from Vegard's rule i.e. from the linearity between the lattice constants of the end-point structures Si and Ge. The departure Δ from Vegard defined by

$$\Delta = a_{SiGe} - [a_{Si}(1-x) + a_{Ge}x] \tag{1}$$

is also listed in TABLE 1. Δ is negative throughout the system with a broad maximum in the middle of the system. This experimentally determined deviation from Vegard's rule has also been found theoretically by Monte Carlo simulations on $Si_{1-x}Ge_x$ alloys [2,3]. By using the values given in [1] for x = 0%, x = 25% and x = 100% a parabolic relation for the $Si_{1-x}Ge_x$ lattice parameter as a function of composition x:

$$a(x) = 0.002733x^2 + 0.01992x + 0.5431 \text{ (nm)} \tag{2}$$

TABLE 1 Lattice parameter a(x) of $Si_{1-x}Ge_x$ alloys for x from 0 to 100 at. % in 5% steps after [1]. The right column gives the deviation parameter Δ.

x (at. % Ge)	a(x) (nm)	Δ
0	0.54310	.....
5	0.54410	-0.0004
10	0.54522	-0.0014
15	0.54624	-0.0026
20	0.54722	-0.0041
25	0.54825	-0.0051
30	0.54928	-0.0062
35	0.55038	-0.0065
40	0.55149	-0.0067
45	0.55261	-0.0068
50	0.55373	-0.0069
55	0.55492	-0.0063
60	0.55609	-0.0060
65	0.55727	-0.0055
70	0.55842	-0.0053
75	0.55960	-0.0048
80	0.56085	-0.0027
85	0.56206	-0.0023
90	0.56325	-0.0019
100	0.56575	.....

can be derived which approaches the experimental data with a maximum deviation of 10^{-4} nm. The lattice parameter of Si of $a_{Si} = 0.5431$ nm is confirmed by recent high precision measurements on pure single crystal Si [4]. The published room temperature data on undoped Ge single crystals range from $a_{Ge} = 0.56573$ nm [5] to 0.56579 nm [6].

Concerning the composition dependency of other structural parameters covering the whole composition range the number of publications in the literature is rather limited. The elastic constants, for example, are required if the lattice parameters of heteroepitaxial $Si_{1-x}Ge_x$ films have to be corrected for elastic strain. Bublik et al [7] investigated three different $Si_{1-x}Ge_x$ alloys (x = 0.36, 0.46 and 0.72) and found experimental values of the elastic constants c_{ij} which are larger than those calculated from a linear combination of the c_{ij} values of the pure components. From Raman measurements on $Si/Si_{0.52}Ge_{0.48}$ strained layer superlattices on Si Zhang et al [8] obtained for the Si layer a smaller sound velocity and density compared with the Si bulk values, and for the SiGe alloy layers a higher sound velocity and higher density than the ones deduced by linear interpolation. However, these changes are possibly due to strain introduced during formation of the superlattice. Extended X-ray-absorption fine-structure spectroscopy (EXAFS) and X-ray diffraction (XRD) have been used to analyse the bond length of strained SiGe layers on Si substrates [9-11]. The results indicate that the Ge-to-Si, Ge-to-Ge, and Si-to-Si nearest-neighbour distances are 2.38 ± 0.02 Å, 2.42 ± 0.02 Å, and 2.35 ± 0.02 Å, respectively, close to the sum of their constituent-element covalent radii and independent of the SiGe composition, while the lattice constant varies monotonically with the Ge-content. Therefore, the structural change occurs in bond angles rather than in bond length.

D LIQUIDUS-SOLIDUS CURVE

The phase diagram in FIGURE 2 was established by elaborate thermal and X-ray analysis [12]. No phase changes or decompositions were detected roentgenographically after annealing homogeneous solid solution crystals for several months at temperatures in the range from 177°C to 925°C. The SiGe system is a typical representative of a system with strong segregation, i.e. a solid solution in which the solid and liquid phases are separated by a large

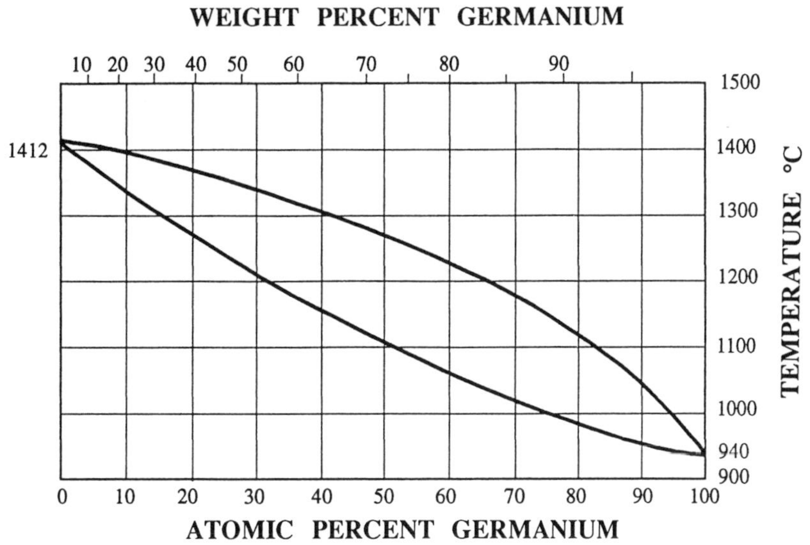

FIGURE 2 Liquidus-solidus curve of the $Si_{1-x}Ge_x$ system after [12].

regime of coexistence. From this it is evident that the preparation of a homogeneous solid solution from Si and Ge requires some effort, because during solidification from the molten phase the Si component strongly segregates and thus quasi-decomposition occurs. Under the assumption that Si and Ge form ideal liquid and solid solutions Thurmond [13] calculated the liquidus and solidus curves and found the latter to coincide with that in FIGURE 2. The liquidus lies slightly below the experimental curve.

E CONCLUSION

We have reviewed the crystal structure, the lattice parameters and the liquidus-solidus curve of the $Si_{1-x}Ge_x$ system. The deviation of the lattice parameters from Vegard's rule is discussed. Probably due to difficulties in preparing homogeneous $Si_{1-x}Ge_x$ bulk crystals there is still a deficiency in material data covering the whole composition range from x = 0 to 1. In such cases where exact values are missing Vegard's rule is usually used.

REFERENCES

[1] J.P. Dismukes, L. Ekstrom, R.J. Paff [*J. Phys. Chem. (USA)* vol.68 (1964) p.3021-7]

[2] R. Fabbri, F. Cembali, M. Servidori, A. Zani [*J. Appl. Phys. (USA)* vol.74 (1993) p.2359-69]

[3] S. de Gironcoli, P. Giannozzi [*Phys. Rev. Lett. (USA)* vol.66 (1991) p.2116-9]

[4] D. Windisch, P. Becker [*Phys. Status Solidi A (Germany)* vol.118 (1990) p.379-88]

[5] O. Brümmer, V. Alex, G. Schulze [*Ann. Phys. (Germany)* vol.28 (1972) p.118-34]

[6] J.F.C. Baker, M. Hart [*Acta Crystallogr. A (Denmark)* vol.31 (1975) p.364-7]

[7] V.T. Bublik, S.S. Gorelik, A.A. Zaitsev, A.Y. Polyakov [*Phys. Status Solidi B (Germany)* vol.66 (1974) p.427-32]

[8] P.X. Zhang, D.J. Lockwood, H.J. Labbé, J.-M. Baribeau [*Phys. Rev. B (USA)* vol.46 (1992) p.9881-4]

[9] J.C. Woicik et al [*Phys. Rev. B (USA)* vol.43 (1991) p.2419-22]

[10] M. Matsuura, J.M. Tonnerre, G.S. Cargill III [*Phys. Rev. B (USA)* vol.44 (1991) p.3842-9]

[11] H. Kajiyama, S. Muramatsu, T. Shimada, Y. Nishino [*Phys. Rev. B (USA)* vol.45 (1992) p.14005-10]

[12] H. Stöhr, W. Klemm [*Z. Anorg. Allg. Chem. (Germany)* vol.241 (1939) p.305-23]

[13] C.D. Thurmond [*J. Phys. Chem. (USA)* vol.57 (1953) p.827-30]

2.2 Ordering in SiGe alloys

W. Jäger

June 1994

A INTRODUCTION

SiGe alloys crystallise in the diamond structure which consists of two face-centred cubic (fcc) sublattices shifted by one quarter of the body diagonal, R = 1/4 <111>. According to the phase diagram calculated in the regular solution model with a positive enthalpy of mixing [1-6], the alloy tends to segregate into pure components at sufficiently low temperatures, forms a solid solution at higher temperatures, and does not have any ordered phases. No ordering has been observed in bulk SiGe alloy crystals of different compositions which were grown from the melt by the vertical Bridgman technique [7,8].

Observations of long-range ordering (LRO) of Si and Ge have been made, however, first in alloy layers [9] and subsequently also at interfaces of superlattices (see references in TABLE 1) prepared by MBE at medium temperatures and for various compositions. These results are unexpected since systems with positive mixing enthalpies are unable to form alloys or ordered structures at low temperature. The origin of ordering has been attributed to processes at the growing surface e.g. reconstruction-induced growth kinetics [10-14], although the occurrence of order with a reversible phase transition at ~800°C has been reported [14] indicating the presence of an equilibrium phase.

This Datareview describes possible atomic structures of ordered phases and their occurrence in epitaxially grown SiGe layers and discusses possible origins of ordering.

B PHASE DIAGRAM

The equilibrium phase diagram for the Si-Ge system [4,15] shows that Ge and Si are completely miscible as liquid and as solid. Stable phases formed by Si and Ge are the liquid and a cubic diamond-type substitutional solid solution. The Ge-Si solid solution may transform into a two-phase mixture at low temperature (below ~170 K) (FIGURE 1) [1,3,6]. At high pressures (above ~150 kbar) the bct structure of a Ge-Si solid solution is more stable than the diamond structure [16]. The lattice parameter increases nearly linearly with increasing Ge concentration x but shows deviations from Vegard's law in the regime of medium compositions [15,17]. Near the melting temperature the calculated phase diagram based on regular solid and liquid solutions can be fitted well to the measured diagram by an appropriate choice of enthalpy of mixing functions [4] (FIGURE 1).

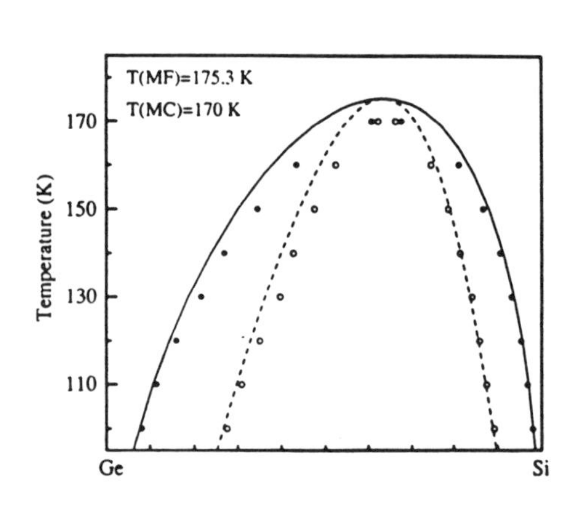

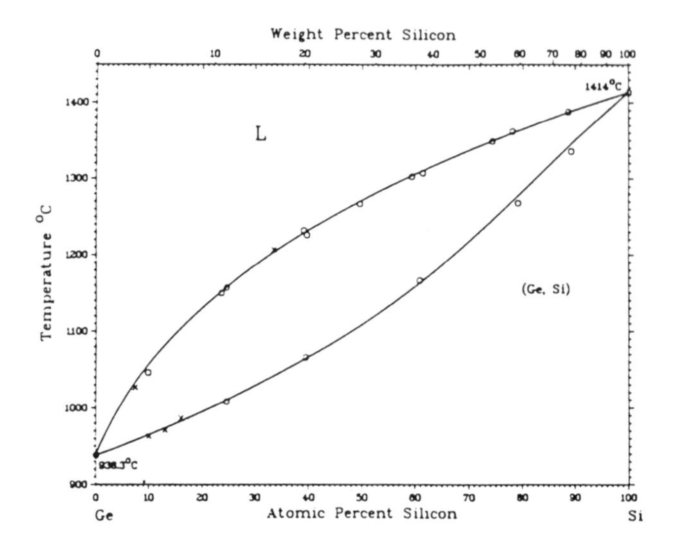

FIGURE 1 Phase diagrams at low temperatures [6] (left) and near the melting point [4] (right). Left: dots are Monte Carlo predictions for the miscibility gap (solid) and spinodal line (open); solid line: mean field prediction for the miscibility gap; dashed line: mean field prediction for the spinodal line. Right: points are experimental data [4].

C ORDER STRUCTURES OF SiGe

The LRO occurring in SiGe has been detected through the appearance of superstructure reflections in selected-area electron diffraction patterns and of dark-field imaging in TEM ([9] and references in TABLE 1). Direct images of ordering on an atomic scale have been obtained by Z-contrast imaging in STEM [10,11]. It was shown recently that different types of order can be detected and separated from each other in <110> HRTEM lattice fringe images taken under optimised conditions [18].

Different order structures were suggested on the basis of the experimental investigations and are depicted in FIGURES 2(a), 2(b) and 2(c). The initial experimental results indicated a bi-layer stacking of Si and Ge atoms on adjacent {111} planes, corresponding to a doubling of the unit cell in the <111> directions for which two rhombohedral structure models (RS1 and RS2) have been proposed [9,19]. For both structures the stoichiometry is $Si_{0.5}Ge_{0.5}$. A general structure model (RS3) of ordered SiGe allowing for compositional differences on specific sites has been suggested recently [12]:

(i) RS1: the widely spaced {111} planes are occupied by the same atom type;
(ii) RS2: the closely spaced {111} planes are occupied by the same atom type;
(iii) RS3: compositional differences between projected sites or columns are allowed, with compositions α, β, γ and δ corresponding to specific Ge-rich and Si-rich sites.

The rhombohedral structures (CuPt ordering, R3m) are among the possible ordered phases of adamantine compounds (i.e. tetrahedral structures with close-packed cubic sublattices) that are likely to be the most stable ones according to the Landau-Lifshitz theory of phase transformations [20]. The remarkable properties of these structures are that the order-disorder transformation can (but need not) be of second order, and that they can exist over a wide concentration range [21]. Among the adamantine phases only the zinc blende and rhombohedral structures possess sufficient structural degrees of freedom to make bond angles and bond lengths ideally tetrahedral.

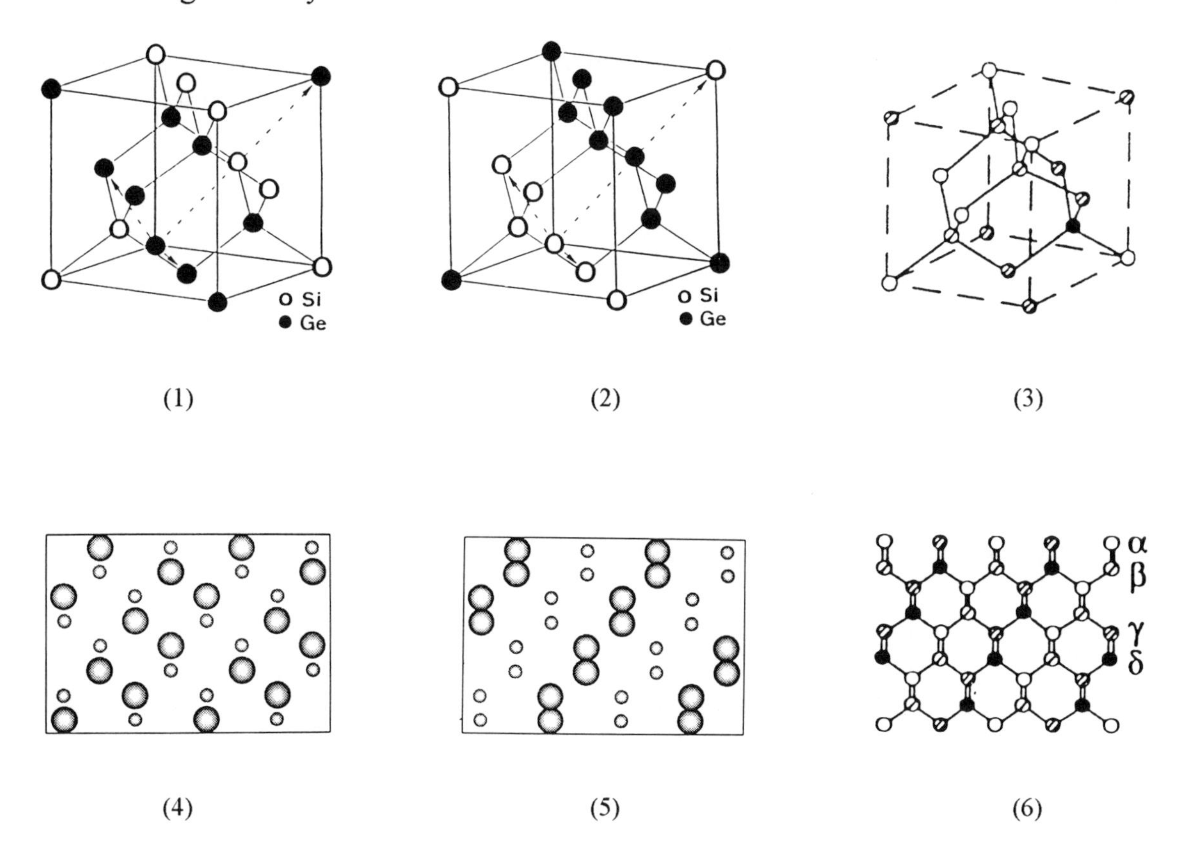

FIGURE 2 Ordered Si-Ge structure with bi-layer stacking on {111} planes as suggested by model RS1 (left, 1, 4), model RS2 (centre, 2, 5) and model RS3 (right, 3, 6). Unit cells are shown above, the <110> projections are shown below. Large (small) dot-shaded circles represent Ge (Si) (4, 5). The four projected compositions α, β, γ and δ correspond to specific Ge-rich and Si-rich sites (right, 3, 6). In this case the cubic unit cell is generated from eight of the structural units shown above (3) but with atom site occupancies in each adjacent unit transformed by the relation ($\alpha \leftrightarrow \gamma$, $\beta \leftrightarrow \delta$).

Bragg reflections in the electron diffraction patterns from the diamond structure have unmixed hkl indices (i.e. h, k and l are either all even or all odd). In addition to structure reflections, reflections resulting from double diffraction may occur. The presence of ordered domains leads to additional superstructure reflections at half-integer positions. For the RS1 structure the structure factor can be calculated as follows:

$$F = 2 \cos \{2\pi \mathbf{g}\,(a/8)\,[111]\}\,\{f_{Si} + f_{Ge} \exp(2\pi i\,\mathbf{g}\,(a/2)\,[101])\}$$

$F = 0$ for all reflections (hkl) with $h + k + l = 4n + 2$ and proportional to $(f_{Si} - f_{Ge})$ for superstructure reflections. Superstructure reflections to be observed from the order structures of the RS1 or RS2 type (FIGURE 2) are of the type 1/2(111), 1/2(311), 1/2(331), ... In the kinematic theory of diffraction the intensity I of superstructure reflections is given by $I = |F|^2 \propto S^2(F_{Ge} - F_{Si})^2$, with a long-range order parameter $S = (r_{Ge} - x)/(1 - x)$ (r_{Ge} = fraction of Ge sites occupied correctly, x = fraction of Ge atoms in the alloy, S = 1 perfect and complete order, S = 0 random alloy). Hence, the degree of long-range order can be quantitatively deduced from the intensity of superstructure reflections. However, because of the presence of defects, such as antiphase boundaries, of superposition of various domains, and of the effect of multiple scattering the observed intensity can deviate substantially from the kinematic value so that quantitative determinations from SAD intensities are obscured.

The type of ordering (RS1, RS2) can be determined from a semi-quantitative evaluation of electron diffraction spot intensities [14] and from [110] STEM Z-contrast [10] or HRTEM lattice fringe [18] images. Generally, the observed diffraction pattern is a superposition of contributions from the four independent domain orientations, each of them defined by a doubling of the unit cell along one of the four equivalent <111> axes. The presence of domains of all of these four equivalent orientations can be deduced from the presence of 1/2 {311} reflections in diffraction patterns taken along the [130] and the [1$\bar{3}$0] direction of the same sample [14]. Intensity distributions in the form of streaks going through half-integer reflections can be explained by shape effects of layer domains of thicknesses of a few nm and below [22,23]. Projected domain sizes can be determined from dark-field images [12,24,25] for the case of 3-dimensional bulk domains or from <110> zone-axis STEM [10] and HRTEM [18] images for the case of 2-dimensional order at interfaces.

D OCCURRENCE AND STABILITY

Ordered SiGe phases have been observed in the bulk of thick alloy layers and at interfaces of superlattices and heterostructures which were grown by MBE at growth temperatures between 300°C and 800°C on Si, Ge and $Si_{1-y}Ge_y$ substrates of (100) orientation with different compositions y (TABLE 1). Several aspects of the occurrence of order have been investigated.

(i) Bulk and interface ordering: Ordering has been observed in the bulk of alloy layers and at interfaces. Order is found to be associated with the growth mode of island growth where it seems to occur preferentially at step edges [10-12].

(ii) Domain structure: All equivalent orientation variants along <111> have been observed in alloy layers and in SL on (100) substrates for the as-grown state [12,14,25]. Projected lateral domain sizes are of the order of 20 nm [10,26]. For bulk layers randomly shaped order domains with extensions of up to 0.5 μm are reported [12,25].

(iii) Stability during annealing: The stability of order during annealing up to the dissociation temperature T_d and the reversibility of the transition have been investigated for RS1 and RS2 in SL structures and for bulk layers [14]. For RS1 $T_d \sim 800$°C, and the transition is found to be reversible. For RS2 $T_d \sim 570$°C (this is close to the value of 650°C reported by [13]), and the transition is irreversible. Disproportionation of the SL occurs at temperatures above ~570°C.

(iv) Influence of surface reconstruction: Ordering during growth is suppressed on (100)-oriented crystal surfaces when Sb surfactants are applied changing the surface reconstruction [13]. Ordering is absent for growth on Si (111) substrates [13].

(v) Influence of composition: Order has been observed over a wide range of layer compositions both for thick alloy layers and for thin layers of SL. For Ge-rich systems RS1 and RS2 are present in the as-grown state whereas for Si-rich systems only RS2 is observed [13,14,26].

(vi) Influence of strain: The occurrence of order has been observed for SL grown on different substrates and with different layer sequences corresponding to individual layers under different amounts of strain [14,22-24,27,28]. Whereas the occurrence of RS2 seems to be independent of the strain distribution in layers, the RS1 phase seems to be associated with the presence of strain [14].

(vii) Growth under near-equilibrium conditions: No observations of ordering are reported for Si-rich and Ge-rich SiGe bulk crystals with different compositions or for $Si_{0.5}Ge_{0.5}$ bulk crystals grown from the melt [7,8].

TABLE 1 Type and occurrence of Si-Ge order phases observed in $Si_{1-x}Ge_x$ crystal layers grown on Si, Ge or $Si_{1-y}Ge_y$ substrates of (100) orientation by MBE.

Substrate	System x/thickness (nm)	T_g (°C)	Technique	Order type/strain	T_d (°C)	Refs
Si (100)	$Si_{1-x}Ge_x/Si$ SL 0.4/7.5	550	SAD	bulk <111> RS1/2		[9]
Si (100)	$Si_{1-x}Ge_x/Si$ SL 0.33/20 0.2/1000	 650 800	DF (SAD)	bulk <111> RS1/2 strained/relaxed		[24]
Si (100)	$Si_{1-x}Ge_x$ layer 0.5/500	390-475	DF, SAD	bulk <111> RS2		[25]
Si (100)	$Si_{1-x}Ge_x$ layer 0.5/750	390-625	SAD, LEED	bulk <111> RS2	650 irr	[13]
Si, Ge (100) $Si_{1-y}Ge_y$(100)	Si_mGe_n SL var m,n/ultrathin	420-470	SAD <130>	interface RS1/var strain		[22,23]
Si, Ge (100) $Si_{1-y}Ge_y$(100)	$Si_{1-x}Ge_x$ layer 0.35-0.7/1000,250	500-700	SAD <130>	bulk RS1/2 relaxed, strained	800 RS1, rev 570 RS2, irr	[14]
Si, Ge (100) $Si_{1-y}Ge_y$(100)	Si_mGe_n SL var m,n/ultrathin	420-450	SAD <130>	interface RS1/2 var strain	800 RS1, rev 570 RS2, irr	[14]
Ge (100)	Si_mGe_n SL m=4, n=8/ultrathin	350	STEM <110>	interface <111> 3 types/Ge segr		[26]
Si (100)	$Si_{1-x}Ge_x$ layer 0.4/350	350	STEM <110>	bulk <111> relaxed		[10]
Si (100)	$Si_{1-x}Ge_x$ layer 0.5/>90	400	DF, SAD STEM, XRD	bulk <111> RS3 islands		[12]
Si (100) $Si_{1-y}Ge_y$(100)	Si_mGe_n SL var m,n/ultrathin	300-400	SAD <110> SAD <130>	interface <111> RS1/2 var strain		[27,28]

E DISCUSSION

Observations of the long-range order of Si and Ge have created interest both in the ordering mechanism and in the technological implication of tuning the alloy bandgaps through ordering. The origin of ordering of Si and Ge is still a matter of controversy. Theoretical studies were conducted for 3-dimensional bulk ordered structures and alloys [19,29-31] and for 3-dimensional epitaxial SL and alloys coherent with a substrate [19,29,31-33]. These studies were concerned with the determination of energies of different configurations, the segregation behaviour of surfaces, the influence of epitaxial strain and predictions for the temperature of the order-disorder transition.

Energy considerations of equilibrium properties favour the RS1 structure over the RS2 structure, which is microscopically strained and therefore energetically less favourable [19,29,32,33]. The RS1 phase is accepted as metastable. Its stability is increased for epitaxially strained $Si_{1-x}Ge_x$ layers. The theoretical order-disorder transition temperature is predicted to be below 150 K [19,33] i.e. far below the experimental values for growth temperatures (TABLE 1).

The formation of ordered Si-Ge phases was suggested to be due to the growth mechanism at the reconstructed surface instead of being an equilibrium property [32,33]. Two models based on growth kinetics were suggested [13,26] in view of the experimental evidence for the occurrence of RS2 as the main structure [13,26] instead of the intrinsically strain-free RS1. The models developed for alloy layers [13] and for strained Si_mGe_n SL [26] are closely related but one of them [13] requires a bi-layer growth mode for which experimental evidence is lacking. They explain the formation of RS2 by Si (Ge) segregation during growth due to the strong compressive and tensile strains in the uppermost subsurface layers of a (2x1) reconstructed surface. Stresses can strongly enhance the diffusion of atoms in the uppermost layers while bulk diffusion is negligible at growth temperatures [34]. Oscillatory segregation at the Si-Ge (100) 2x1 reconstructed surface and lateral ordering driven by local stress fields is predicted also by simulations [14,31]. The RS2 structure is metastable and irreversibly destroyed after dissociation above a specific temperature, in agreement with experiments ([14], TABLE 1). A recent model attributes the atomic-scale compositional ordering to Ge segregation at kinks of step edges during the formation of coherent islands during Stranski-Krastanov growth [12]. This agrees with recent investigations of $Si_{1-x}Ge_x$ surface reconstruction showing that Ge indeed segregates to the exposed surface during growth [35]. The general ordered phase RS3 suggested on the basis of this kink model (FIGURE 2) describes variants of rhombohedral ordered structures to be considered in alloy films. The prevention of formation of the RS2 structure may be possible by suppression of the 2x1 and 1x2 surface reconstruction by applying surfactants during MBE growth [13]. The models cannot explain the presence of RS1 and the occurrence of a reversible phase transformation in Ge-rich samples and, at high temperatures, in Si-rich samples ([14], TABLE 1).

Optical and electrical properties still remain to be explored. For instance, ordered Si-Ge alloy layers at the interfaces of superlattices are predicted to have considerably increased oscillator strengths of the optical transitions as compared to random alloy interfaces [36,37].

Long-range order along <111> directions (CuPt-type) is not only a phenomenon observed for the Si-Ge system but was found also in many III-V compound semiconductor alloys ([38], refs in [39,40]). Further ordering patterns observed in such alloys are the chalcopyrite structure and the CuAuI structure (refs in [39,40]).

F CONCLUSION

Order of Si and Ge is found in the bulk and at interfaces of SiGe layers deposited by MBE on substrates of (100) orientation at growth temperatures between 300°C and 800°C. Formation of order phases seems to be suppressed for growth on (111) substrates and upon the use of surfactants changing the (100) surface reconstruction during growth. No observations of ordered phases exist for bulk alloy crystals grown from the melt.

The types of ordered atomic arrangements observed show a bi-layer stacking along <111> directions. Differences in the temperature stability of different order phases are reported. However, both the stability regime and the reversibility of the order-disorder transition still remain to be explored.

These results are unexpected from the viewpoint that all systems with positive mixing enthalpies are unable to form alloys or ordered structures at low temperature. The detailed mechanisms for ordering are still not clear. The present state of experimental work indicates that ordering is a growth-induced rather than an equilibrium phenomenon.

ACKNOWLEDGEMENT

The critical reading of the manuscript by Dr. R. Butz, Prof. K. Schroeder, Dr. D. Stenkamp and Prof. H. Wenzl is gratefully acknowledged.

REFERENCES

[1] V.T. Bublik, S.S. Gorelik, A.A. Zaitsev, A.Y. Polyakov [*Phys. Status Solidi B (Germany)* vol.65 (1974) p.K79-K84]
[2] G.B. Stringfellow [*J. Phys. Chem. Solids (UK)* vol.34 (1973) p.1749]
[3] T. Soma [*Phys. Status Solidi B (Germany)* vol.95 (1979) p.427 and *Phys. Status Solidi B (Germany)* vol.98 (1980) p.637]
[4] R.W. Olesinski, G.J. Abbaschian [*Bull. Alloy Phase Diagrams (USA)* vol.5 (1984) p.180-3]
[5] L.G. Ferreira, S.-H. Wei, A. Zunger [*Phys. Rev. B (USA)* vol.40 (1989) p.3197]; S.-H. Wei, L.G. Ferreira, A. Zunger [*Phys. Rev. B (USA)* vol.41 (1990) p.8240]
[6] S. Baroni, S. de Gironcoli, P. Giannozzi [in *Structural and Phase Stability of Alloys* Eds J.L. Moran-Lopez et al (Plenum Press, New York, 1992) p.133]
[7] D. Stenkamp, W. Jäger [*Philos. Mag. A (UK)* vol.65 (1992) p.1369-82]
[8] J. Schilz, V.N. Romanenko [to be published in J. Mater. Sci., Mater. Electron. (UK) (1994)]
[9] A. Ourmazd, J.C. Bean [*Phys. Rev. Lett. (USA)* vol.55 (1985) p.765]
[10] D.E. Jesson, S.J. Pennycook, J.-M. Baribeau, D.C. Houghton [*Phys. Rev. Lett. (USA)* vol.68 (1992) p.2062]
[11] D.E. Jesson, S.J. Pennycook, J.-M. Baribeau, D.C. Houghton [*Thin Solid Films (Switzerland)* vol.222 (1992) p.98-103]
[12] D.E. Jesson, S.J. Pennycook, J.Z. Tischler, J.D. Budai, J.-M. Baribeau, D.C. Houghton [*Phys. Rev. Lett. (USA)* vol.70 (1993) p.2293]
[13] F.K. LeGoues, V.P. Kesan, S.S. Iyer, J. Tersoff, R. Tromp [*Phys. Rev. Lett. (USA)* vol.64 (1990) p.2038]
[14] E. Müller, H.-U. Nissen, K.A. Mäder, M. Ospelt, H. von Känel [*Philos. Mag. Lett. (UK)* vol.46 (1991) p.183]; E. Müller [Dissertation No.9848, ETH Zürich (1992)]
[15] Landolt-Börnstein [*Numerical Data and Fundamental Relationships in Science and Technology* vol.17a (1982) and vol.17c (1984) (Springer Verlag, Berlin)]
[16] T. Soma [*Phys. Status Solidi B (Germany)* vol.111 (1982) p.K23-K26]
[17] J.P. Dismukes, L. Ekstrom, R.J. Paff [*J. Phys. Chem. (USA)* vol.68 (1964) p.3021]
[18] D. Stenkamp, W. Jäger [*Ultramicroscopy (Netherlands)* vol.50 (1993) p.321-54]
[19] P.B. Littlewood [*Phys. Rev. B (USA)* vol.34 (1986) p.1363]
[20] L.D. Landau, E.M. Lifshitz [*Statistical Physics* (Pergamon, Oxford, 1969) ch.14]
[21] A.G. Khachaturyan [*Theory of Structural Transformations in Solids* (Wiley, New York, 1983)]

[22] E. Müller, H.-U. Nissen, M. Ospelt, H. von Känel [*Phys. Rev. Lett. (USA)* vol.63 (1989) p.1819]
[23] M. Ospelt, J. Henz, E. Müller, H. von Känel [*Mater. Res. Soc. Symp. Proc. (USA)* vol.198 (1990) p.485]
[24] D.J. Lockwood, K. Rajan, E.W. Fenton, J.-M. Baribeau, D.W. Denhoff [*Solid State Commun. (USA)* vol.61 (1987) p.465]
[25] F.K. LeGoues, V.P. Kesan, S.S. Iyer [*Phys. Rev. Lett. (USA)* vol.64 (1990) p.40]
[26] D.E. Jesson, S.J. Pennycook, J.-M. Baribeau [*Phys. Rev. Lett. (USA)* vol.66 (1991) p.750]
[27] W. Jäger, K. Leiger, P. Ehrhart, E. Kasper, H. Kibbel [*Mater. Res. Soc. Symp. Proc. (USA)* vol.220 (1991) p.167 and unpublished results]
[28] W. Jäger et al [*Thin Solid Films (Switzerland)* vol.222 (1992) p.221-6 and unpublished results]
[29] J.L. Martins, A. Zunger [*Phys. Rev. Lett. (USA)* vol.56 (1986) p.1400]
[30] A. Qteish, R. Resta [*Phys. Rev. B (USA)* vol.37 (1988) p.1308-14 and *Phys. Rev. B (USA)* vol.37 (1988) p.6983-90]
[31] P.C. Kelires, J. Tersoff [*Phys. Rev. Lett. (USA)* vol.63 (1989) p.1164]
[32] S. Ciraci, I.P. Batra [*Phys. Rev. B (USA)* vol.38 (1988) p.1835]
[33] B. Koiller, M.O. Robbins [*Phys. Rev. B (USA)* vol.40 (1989) p.12554]
[34] S.S. Iyer, F.K. LeGoues [*J. Appl. Phys. (USA)* vol.65 (1989) p.4693]
[35] R. Butz, S. Kampers [*Thin Solid Films (Switzerland)* vol.222 (1992) p.104]
[36] M. Jaros, A.W. Beavis, E. Corbin, J.P. Hagon, R.J. Turton, K.B. Wong [*J. Vac. Sci. Technol. B (USA)* vol.11 (1993) p.1689]
[37] R.J. Turton, M. Jaros [*Semicond. Sci. Technol. (UK)* vol.8 (1993) p.2003-9]
[38] A.G. Norman [*NATO ASI Ser. B, Phys. (USA)* vol.203 (1988) p.233-53]
[39] R. Osorio, J.E. Bernard, S. Froyen, A. Zunger [*Phys. Rev. B (USA)* vol.45 (1992) p.11173-91]
[40] T.S. Kuan [in *Properties of Aluminium Gallium Arsenide* Ed. S. Adachi, EMIS Datareviews Series No.7 (INSPEC, IEE, 1993) p.7-9]

2.3 The Si/Ge interface: structure, energy and interdiffusion

G. Theodorou and P.C. Kelires

January 1994

A INTRODUCTION

It is well known that the lattice constants of bulk Si and Ge differ by 4.2%. The epitaxy of lattice-mismatched materials results either in a strained-layer configuration, if the layers are sufficiently thin, or in strain relaxation basically by a misfit dislocation network [1,2]. In the strained-layer configuration the thin layer lattice constant parallel to the interface, $a_{\|}$, is equal to that of the substrate, which is assumed to be rigid. We discuss here several aspects related to Si/Ge interfaces such as the elastic energy (Section B), formation enthalpy (Section C), interlayer distances (Section D) and interdiffusion (Section E).

B ELASTIC ENERGY

The elastic energy stored in the layer grown on a substrate with lattice constant a_s is composed of the energy E_h of a homogeneously strained layer and the energy nE_d of the misfit dislocation network [3-5]:

$$E = E_h + nE_d \tag{1}$$

where n is the number of dislocation sets. When the layer is grown on a (001) surface of a diamond structure semiconductor, there are two sets of misfit dislocations running perpendicular to each other and the value of n is 2 [6]. In the elastic continuum theory, the energy E_h due to strain in the epilayer is well known and is given by [5]:

$$E_h = 2\mu \frac{1+\nu}{1-\nu} \varepsilon^2 h \tag{2}$$

where μ is the shear modulus, ν the Poisson's ratio, ε the partially relaxed strain and h the epilayer thickness. The strain ε is related to the misfit parameter $f_m = (a - a_s)/a_s$ and the average number of dislocations present at the interface by the following relation:

$$f_m = -\varepsilon + b'/p \tag{3}$$

where b' is the active component of the Burgers vector, p the average distance between dislocations, a the bulk lattice constant of the epilayer and a_s the substrate lattice constant. The Poisson's ratio ν is given by the relation:

$$\nu = c_{12}/(c_{11} + c_{12}) \tag{4}$$

with c_{11} and c_{12} the elastic constants of the material.

C FORMATION ENTHALPY

The energetics of Si/Ge interfaces can be studied by considering the individual contributions to the total formation enthalpy ΔH of Si-Ge superlattices (SLs). In such an approach, Bernard and Zunger [6] expressed ΔH as a sum of bulk-like and interface-like terms:

$$\Delta H = \Delta E_{cs} + [\Delta E_{sr} + \Delta E_{chem}]$$

The bulk contribution is the 'constituent strain' energy ΔE_{cs}, which represents deformations of the equilibrium structures of the constituents, and it is therefore a positive quantity $\Delta E_{cs} > 0$. The interface-like term (in square brackets) consists of a 'strain-relief' energy $\Delta E_{sr} < 0$, which corresponds to relaxations of strained bonds at the interface, and a chemical energy $\Delta E_{chem} > 0$. This can be thought of as arising from charge transfer at the interface and the formation of Si-Ge bonds.

An extensive analysis shows [6] that the relative stability of long-period superlattices (isolated interfaces) is mostly controlled by ΔE_{cs}. Typical values of this quantity in this case are ~12 meV/atom. For example, the (100) Si/Ge SL has a ΔE_{cs} of 11.7 meV/atom. For short-period SLs (interacting interfaces) the interface-like terms are important. We distinguish between attractive interfaces ($\Delta E_{sr} + \Delta E_{chem} < 0$) or repulsive ones ($\Delta E_{sr} + \Delta E_{chem} > 0$) (see [6] for a listing of ΔE_{sr} and ΔE_{chem} values, which depend on SL orientation, repeat period and substrate lattice parameters).

A similar approach is followed by Ciraci and Batra [7] who consider the formation enthalpy as the sum of strain energy and interfacial energy, which is approximately equal to the difference in energy between the heteropolar and homopolar average. They estimate the interfacial energy to be 7 meV/atom (on the average).

D INTERLAYER DISTANCE

The choice of a substrate with a proper lattice constant can also be achieved by growing a partly relaxed $Si_{1-x}Ge_x$ buffer layer between the strained-layer configuration and the Si substrate [8]. The quality of this buffer layer and thereby that of the subsequent strained-layer configuration is strongly influenced by elastic-strain-driven phenomena, dislocation generation and surface waviness. For Si and Ge layers grown on a $Si_{1-x}Ge_x(001)$ buffer, with thickness smaller than the critical thickness, the lattice constant of the strained layer in the growth plane, $a_{\|}$, is equal to that of the buffer layer, while perpendicular to the growth plane, $a_{\perp}$ is given to a very good approximation by the elastic theory and is equal to [9]:

$$a_{i\perp} = a_i\,[1 - D^i(a_{\|}/a_i - 1)] \tag{5}$$

with $D^i = 2c_{12}^i/c_{11}^i$, a_i the bulk lattice constant and c_{11}^i and c_{12}^i the elastic constants of the i (Si or Ge) material. The room temperature values of the elastic constants are given in TABLE 1.

TABLE 1 Room temperature elastic constants of Si and Ge (in Mbar) [10].

	c_{11}	c_{12}
Si	1.675	0.650
Ge	1.315	0.494

Valence-force-field calculations and total energy calculations [11-13] have shown that the elastic theory predicts very accurately the elastic constants of strained Si and Ge. Also for growth on the same substrate, the interlayer distance between Si and Ge layers at the interface is very close to the mean value of the interlayer distance in the distorted materials [11-13]. A listing of typical interlayer Si-Ge distances is given in TABLE 2 [13].

TABLE 2 Interfacial Si-Ge distances (in Å) for the epitaxial $(Si)_5/(Ge)_5$ superlattice grown pseudomorphically on various substrates. Respective distances in the bulk constituents coh-Si and coh-Ge (coherently grown on the substrate) are also given for comparison.

	$(Si)_5/(Ge)_5$	coh-Si	coh-Ge
α_{Si}	2.381	2.352	2.408
$\alpha_{Si_{0.56}Ge_{0.44}}$	2.97	2.366	2.425
α_{Ge}	2.419	2.385	2.449

E INTERDIFFUSION

Finally, we consider the question of interdiffusion at the Si/Ge interface. Experimental studies [14] showed that substantial intermixing occurs even for short anneals. A diffusion constant of 10^{-24} m^2/s was measured [15]. Interdiffusion is rapid only at the very early annealing stages, after which it diminishes. Obviously, the driving force behind intermixing is strain relaxation. It was found [14] that the amount of intermixing at the two Si-Ge interfaces (per unit cell) is $<20\%$ (assuming that both interfaces mix equally).

The issue of interfacial stability, and the more subtle question of whether intermixing leads to randomised or to ordered interface layers, is debated among theoretical studies (for a discussion of ordering in Si-Ge systems see [6]). Mader et al [16] find a tendency for thin ordered layers (they have not considered random layers) to be more stable than the abrupt interface. Bernard and Zunger [6] find no such tendency (even for random interface layers) for any of the substrate lattice parameters studied. Recent empirical-potential calculations [17] found that both ordered and randomised layers are more stable than the abrupt interface for most of the substrate parameter values. Relevant enthalpies of formation ΔH of the $(Si)_4/(Ge)_4$ SL are given in TABLE 3. These results might have implications for the optical properties of Si-Ge SLs.

TABLE 3 Enthalpy of formation ΔH (in meV/atom) of the $(Si)_4/(Ge)_4$ superlattice constrained on various substrates. The number of ordered monolayers is denoted by q; r denotes randomised layers.

	$\alpha_s = \alpha_{Si}$	$\alpha_s = \alpha_{SL}$	$\alpha_s = \alpha_{Ge}$
q = 0	17.5	7.9	14.4
q = 2	16.2	7.3	14.7
r = 2	16.8	7.7	15.0

F CONCLUSION

In this Datareview we discussed the structure of a Si/Ge interface. The main conclusions are that thin epilayers form a strain-layer configuration with in-plane lattice constant equal to that of the substrate. The perpendicular lattice constant in this case is well described by the elastic theory. The critical thickness for strain-layer configuration is presented. Formation energies and interdiffusion are also discussed.

ACKNOWLEDGEMENT

This work has been supported in part by the ESPRIT Basic Research Action No. 7128.

REFERENCES

[1] E. Kasper [*NATO ASI Ser. B, Phys. (USA)* vol.170 (1987)]
[2] S.C. Jain, J.R. Willis, R. Bullough [*Adv. Phys. (UK)* vol.39 (1990) p.127]
[3] J.H. Van der Merwe [*Surf. Sci. (Netherlands)* vol.31 (1972) p.198]
[4] J.W. Mathews, A.E. Blakeslee [*J. Cryst. Growth (Netherlands)* vol.27 (1974) p.118]
[5] E. Kasper, H.J. Herzog [*Thin Solid Films (Switzerland)* vol.44 (1977) p.357]
[6] J.E. Bernard, A. Zunger [*Phys. Rev. B (USA)* vol.44 (1991) p.1663]
[7] S. Ciraci, I.P. Batra [*Phys. Rev. B (USA)* vol.38 (1988) p.1835]
[8] E. Kasper, H.J. Herzog, H. Jorke, G. Abstreiter [*Superlattices Microstruct. (UK)* vol.3 (1987) p.141]
[9] C.G. Van de Walle, R.M. Martin [*Phys. Rev. B (USA)* vol.34 (1986) p.5621]
[10] E. Anastassakis [in *Light Scattering in Semiconductor Structures and Superlattices* Eds D.J. Lockwood, J.F. Young (Plenum, New York, 1991)]
[11] S. Ciraci, A. Baratoff, I.P. Batra [*Phys. Rev. B (USA)* vol.41 (1990) p.6069]
[12] S. Froyen, D.M. Wood, A. Zunger [*Phys. Rev. B (USA)* vol.36 (1987) p.4547 and *Phys. Rev. B (USA)* vol.37 (1988) p.6893]
[13] P.C. Kelires, C. Tserbak, G. Theodorou [*Phys. Rev. B (USA)* vol.48 (1993) p.14238]
[14] T.E. Jackman, J.-M. Baribeau, D.J. Lockwood, P. Aebi, T. Tyliszczak, A.P. Hitchcock [*Phys. Rev. B (USA)* vol.45 (1992) p.13591]
[15] J.-M. Baribeau, R. Pascual, S. Saimoto [*Appl. Phys. Lett. (USA)* vol.57 (1990) p.1502]
[16] K.A. Mader, H. von Kanel, A. Baldereschi[*Superlattices Microstruct. (UK)* vol.9 (1991) p.15]
[17] P.C. Kelires [*Phys. Rev. B (USA)* vol.49 (1994) p.11496-9]

CHAPTER 3

THERMAL, MECHANICAL AND LATTICE VIBRATIONAL PROPERTIES

3.1 Elastic stiffness constants of SiGe

S.P. Baker and E. Arzt

May 1994

A INTRODUCTION

No data from reliable experimental measurements (e.g. ultrasonic measurements of bulk single crystals) of the elastic constants of Si-Ge alloys have been published to date. However, the data that do exist suggest that the elastic constants can be adequately determined using simple linear rule-of-mixtures (ROM) interpolations from the elastic constants of pure Si and Ge, which are very well documented. Furthermore, there is a variety of indirect evidence, both experimental and theoretical, which suggests that deviations from ROM values are small.

This possibility is not surprising considering the close similarity of Si-Si and Ge-Ge bonds and the very low ionicity of Si-Ge bonds. For covalently bonded materials, the stiffness can be related to bond length. The lattice parameters of Si and Ge differ by only 4% and the average lattice parameters of Si-Ge alloys follow the ROM prediction with only small deviations [1].

Si-Ge alloys crystallise in the diamond cubic form. Thus the elastic stiffness tensor, c_{ij}, is reduced to the three independent stiffness constants c_{11}, c_{12} and c_{44}. TABLE 1 shows typical values of these constants for pure Si and Ge.

TABLE 1 Elastic constants of Si (p-type, $\rho = 410\ \Omega$ cm) and Ge (n-type, $\rho = 45\ \Omega$ cm) at 298 K, 1 atm, in GPa (see [2] and references therein).

	Si	Ge
c_{11}	165.8	128.5
c_{12}	63.9	48.3
c_{44}	79.6	66.8

The rule of mixtures for the alloy elastic constants can then be simply written as

$$c_{ij} = c_{ij,Ge}x_{Ge} + c_{ij,Si}(1-x_{Ge}) \tag{1}$$

where c_{ij}, $c_{ij,Ge}$ and $c_{ij,Si}$ are the elastic stiffness constants of the alloy, pure germanium and pure silicon, respectively, and x_{Ge} is the germanium concentration.

B EXPERIMENTAL RESULTS

We have found only two reports of direct measurements of the elastic constants of Si-Ge alloys [3,4], of which only one appears to be reliable [5]. Mendik et al determined the elastic constants of a relaxed $Si_{0.49}Ge_{0.51}$ film by Brillouin light scattering and obtained c_{ij}'s which were identical (within experimental error) to ROM averages of the c_{ij}'s obtained in the same way from pure Si and Ge films.

A few studies support the use of ROM values indirectly. When an (001) $Si_{1-x}Ge_x$ film is grown epitaxially on an (001) silicon substrate, the film is biaxially compressed in the film plane to adopt the lattice parameter of the substrate and expands in the direction perpendicular to the film. This tetragonal distortion can be described by

$$\varepsilon_{\perp} = -[2\nu(1-\nu)]\varepsilon_{\parallel} = -2(c_{12}/c_{11})\varepsilon_{\parallel} \qquad (2)$$

where $\varepsilon_{\perp}$ and $\varepsilon_{\parallel}$ are the strain perpendicular and parallel to the film plane, respectively, ν is Poisson's ratio and the c_{ij}'s are the appropriate values for the alloy film. In studies of epitaxial films with $x_{Ge} = 0.05$ [6], $0.16 \leq x_{Ge} \leq 0.25$ [7] and $x_{Ge} = 0.31$ [8], the composition (and thus the unstrained lattice parameter) and some measure of the tetragonal distortion were determined independently and were found to be consistent with ROM values at least for the c_{ij}'s in Eqn (2). In another study [9], X-ray diffraction was used to obtain ε, and ROM c_{ij}'s were used along with published Raman stress coefficients to predict Raman shifts in rough agreement with experimentally measured values in epitaxial films having $0.04 \leq x_{Ge} \leq 0.23$. Furthermore, such diverse quantities as atomic volume, stacking fault energy, and some Raman strain shift coefficients, which all depend in some way on the elastic stiffness tensor, have been observed to follow ROM behaviour with x_{Ge}.

C CALCULATED RESULTS

In contrast to experimental results, the elastic constants of at least $Si_{0.5}Ge_{0.5}$ alloys have been calculated using both empirical and theoretical models. Calculations using empirical interatomic potentials [10] and using the theoretical local (electron) density approximation [11] for the zinc-blende (ordered) structure accurately reproduce the elastic constants of pure Si and Ge and predict ROM values for the alloy elastic constants. Elastic network relaxation calculations which assume ROM elastic constants for a random alloy result in ROM lattice parameters.

D CONCLUSION

Reports of direct experimental studies of the elastic constants of Si-Ge alloys are not currently available. However, both experimental evidence and atomic-scale calculations suggest that these values for the alloy closely follow a simple rule of mixtures formulation for germanium compositions up to 50%. The accuracy of the indirect correlations from strained layers also suggests that the third-order elastic constants have only a small effect up to about 2% elastic strain.

REFERENCES

[1] J.P. Dismukes, L. Ekstrom, R.J. Paff [*J. Phys. Chem. (USA)* vol.68 (1964) p.3201]
[2] Landolt-Börnstein [*Numerical Data and Functional Relationships in Science and Technology* New Series, Group III, vol.17a (Springer, Berlin, 1982)]
[3] V.T. Bublik, S.S. Gorelik, A.A. Zaitsev, A.Y. Polyakov [*Phys. Status Solidi B (Germany)* vol.66 (1974) p.427]

[4] M. Mendik, M. Ospelt, H. von Känel, P. Wachter [*Appl. Surf. Sci. (Netherlands)* vol.50 (1991) p.303]

[5] In the work by Bublik et al a positive deviation from the ROM predictions to values greater than those of Si was reported for the c_{ij}'s. However, these authors reported some difficulty in formulating their alloys and this work has not been substantiated.

[6] D.M. Maher, H.L. Fraser, C.J. Humpfreys, R.V. Knoell, J.C. Bean [*Appl. Phys. Lett. (USA)* vol.50 (1987) p.574]

[7] N.R. Parikh et al [*Thin Solid Films (Switzerland)* vol.163 (1988) p.455]

[8] J.C. Woicik et al [*Phys. Rev. B (USA)* vol.43 (1991) p.2419]

[9] B. Dietrich et al [*J. Appl. Phys. (USA)* vol.74 (1993) p.3177]

[10] K.E. Khor, S. Das Sarma [*J. Vac. Sci. Technol. B (USA)* vol.10 (1992) p.1994]

[11] S. Wei, D.C. Allan, J.W. Wilkins [*Phys. Rev. B (USA)* vol.46 (1992) p.12411]

3.2 Thermal properties of SiGe

K.L. Wang and X. Zheng

June 1994

A INTRODUCTION

Presented here is a review of the thermal properties of Ge-Si alloys. The review mainly covers the experimental data spanning more than three decades. Some important theoretical work is also included.

Ge-Si is important not only as an electronic material, but also for thermoelectric power applications. Recent developments in research into the alloys also show that $Ge_{1-x}Si_x$ in the form of layered material may be important for microsensor and micromachine applications. The thermal, mechanical and lattice vibrational data on Ge-Si alloys, especially those on layered Ge-Si material, are rather incomplete. It is hoped that this Datareview will serve the purpose of further promoting research in these areas.

B THERMAL CONDUCTIVITY

Systematic work on thermal conductivity of $Ge_{1-x}Si_x$ alloys was performed by the RCA Princeton Laboratory in the early 1960s [1]. Shown in FIGURE 1 is the room temperature thermal resistivity ($1/\kappa$) of undoped and doped $Ge_{1-x}Si_x$ alloys as a function of alloy composition, measured by an absolute thermal conductivity apparatus [2]. The thermal conductivity is the largest for undoped samples; as the doping density increases the thermal conductivity decreases. Heavily doped samples, both n- and p-type, have significantly larger thermal resistivity than undoped samples, due to added scattering of phonons by dopants.

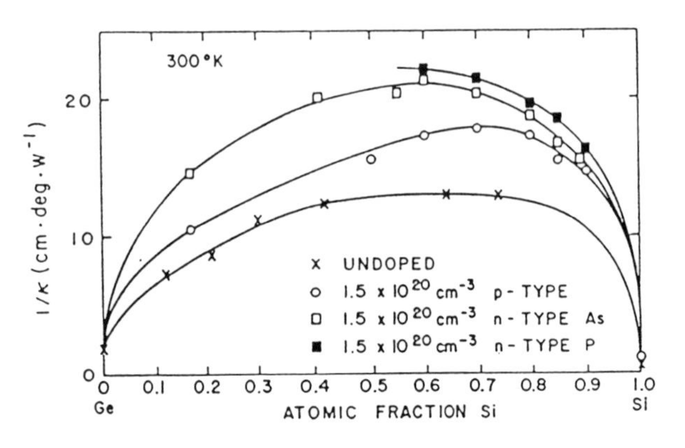

FIGURE 1 Compositional dependence of the thermal resistivity of $Ge_{1-x}Si_x$ at 300 K for 'undoped' material ($n \sim 2 \times 10^{18}$ cm^{-3}) and material doped to 1.5×10^{20} cm^{-3} with B, As and P [1].

For extremely heavily doped p-type samples, the significant contribution of free carriers to the thermal conductivity offsets the effect of the scattering, leading to a decrease in thermal resistivity in the very high carrier concentration regime. FIGURES 2 and 3 show the thermal resistivity of p- and n-type $Ge_{1-x}Si_x$ alloys as a function of carrier concentration obtained by measurement of thermal diffusivity κ/C_P [1]. The maximum thermal resistivity of p-type samples is shown to be approximately 3×10^{20} cm^{-3} at 300 K. At elevated temperatures, the maximum shifts to slightly lower concentrations. The thermal resistivity of n-type samples exhibits no peak due to the fact that extremely high doping density was not achieved.

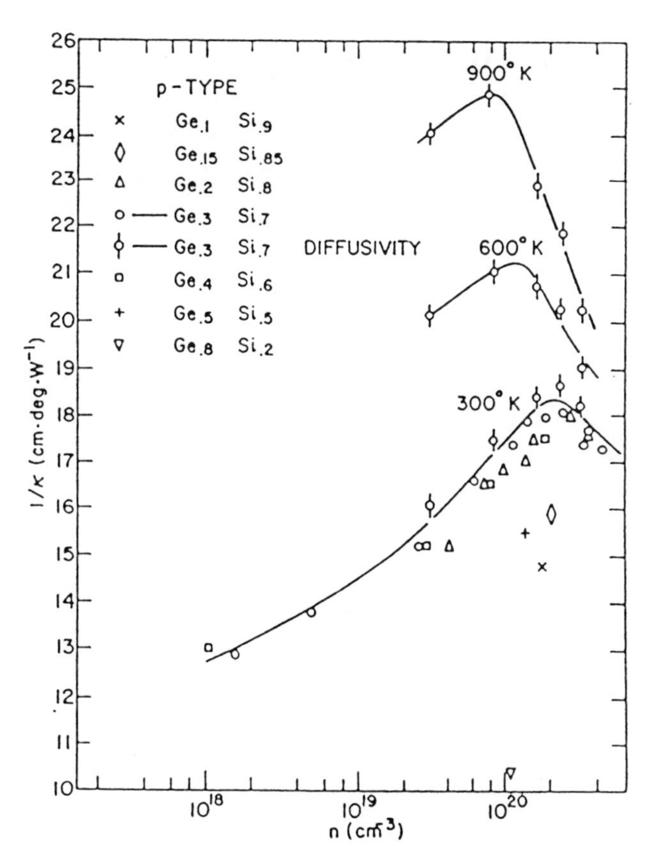

FIGURE 2 Thermal resistivity of p-type $Ge_{1-x}Si_x$ alloys as a function of carrier concentration with temperature and alloy composition as parameters [1].

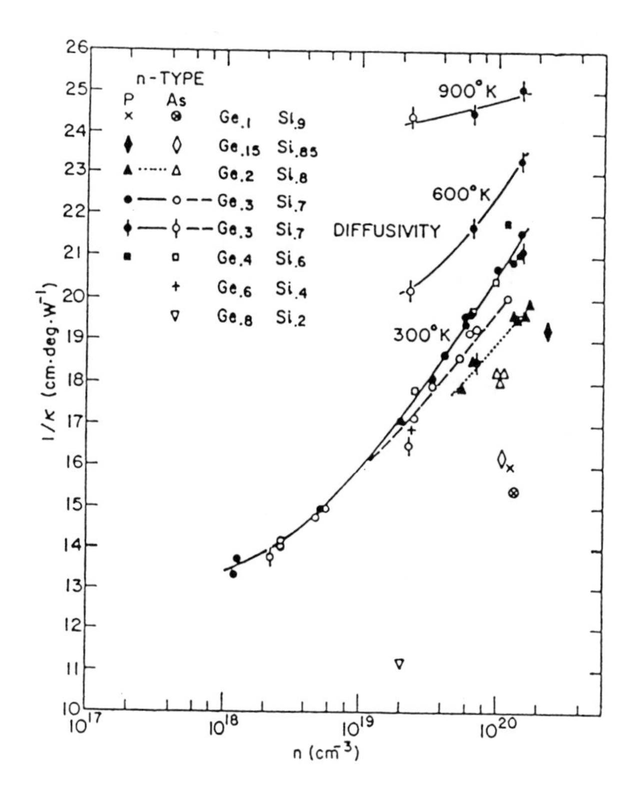

FIGURE 3 Thermal resistivity of n-type $Ge_{1-x}Si_x$ alloys as a function of carrier concentration with temperature, alloy composition and doping species as parameters [1].

One method of achieving low thermal conductivity is by the use of the 'fine grain GeSi' technique, in which a thermal conductivity decrease of up to 50% was attained by increasing grain boundary scattering of phonons [3,4]. Shown in FIGURE 4 is the normalised thermal conductivity of a sintered p-type $Ge_{0.2}Si_{0.8}$ alloy as a function of reciprocal particle size [4]. An observable trend of the grain size effect on thermal conductivity is evident from the figure.

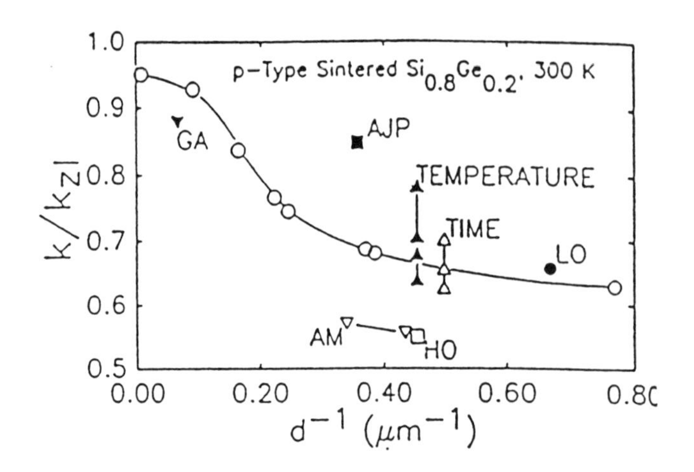

FIGURE 4 Thermal conductivity of sintered p-type $Ge_{0.2}Si_{0.8}$ normalised to zone levelled material as a function of reciprocal particle size d^{-1}. The labels indicate processing variations discussed in [4].

C THERMAL EXPANSION, GRÜNEISEN CONSTANT, SPECIFIC HEAT

In an early work, the linear thermal expansion coefficient α of $Ge_{1-x}Si_x$ alloys at 273 K and 473 K was measured by Zhdanova et al [5]. The data are shown in FIGURES 5 and 6. The value α is shown to decrease monotonically from Ge to Si in FIGURE 5 and to increase with temperature in FIGURE 6. It is interesting to note that there is a sudden change of the slope which coincides with the transition of the band structure from Ge-like to Si-like.

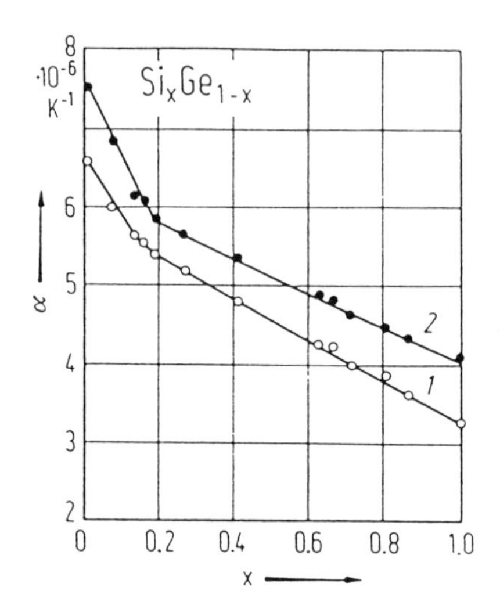

FIGURE 5 Compositional dependence of the linear thermal expansion coefficient of $Ge_{1-x}Si_x$ at (1) 473 K and (2) 773 K [5].

Theoretical work on the thermal expansion coefficient, α, Grüneisen constant, γ, and specific heat at constant volume, C_Ω, is described in [6]. In the treatment a simplified model was used where the lattice vibrations of Si and Ge atoms in the $Ge_{1-x}Si_x$ solid solution were assumed to be those of pure Si and Ge crystals except that the lattice constant of the alloy was used. Considering the volume effect on the force constant of the pure constituent, the phonon dispersion curves of local and band modes were obtained and used for the calculations of α, γ and C_Ω. The results are discussed below. Generally, there is good agreement between the calculations and experiments. Shown in FIGURE 7 is the calculated linear expansion coefficient α of $Ge_{1-x}Si_x$ alloys from 0 K to 500 K [6]. Also shown are the experimental data for pure Ge and pure Si [7-9]. The negative thermal expansion at low temperature is related to the softening of the transverse-like acoustic phonons of the crystal under pressure [10-12].

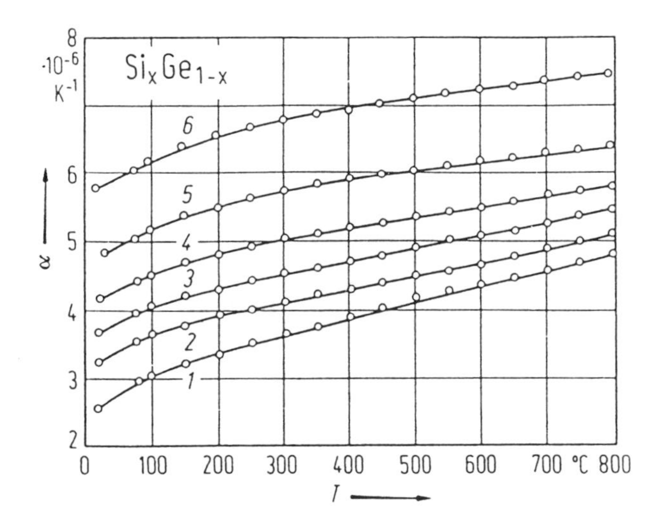

FIGURE 6 Temperature dependence of the linear thermal expansion coefficient. (1: Si, 2: 79.7 at. % Si, 3: 64.9 at. % Si, 4: 40.2 at. % Si, 5: 15 at. % Si, 6: Ge) [5].

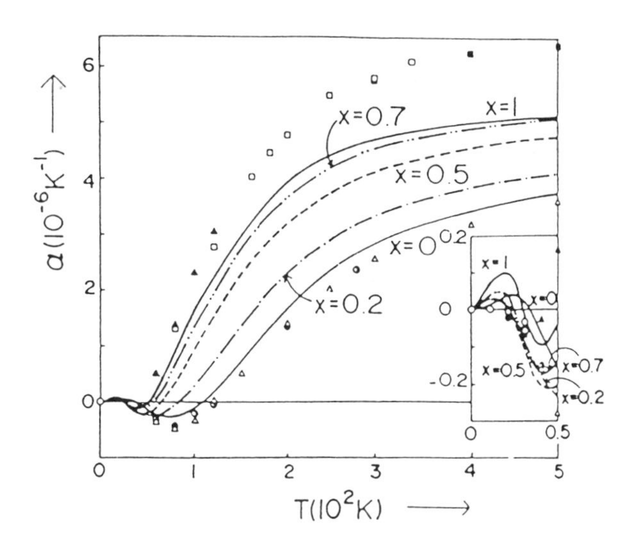

FIGURE 7 Linear expansion coefficient α of Si, $Ge_{0.2}Si_{0.8}$, $Ge_{0.5}Si_{0.5}$, $Ge_{0.7}Si_{0.3}$ and Ge vs. temperature. The parameter x is the germanium fraction [6]. The points are observed data; the upper curve is for Ge and the lower one is for Si [7-9].

Shown in FIGURE 8 is the calculated Grüneisen constant γ of $Ge_{1-x}Si_x$ alloys [6] together with the experimental data for $Ge_{0.2}Si_{0.8}$ obtained directly from measurements of inertia thermoelastic stress by pulse-heating a portion of the sample [13]. The near constant γ at high temperature is in good agreement with the experimental data. At low temperature, there is a negative maximum value for x = 0.2, and a negative minimum value for x = 0.7.

Zhifeng et al [14] have reported the alloy composition dependence of the Grüneisen constant associated with Ge-Ge, Ge-Si and Si-Si bonds at ambient temperature for a polycrystalline $Ge_{0.45}Si_{0.55}$ sample. In that work, the Raman shifts for different modes were measured for samples under hydrostatic pressure up to 8 kbar in order to evaluate γ, since the value of γ depends on the differential of the phonon frequency with respect to the volume, dω/dV. To

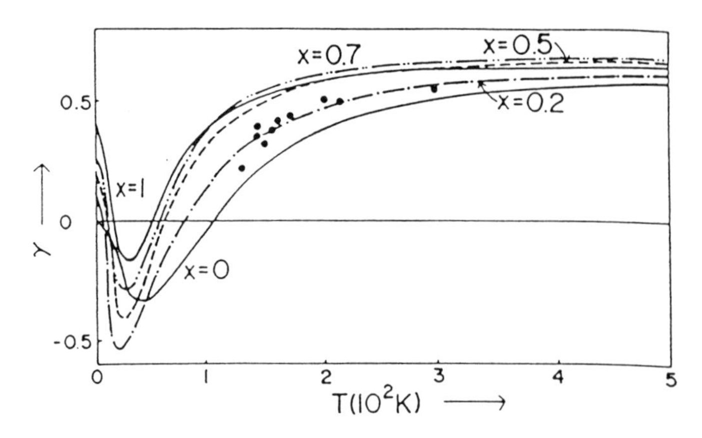

FIGURE 8 Grüneisen constant γ of Si, $Ge_{0.2}Si_{0.8}$, $Ge_{0.5}Si_{0.5}$, $Ge_{0.7}Si_{0.3}$ and Ge vs. temperature. The parameter x is the germanium fraction [6]. The points are observed data for $Ge_{0.2}Si_{0.8}$ alloy [13].

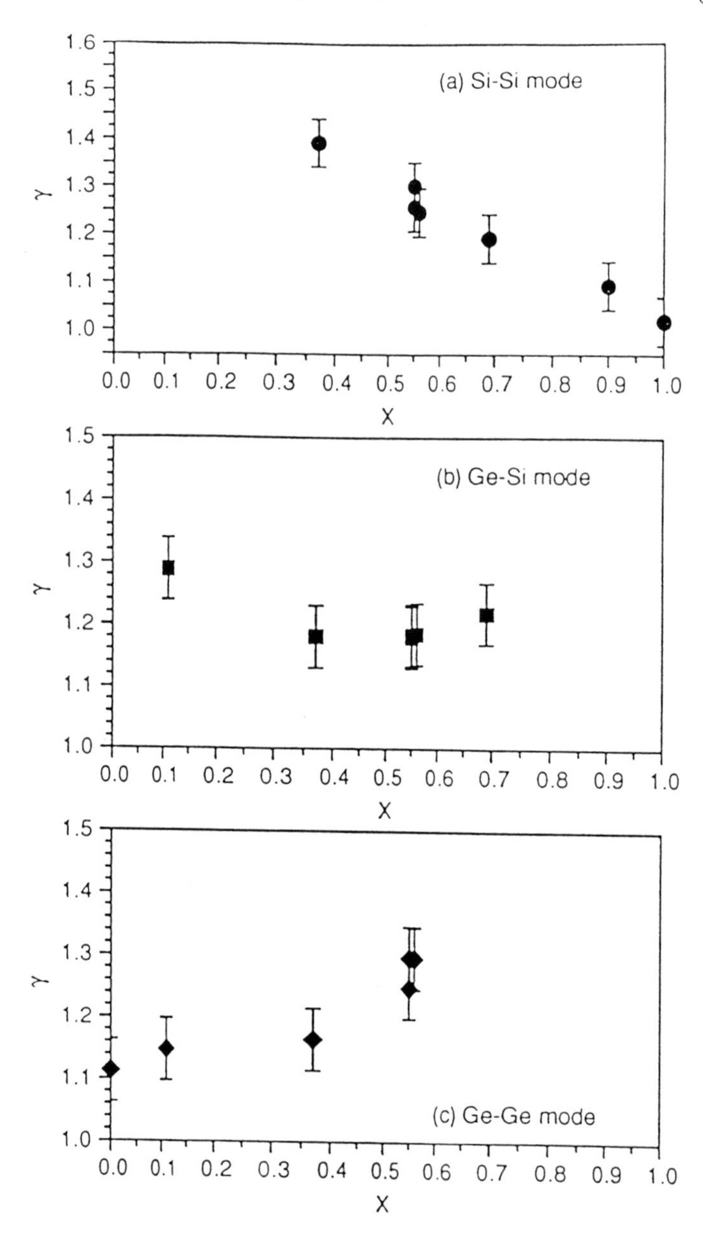

FIGURE 9 Grüneisen constant γ as a function of the Si fraction x for the three modes [14].

evaluate the Grüneisen constant, the linear interpolation of the bulk moduli between the values of pure Si and pure Ge was used. The result for γ is shown in FIGURE 9 for the three different modes.

FIGURE 10 shows the calculated specific heat, C_Ω, of $Ge_{1-x}Si_x$ alloys from 0 K to 500 K [6]. Also exhibited are the experimental data (unfilled circles) for the specific heat of pure Si and pure Ge for constant pressure, C_P [15]. The difference $C_P - C_\Omega$ is negligible at low temperature, but increases by less than 3 - 3.5% at 500 K. The calculated and experimental results are in reasonable agreement. Another set of data, C_P of the $Ge_{0.3}Si_{0.7}$ alloy, is given in [1] and is shown in FIGURE 11.

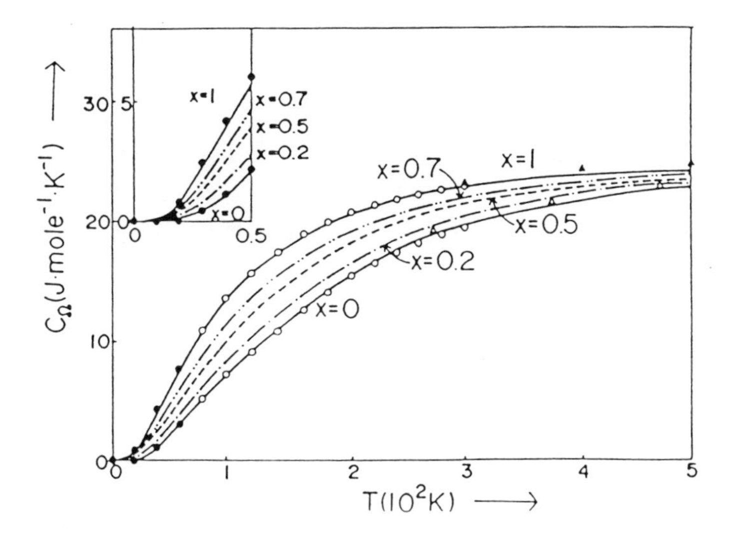

FIGURE 10 Specific heat at constant volume C_Ω of Si, $Ge_{0.2}Si_{0.8}$, $Ge_{0.5}Si_{0.5}$, $Ge_{0.7}Si_{0.3}$ and Ge vs. temperature. The parameter x is the germanium fraction [6]. The points are the observed data for Si and Ge [15].

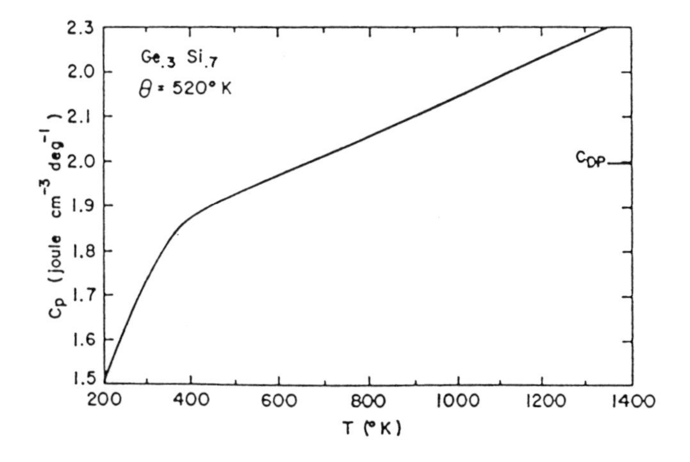

FIGURE 11 Specific heat of $Ge_{0.3}Si_{0.7}$ alloy as a function of temperature [1].

D THERMOELECTRIC POWER MATERIAL - SEEBECK COEFFICIENT

Ge-Si alloy is useful for thermoelectric power applications. It is particularly suitable for space power applications. There are some advantageous properties. First, Ge-Si alloys are competitive in efficiency with the best available high temperature thermoelectric materials.

Further, the thermal and electrical properties of the n-type and p-type alloys are closely matched, which is a considerable advantage in device design. The most important point is that the chemical stability and mechanical strength of Ge-Si alloys at elevated temperatures allow them to be operated both in air and in vacuum without an appreciable change in efficiency [1].

The figure of merit Z, usually defined as $Z = Q^2/\rho\kappa$ (ρ and κ are the electric resistivity and thermal conductivity, respectively), is a critical parameter for thermoelectric power application, where Q is the Seebeck coefficient [1]. Shown in FIGURES 12 and 13 respectively are the Seebeck coefficients of p-type and n-type Ge-Si alloys as a function of carrier concentration parameter. There are two main approaches for improving the figure of merit Z: (1) reduce the thermal conductivity, and (2) improve the electric properties. In addition to the approach of 'fine grain GeSi' mentioned in Section B, another way is to incorporate other 'impurities' such as gallium phosphide into GeSi [16]. The dimensionless figure of merit ZT of an n-type GeSi sample calculated by Vining is shown in FIGURE 14 [17]. Also exhibited in it are experimental results for GaP-incorporated GeSi samples.

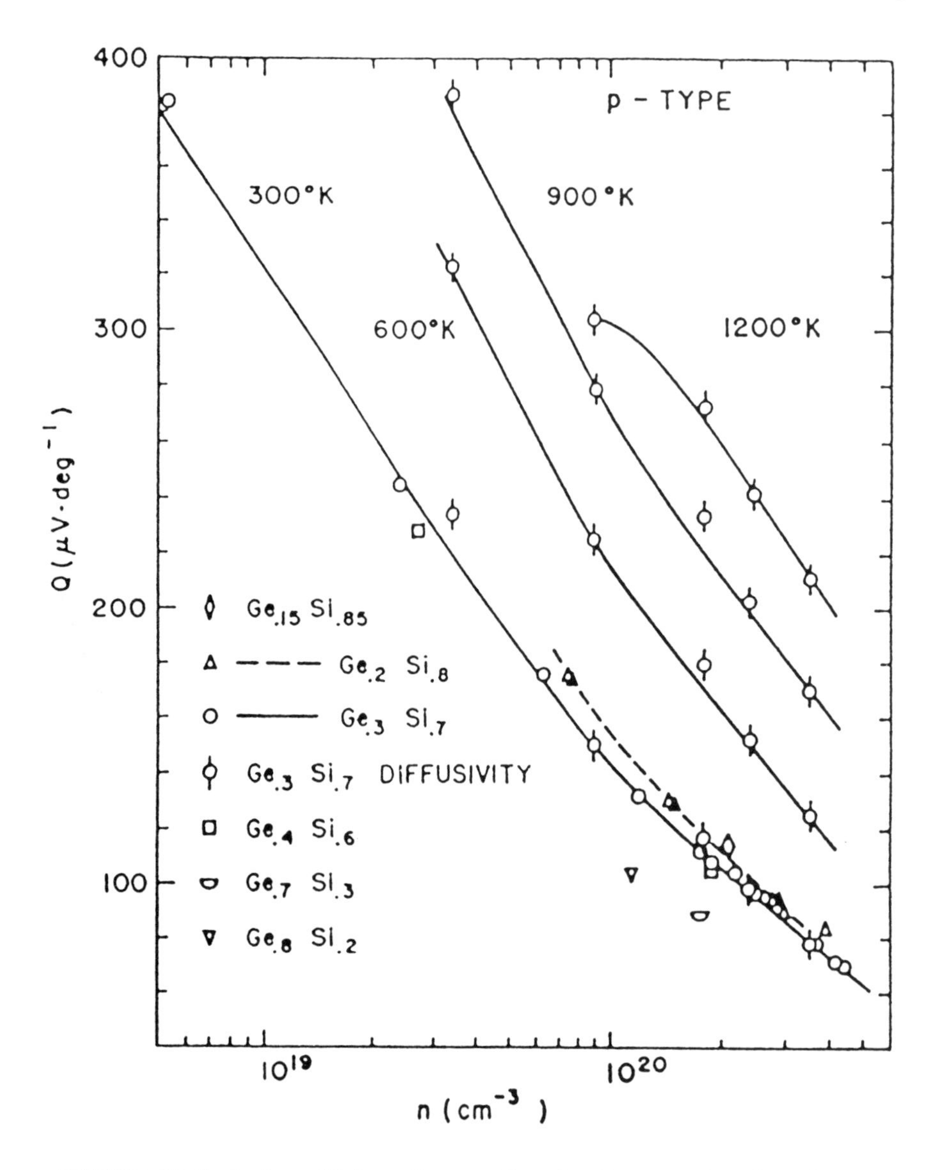

FIGURE 12 Seebeck coefficient of p-type $Ge_{1-x}Si_x$ alloys as a function of carrier concentration with temperature and alloy composition as parameters [1].

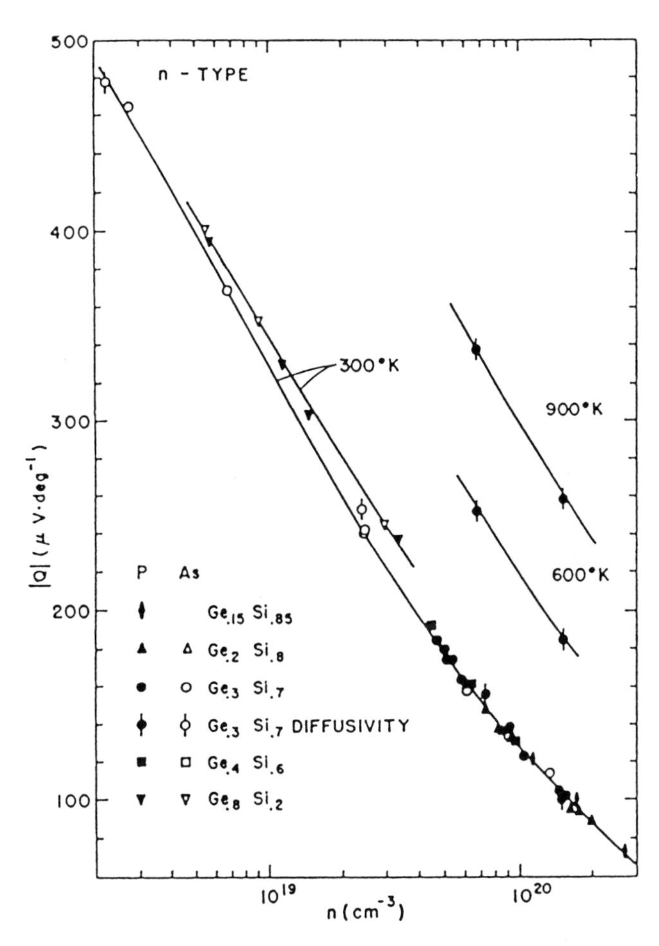

FIGURE 13 Seebeck coefficient of n-type $Ge_{1-x}Si_x$ alloys as a function of carrier concentration with temperature, alloy composition and doping species as parameters [1].

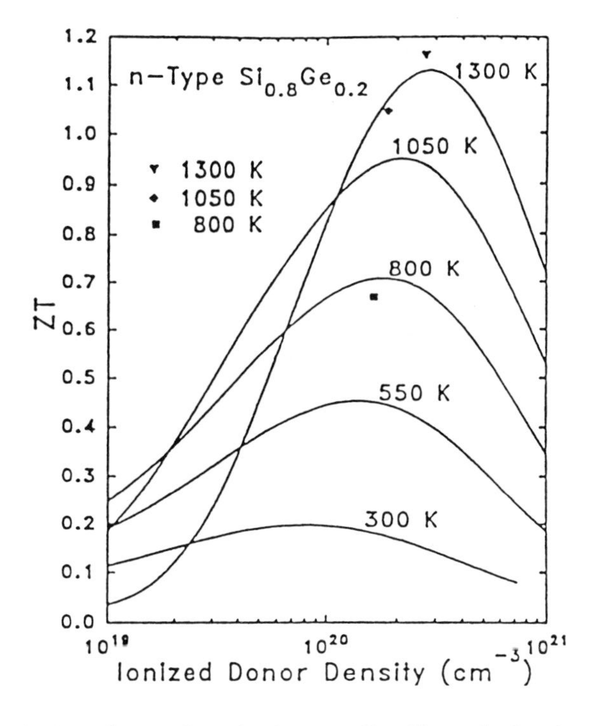

FIGURE 14 Dimensionless figure of merit of n-type $Ge_{0.2}Si_{0.8}$ calculated as described in [17]. The solid points represent experimental results for a sample of GeSi/GaP.

REFERENCES

[1] J.P. Dismukes, L. Ekstrom, E.F. Steigmeier, I. Kudman, D.S. Beers [*J. Appl. Phys. (USA)* vol.35 (1964) p.2899]

[2] F.D. Rosi, B. Abeles, R.V. Jensen [*J. Phys. Chem. Solids (UK)* vol.10 (1959) p.191]

[3] D.M. Rowe [*J. Phys. D, Appl. Phys. (UK)* vol.7 (1974) p.1843]

[4] C.B. Vining, W. Laskow, J.O. Hanson, R.R. Van der Beck, P.D. Gorsuch [*J. Appl. Phys. (USA)* vol.69 (1991) p.4333]

[5] V.V. Zhdanova, M.G. Kakna, T.Z. Samadashvili [*Izv. Akad. Nauk. SSSR Neorg. Mater. (Russia)* vol.3 (1967) p.1263]

[6] H.-Matsuo Kagaya, Y. Kitani, T. Soma [*Solid State Commun. (USA)* vol.58 (1986) p.399]

[7] Y.S. Touloukian, R.K. Kirby, R.E. Taylor, T.Y.R. Lee [*Thermophysical Properties of Matter, vol.12/13, Thermal Expansion* (IFI/Plenum, New York, 1975/1977)]

[8] G.A. Slack, S.F. Bartram [*J. Appl. Phys. (USA)* vol.46 (1975) p.89]

[9] T.F. Smith, G.K. White [*J. Phys. C (UK)* vol.8 (1975) p.2031]

[10] B.A. Weinstein [*Solid State Commun. (USA)* vol.24 (1977) p.595]

[11] T. Soma [*Solid State Commun. (USA)* vol.34 (1980) p.927]

[12] T. Soma, J. Satoh, H. Matsuo [*Solid State Commun. (USA)* vol.42 (1982) p.889]

[13] W.B. Gauster [*J. Appl. Phys. (USA)* vol.44 (1973) p.1089]

[14] Zhifeng Sui, H.H. Burke, I.P. Herman [*Phys. Rev. B (USA)* vol.48 (1993) p.2162]

[15] Y.S. Touloukian, R.K. Kirby, R.E. Taylor, T.Y.R. Lee [*Thermophysical Properties of Matter, vol.4, Specific Heat* (IFI/Plenum, New York, 1970)]

[16] R.K. Pisharody, L.P. Garvey [*13th Intersoc. Energy Conv. Eng. Conf.*, San Diego, CA, 1978, p.1969]

[17] C.B. Vining [*J. Appl. Phys. (USA)* vol.69 (1991) p.331]

3.3 Optical and acoustical phonons in SiGe; Raman spectroscopy

R. Schorer

April 1994

A INTRODUCTION

The lattice dynamics of SiGe alloy crystals is a rather complex topic. The average energy difference between the optical phonon bands of bulk Si and Ge is significantly larger than the width of these bands, which prevents the description of the optical phonon modes by the virtual crystal approximation.

In Raman spectra, three main peaks are observed which can be attributed to local Si-Si, Ge-Ge and Si-Ge vibrational modes. Section B will describe the composition dependence of these modes and also the origin of additional weak structures. Section C will review the theoretical approaches to the lattice dynamics of SiGe alloys, whereas experimental findings for resonances and acoustic modes will be presented in Sections D and E. Theoretical and experimental data on the strain-shift of optical phonons for bulk Si and Ge as well as for the alloy will be discussed in Section F.

B THE THREE MODE BEHAVIOUR OF THE OPTICAL PHONONS

The first Raman measurements on polycrystalline SiGe alloys in 1966 [1] were soon followed by studies with a wide systematic variation of x [2-4]. Measurements on liquid-phase-epitaxy (LPE)-grown single crystals are in good agreement with this early work [5].

For a wide composition range three optical modes dominate the Raman spectra, the frequencies of which are close to those of the bulk constituents and their linear average, respectively. They are attributed to local vibrations of Si-Si, Ge-Ge and Si-Ge atom pairs. The localisation is due to the large atomic mass discrepancy of Si and Ge. Their relative intensities are roughly proportional to the relative numbers of the corresponding bond types, i.e. $(1-x)^2$, $2x(1-x)$ and x^2, respectively [2]. Due to the negligible polarity of the Si-Ge bond [6], TO and LO modes are degenerate. The Raman lines are asymmetrically broadened on the low energy side. This effect is attributed to disorder induced scattering from forbidden modes with large k due to compositional disorder induced relaxation of k conservation [2]. The linewidth is strongly composition dependent [2].

FIGURE 1 shows the composition dependence of the frequencies of the optical modes. Typical Raman spectra are found in FIGURE 2. This dependence is linear for the Si-Si peak with $\omega_{Si-Si} = 520 - 70x$ [5], whereas it is monotonic but nonlinear for Ge-Ge (except for $x < 0.02$, where a weak maximum is observed [7]) and cusp-like for Si-Ge. Qualitatively, this behaviour can be explained in a simple cluster model [8]. The optical modes can be regarded as localised in random 3D clusters of the corresponding atom pairs. For increasing Ge content, for instance, the average random cluster size of Si-Si atom pairs is decreasing. Thus

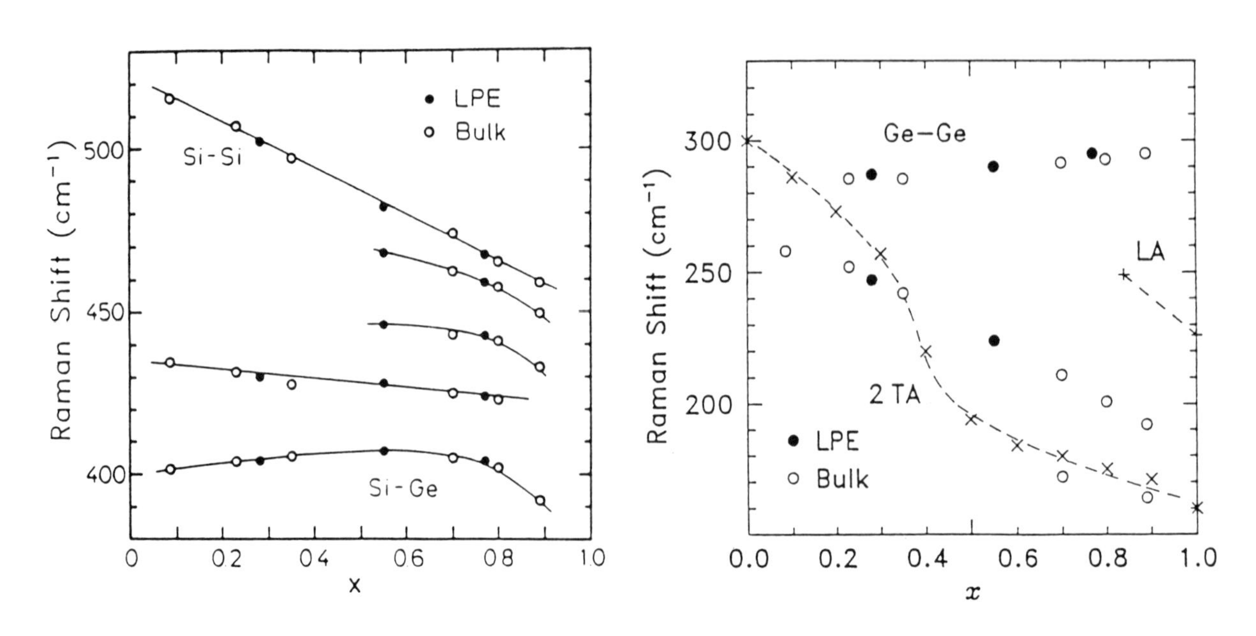

FIGURE 1 Frequencies of Raman active phonon modes in single crystal SiGe alloys measured at room temperature for different values of the Ge concentration x [5].

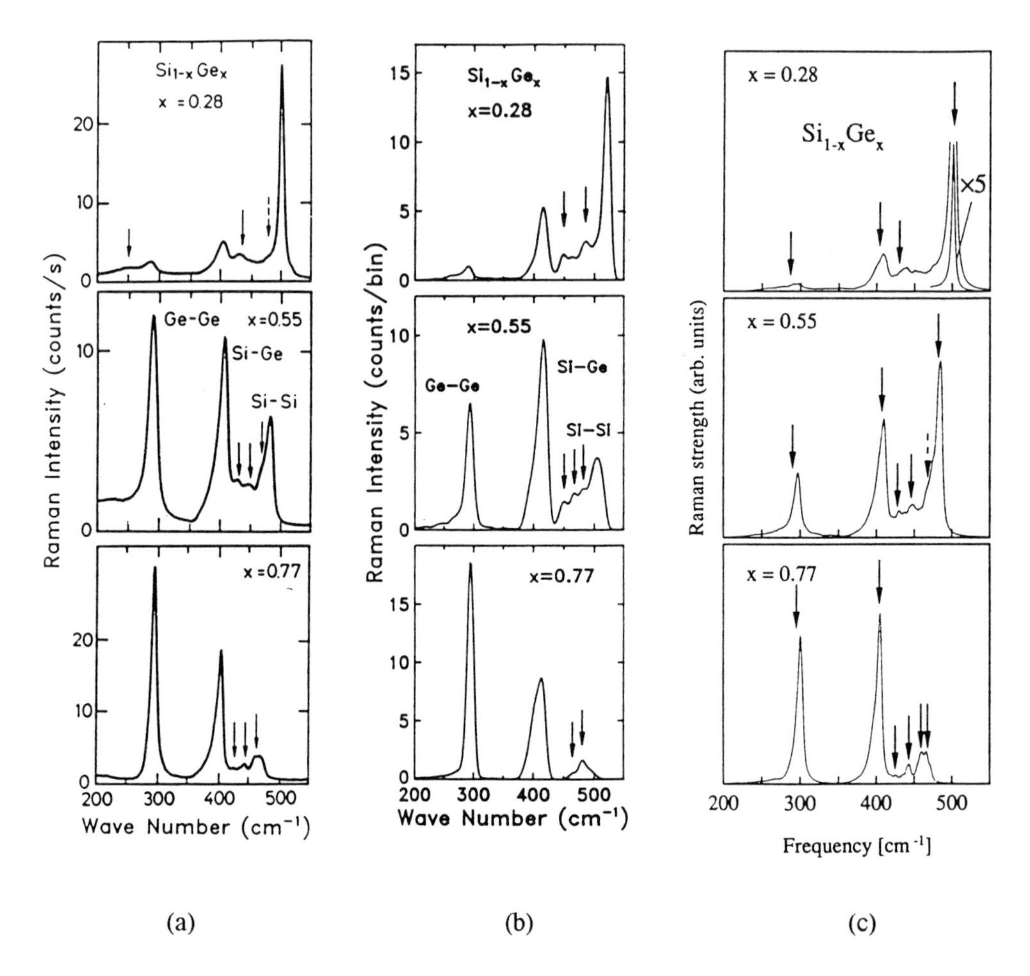

FIGURE 2 Raman spectra of SiGe alloys for different values of the Ge concentration x. (a) Experimental spectra [5]. (b) Spectra obtained from supercell calculations within a Keating-type model [5]. (c) Spectra obtained from supercell calculations with force constants derived from first principles [11].

the corresponding Si-Si frequency is decreasing due to a larger average confinement wave-vector, similarly to a downward shift in short-period Si/Ge or Si/SiGe superlattices (note that the confinement potential is 1D in this case as compared to 3D for the alloy). This effect is superposed with a phonon shift due to the internal strain.

Additionally, four weak peaks between Si-Ge and Si-Si are observed, especially for large x [2-5,9]. It was shown that these modes are not due to ordering or to defects, as proposed in earlier work [9], but can rather be explained in terms of localised Si-Si optical modes surrounded by an increasing number of Ge atoms [5]. A weak peak below 280 cm^{-1} is attributed to a superposition 2TA with disorder induced 1LA contributions [5].

C LATTICE DYNAMICS OF DISORDERED SYSTEMS

Only a small number of theoretical approaches to the lattice dynamics of SiGe alloys have been undertaken yet [5,10-12]. Mean-field approximations such as the coherent-potential-approximation (CPA), which were successfully used for $Al_{1-x}Ga_xAs$, fail in the case of SiGe alloys [13]. The frequencies of the three dominant Raman peaks have been calculated as a function of the composition x within an isodisplacement model [12]. More recently, supercell calculations were performed, where an artificial periodicity is introduced in the crystal. The lattice sites in this supercell (typical size ~200 atoms) are randomly occupied by Si and Ge atoms until the desired composition is achieved. Eigenfrequencies are then obtained by direct diagonalisation and results are averaged over several random configurations. Interatomic force constants were either obtained within a Keating-type model by fitting the bulk dispersions of the constituents [5] or by direct first principle calculations [11]. In the latter case strain effects are included by higher-order interatomic force constants also derived from first principles. Raman spectra were calculated within the bond polarisability model, assuming an additional Lorentzian line broadening. Results obtained from both methods are shown in FIGURE 2 in comparison with the experimental spectra. The qualitative agreement is excellent, especially concerning the weak optical peaks, although some quantitative deviations are still present. A comparison to the calculated phonon density of states (DOS) shows that k-conservation is not completely relaxed, since the maxima of DOS and Raman line often do not coincide [5].

D RESONANCES

Little work has been devoted to the electron-phonon coupling in SiGe alloys yet. In a resonant Raman scattering study on LPE-grown bulk alloys good agreement of the position of the resonance peaks for the E_0, $E_0 + \Delta_0$ and E_1, $E_1 + \Delta_1$ transitions, as determined by ellipsometry, was found [14]. The resonance curves of the main optical modes are rather similar, indicating that there is no pronounced spatial localisation of electronic states. Electronic confinement effects are observed in Si/SiGe superlattices [15].

E ACOUSTICAL PHONONS

Very few data are reported concerning the sound velocity v_s of SiGe alloys. A pronounced nonlinear composition dependence of the elastic constants of bulk alloys has been observed in an X-ray diffraction study [16]. Brugger et al [17] determined v_s for $x = 0.5$ from the Brillouin line and the doublet splitting in Raman spectra from a $Si/Si_{0.5}Ge_{0.5}$ superlattice and obtained a longitudinal $v_s = 7.5 \pm 0.4 \times 10^5$ cm/s which is considerably above the linear average of 6.67×10^5 cm/s. This result is consistent with [16]. Recently, from a fit to the splitting of higher index doublets of Raman data on similar structures, values of v_s much closer to the linear average were obtained [18]. A precise, systematic study on the composition dependence of v_s of unstrained SiGe alloys is still lacking.

F PHONON STRAIN-SHIFT

F1 Bulk Constituents

Reliable phonon strain-shift data are essential for characterising epitaxial SiGe alloy layers due to the presence of biaxial strain for growth on Si or Ge substrates. The first uniaxial stress measurements for bulk Si and Ge were reported in the early 1970s [19-21]. Three phenomenological coefficients p, q, r are obtained from these experiments, which describe the changes in the 'spring constants' for $k = 0$ optical modes with strain. With these, the corresponding strain shifts can be calculated for arbitrary strain directions [22]. The values for Si have been corrected recently by Raman measurements with excitation below the fundamental bandgap [22].

The k-dependence of the phonon shift was investigated theoretically by a Keating-type [24] and by a first principle approach [25]. The effect was found to be weak for LO phonons (<15%), whereas it is pronounced for TO phonons. It should be noted that in principle these results could be proved experimentally on short-period Si/Ge superlattice structures on different substrates. A problem, however, is the presence of interface roughness in these structures [26].

F2 SiGe Alloys

The experimental strain shift parameters for the bulk constituents for $k = 0$ can not be simply transferred to alloy phonon modes due to a relaxation of the k-conservation. There are very few experiments reported concerning the strain shift of SiGe alloy phonon modes [27-29]. Cerdeira et al compared thin pseudomorphic and thick totally relaxed alloy epilayers with the same x, whereas in the work of Lockwood et al thick metastable alloy layers on Si were subsequently relaxed by annealing. In this case strain was monitored by X-ray diffraction.

In the first work good agreement with the respective bulk values was found for Si-Si and Ge-Ge [27]. This is in contrast to the later work, where all three modes were found to shift similarly, with values considerably larger than for the bulk constituents [28]. The results for the Si-Si mode are confirmed in [29]. Lockwood et al attributed the discrepancy to the lack of strain monitoring in [27]. They also found a pronounced composition dependence of the strain shift coefficient. The results for the phonon strain-shift parameter b in the case of biaxial stress are summarised in TABLE 1. More accurate studies are still desirable. Especially uniaxial pressure experiments on SiGe alloys for determining p, q, r are still lacking, although hydrostatic pressure experiments were performed very early [30].

TABLE 1 Strain-shift coefficient b (in cm^{-1}) for the main optical modes of SiGe alloys and the bulk constituents. The phonon frequency shift is given by $\Delta\omega = b\varepsilon$, where ε is the in-plane lattice mismatch between epilayer and substrate:

$$\varepsilon = (a_{sub} - a_{epi})/a_{epi}.$$

	Si-Si	Si-Ge	Ge-Ge
Si[a]	-832		
Si[b]	-723		
Si (extrapolated)[c]	-715 ± 50		
x = 0.20[c]	-850 ± 50	-710 ± 50	-810 ± 50
x = 0.35[c]	-980 ± 50	-990 ± 50	-925 ± 50
x = 0.15 to 0.65[d]		-455	
Ge[e]			-408

[a][23]; [b][19]; [c][28]; [d][27]; [e][20].

REFERENCES

[1] D.W. Feldman, M. Ashkin, J.H. Parker Jr. [*Phys. Rev. Lett. (USA)* vol.17 (1966) p.1209]

[2] M.A. Renucci, J.B. Renucci, M. Cardona [in *Light Scattering in Solids* Ed. M. Balkanski (Flammarion, Paris, 1971) p.326]

[3] W.J. Brya [*Solid State Commun. (USA)* vol.12 (1973) p.253]

[4] J.S. Lannin [*Phys. Rev. B (USA)* vol.16 (1977) p.1510]

[5] M.I. Alonso, K. Winer [*Phys. Rev. B (USA)* vol.39 (1989) p.10056]

[6] U. Schmid, M. Cardona, N.E. Christensen [*Phys. Rev. B (USA)* vol.41 (1990) p.5919]

[7] H.D. Fuchs, C.H. Grein, M.I. Alonso, M. Cardona [*Phys. Rev. B (USA)* vol.44 (1991) p.13120]

[8] J. Menéndez, A. Pinczuk, J. Bevk, J.P. Mannaerts [*J. Vac. Sci. Technol. (USA)* vol.6 (1988) p.1306]

[9] D.J. Lockwood, K. Rajan, E.W. Fenton, J.-M. Baribeau, M.W. Denhoff [*Solid State Commun. (USA)* vol.61 (1987) p.465]

[10] T. Soma, Y. Kitani, H.-Matsuo Kagaya [*Solid State Commun. (USA)* vol.50 (1984) p.1007]

[11] S. de Gironcoli [*Phys. Rev. B (USA)* vol.46 (1992) p.2412]

[12] G.M. Zinger, I.P. Ipatova, A.V. Subashiev [*Fiz. Tekh. Poluprovodn. (Russia)* vol.11 (1977) p.656; *Sov. Phys.-Semicond. (USA)* vol.11 (1977) p.383]

[13] S. de Gironcoli, S. Baroni [*Phys. Rev. Lett. (USA)* vol.69 (1992) p.1959]

[14] M.I. Alonso, H.-P. Trah, E. Bauser, H. Cerva, H.P. Strunk [in *Epitaxy of Semiconductor Layered Structures* Eds R.T. Tung, L.R. Dawson, R.L. Gunshor (Materials Research Society, Pittsburgh, PA, 1988) vol.102 p.419]; F. Cerdeira et al [*Phys. Rev. B (USA)* vol.40 (1989) p.1361]; M.I. Alonso [PhD thesis, Max-Planck-Institut, Stuttgart, 1990]

[15] F. Cerdeira, A. Pinczuk, J.C. Bean [*Phys. Rev. B (USA)* vol.31 (1985) p.1202]

[16] V.T. Bubelik, S.S. Gorelik, A.A. Zaitsev, A.Y. Polyakov [*Phys. Status Solidi B (Germany)* vol.66 (1974) p.427]

[17] H. Brugger, G. Abstreiter, H. Jorke, H.J. Herzog, E. Kasper [*Phys. Rev. B (USA)* vol.33 (1986) p.5928]

[18] P.X. Zhang, D.J. Lockwood, H.J. Labbé, J.-M. Baribeau [*Phys. Rev. B (USA)* vol.46 (1992) p.9881]

[19] E. Anastassakis, A. Pinczuk, E. Burstein, F.H. Pollak, M. Cardona [*Solid State Commun. (USA)* vol.8 (1970) p.133]

[20] F. Cerdeira, C.J. Buchenauer, F.H. Pollak, M. Cardona [*Phys. Rev. B (USA)* vol.5 (1972) p.580]

[21] M. Chandrasekhar, J.B. Renucci, M. Cardona [*Phys. Rev. B (USA)* vol.17 (1978) p.1623]
[22] E. Anastassakis [*NATO ASI Ser. B, Phys. (USA)* vol.273 (1991) p.173]
[23] E. Anastassakis, A. Cantarero, M. Cardona [*Phys. Rev. B (USA)* vol.41 (1990) p.7529]
[24] Jian Zi, Kaiming Zang, Xide Xie [*Phys. Rev. B (USA)* vol.45 (1992) p.9447]
[25] A. Fasolino, E. Molinari, A. Qteish [in *Condensed Systems of Low Dimensionality* Eds J.L. Beeby et al (Plenum Press, New York, 1991) p.495]
[26] R. Schorer, G. Abstreiter, S. de Gironcoli, E. Molinari, H. Kibbel, H. Presting [*Phys. Rev. B (USA)* vol.49 (1994) p.5406]; S. de Gironcoli, E. Molinari, R. Schorer, G. Abstreiter [*Phys. Rev. B (USA)* vol.48 (1993) p.8959]
[27] F. Cerdeira, A. Pinczuk, J.C. Bean, B. Batlogg, B.A. Wilson [*Appl. Phys. Lett. (USA)* vol.45 (1984) p.1138]
[28] D.J. Lockwood, J.-M. Baribeau [*Phys. Rev. B (USA)* vol.45 (1992) p.8565]
[29] M.A.G. Haliwell, M.H. Lyons, S.T. Davey, M. Hockly, C.G. Tuppen, C.J. Gibbings [*Semicond. Sci. Technol. (USA)* vol.4 (1989) p.10]
[30] J.B. Renucci, M.A. Renucci, M. Cardona [*Solid State Commun. (USA)* vol.9 (1971) p.1651]

CHAPTER 4

BAND STRUCTURE

4.1 Energy gaps and band structure of SiGe and their temperature dependence

T. Fromherz and G. Bauer

April 1994

A INTRODUCTION

The binary alloys of silicon and germanium form a continuously variable system with a wide range of energy gaps and thus optical properties. Since the pioneering work of Braunstein [1] it has been known that for the lowest lying conduction bands the crossover from the $\Delta(6)$ states to the L(4) states occurs for a germanium content as high as 85%. Consequently, the conduction band structure of bulk or relaxed SiGe alloys is silicon-like for a wide range of alloy compositions.

For epitaxially grown $Si_{1-x}Ge_x$ films the band structure is drastically altered by the built-in strain which is fixed by the lattice constant of the substrate as long as the growth is pseudomorphic (lattice matched). Since the lattice mismatch between Si and Ge is about 4%, the critical thickness for pseudomorphic growth, which depends also on the growth conditions, decreases quite rapidly with increasing Ge content. For a pseudomorphic SiGe layer grown along the [001] direction, the biaxial strain causes a tetragonal distortion which shifts and splits the valence and conduction band edges and thus alters the energy gap.

In Section B experimental findings on the indirect minimum bandgap are summarised and compared with theoretical results for the bulk alloys and for $Si_{1-x}Ge_x$ films grown on Si substrates. Experimental data on the direct energy gaps are presented in Section C.

B FUNDAMENTAL ABSORPTION EDGE: INDIRECT MINIMUM ENERGY GAP

B1 Unstrained SiGe Alloys (Bulk)

The dependence of the indirect bandgap in $Si_{1-x}Ge_x$ alloys on both the Ge content and the temperature was first determined by Braunstein et al [1] by absorption measurements. The bandgap was determined by fitting the absorption data at low absorption levels to a Macfarlane-Roberts expression [1]. In FIGURE 1 the values of the indirect bandgap obtained by Braunstein et al are indicated by triangles. These data are the result of an extrapolation of the temperature dependence of the indirect energy gap (shown in FIGURE 2 for the entire range of Ge contents) to a temperature of 4.2 K. At this temperature Weber et al [2] measured photoluminescence (PL) more recently. The excitonic bandgap varies smoothly with the Ge content x from the Si gap at 1.155 eV to the gap of Ge at 0.740 eV as shown in FIGURE 1 by the full squares. The data of [1] deviate by up to 40 meV from the data reported by Weber and Alonso [2]. This is probably due to an oversimplification in the Macfarlane-Roberts expression used in the analysis of the absorption line in [1]. At around x = 0.85 the crossover from the Si-like Δ-conduction-band minimum to the Ge-like

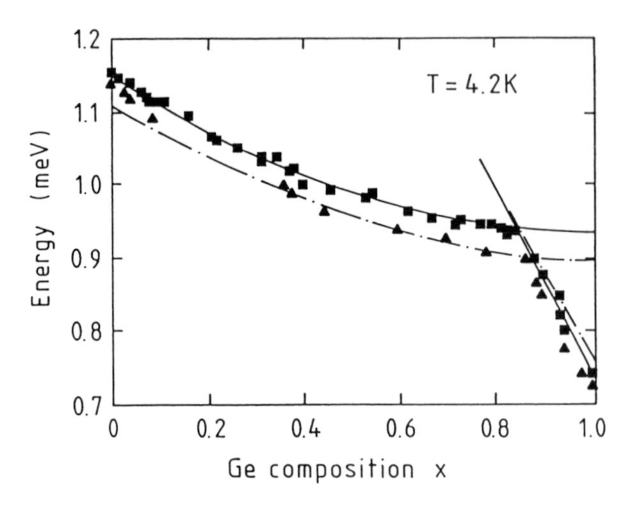

FIGURE 1 Bandgap determined by absorption measurements [1] (triangles) and excitonic bandgap obtained by photoluminescence [2] (full squares) vs. composition for bulk $Si_{1-x}Ge_x$ alloys. The broken line corresponds to the analytical expression given in [3]. The solid line gives the fit to measurements according to Eqns (1) and (2).

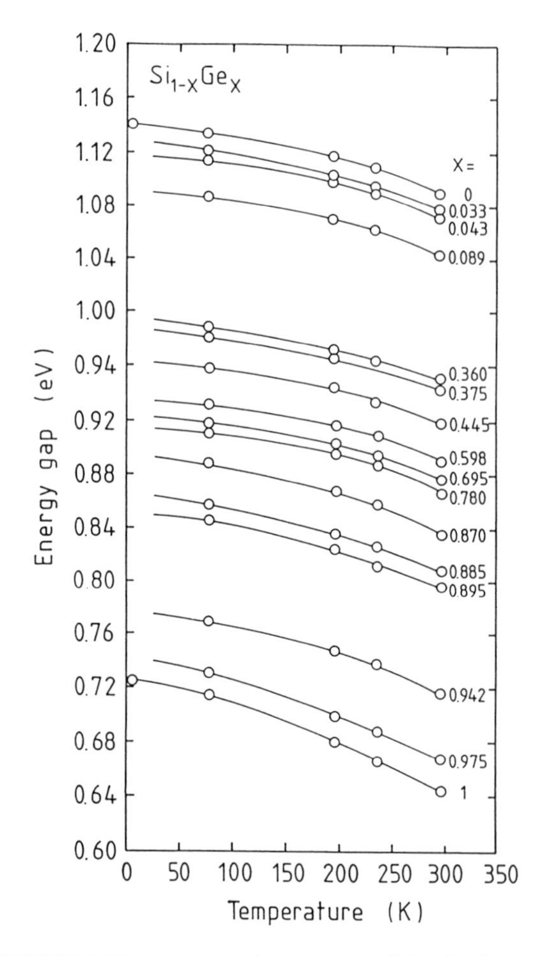

FIGURE 2 Temperature dependence of the fundamental indirect energy gap for $Si_{1-x}Ge_x$ bulk alloys [1].

L-conduction-band minimum occurs. The experimentally observed free-exciton bandgaps can be fitted to the quadratic expressions given below [2] and shown by the solid lines in FIGURE 1.

$$E_g^{(\Delta)}(x) = (1.155 - 0.43x + 0.0206x^2)\ eV \qquad for \quad 0 < x < 0.85 \qquad (1)$$

and

$$E_g^{(L)}(x) = (2.010 - 1.27x)\ eV \qquad for \quad 0.85 < x < 1 \qquad (2)$$

Recent band structure calculations of unstrained SiGe alloys were performed using the molecular coherent potential approximation (MCPA) [3] and nonlocal empirical pseudopotential calculations [4]. The results obtained in [3] are shown in FIGURE 1 by the broken line. Due to the less accurate band structures of the constituent materials involved in the alloy formalism, the results of these theories predict only trends of the bandgap, but not quantitatively accurate results. However, if the effect of alloy disorder is accounted for [3], the calculated bending of the energy gap as a function of the Ge content is in excellent agreement with the experimentally observed one.

B2 Strained SiGe Alloys

Experimental data for the indirect energy gap of strained SiGe alloys only exist for the case of alloys grown pseudomorphically on Si substrates. In this situation, the alloy layer has an in-plane lattice constant equal to that of silicon. Due to the large lattice mismatch, $\Delta a/a = 4\%$ for pseudomorphic SiGe growth, the critical layer thickness decreases drastically with increasing Ge content [5]. The SiGe alloy film is under biaxial compressive strain for any Ge concentration and nearly the entire band offset between Si and the alloy occurs in the valence band. Lang et al [6] have determined the fundamental indirect energy gap of strained SiGe alloys by photo-current measurements at T = 90 K on p-i-n diode structures, in which the intrinsic region was formed by twenty periods of 250 Å Si and 75 Å $Si_{1-x}Ge_x$. The Ge content of the alloy layers varied from 0 to 70% for the different samples. The results of Lang et al are indicated by the full symbols in FIGURE 3.

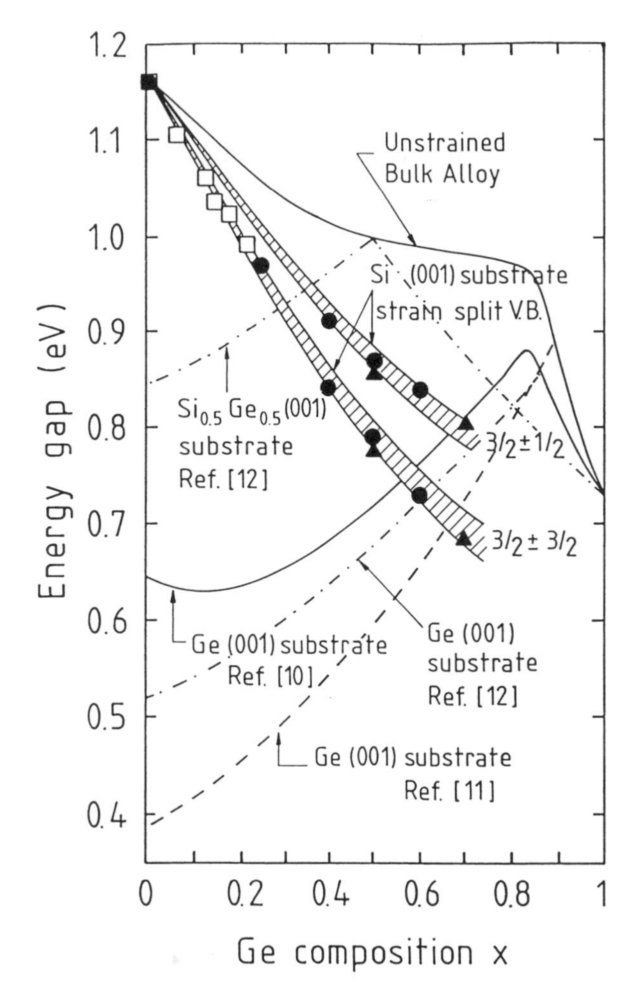

FIGURE 3 Fundamental indirect bandgap of strained $Si_{1-x}Ge_x$ alloys in comparison with the bulk alloy. The full (open) symbols are the results of photocurrent [6] (photoluminescence [8]) measurements at T = 90 K for $Si_{1-x}Ge_x$ pseudomorphic films on Si (001) substrates. The hatched areas between the full lines show the theoretical results of People et al [10] for transitions involving heavy holes (±3/2) and light holes (±1/2). Also shown are the calculated data for pseudomorphic films on either $Si_{0.5}Ge_{0.5}$ substrates [12] or on Ge substrates [10-12].

Photoluminescence data on strained alloy layers exist for samples with a Ge content between 0% and 24% [7,8].

Dutartre et al [8] performed PL measurements at two different temperatures (6 K, 90 K). The values of the energy gap determined by photoluminescence at T = 90 K (indicated by the open symbols in FIGURE 3) are in good agreement with the data of Lang et al [6]. It should be noted that the values given by Lang et al reflect the experimental bandgap which is the sum of the fundamental bandgap and the average phonon energy of the phonons involved in the absorption process [1]. The energy gap obtained by the low temperature (T = 6 K) PL measurements of Dutartre et al [8] are fitted to the following quadratic expression:

$$E_g = (1.171 - 1.01x - 0.835x^2)\ \text{eV} \qquad (x < 0.25) \qquad (3)$$

The energy gap E_g is determined by adding the binding energy of the free exciton, which is estimated by linear interpolation between the values for Si (14.7 meV) and Ge (4.15 meV), to the experimentally determined no-phonon free-exciton luminescence line.

The PL measurements of Robbins et al [7] were performed at T = 4.2 K on strained alloy layers with a Ge content between 0% and 22%. The binding energy E_b of the excitons is calculated using the perturbation theory of Lipari and Baldereschi [9] and the results of these calculations are fitted to the quadratic expression:

$$E_b = (0.0145 - 0.022x + 0.020x^2)\ \text{eV} \qquad (x < 0.25) \qquad (4)$$

Adding E_b to the measured energy of the free exciton line and fitting the results to a quadratic form yields:

$$E_g = (1.17 - 0.896x + 0.396x^2)\ \text{eV} \qquad (5)$$

Eqns (3) and (5) give values for E_g which differ by less than 7 meV for $x < 0.3$, i.e. for the range of alloy compositions investigated in [7,8].

Theoretical calculations of the bandgap of arbitrarily strained alloy layers have been performed by several authors [4,10-12]. For SiGe alloys on a Si substrate, the results of [10-12] are in good agreement with the experimental data of [6,8] (see FIGURE 3). For pseudomorphically strained $Si_{1-x}Ge_x$ alloys on $Si_{1-y}Ge_y$ substrates up to now only theoretical data exist. In FIGURE 3, the fundamental energy gaps of pseudomorphically strained $Si_{1-x}Ge_x$ alloys on a $Si_{0.5}Ge_{0.5}$ [12] and on a Ge [10-12] substrate are shown. According to the change of the biaxial strain from a compressive to a tensile one, the minimum gap changes from an lh-Δ(2) transition (increasing gap with increasing Ge content) to an hh-Δ(4) transition (decreasing gap with increasing Ge content). For higher Ge content in the substrate and in the SiGe alloy the latter transition finally becomes a Γ-L transition. It is important to note that for any composition of the substrate the fundamental bandgap of a strained SiGe alloy is smaller than the bandgap of the unstrained one. The theoretical values for the fundamental bandgaps of strained SiGe alloys reported in the literature differ quite substantially as is shown in FIGURE 3 for the case of strained SiGe alloys on a Ge substrate.

This discrepancy is due to the different methods of calculating the fundamental energy gap used by the authors of [10-12]: People et al [10] calculated the strain induced shift and splitting of the valence and conduction band using the deformation potential parameters of pure Si for $x < 0.85$ and those of pure Ge for $x > 0.85$, as at $x = 0.85$ the bandgap changes from the Si-type to the Ge-type. Using the experimental data for the bandgap of unstrained SiGe alloys obtained by Braunstein et al [1], they calculate the results shown by the solid line in FIGURE 3. Van de Walle et al [11] performed calculations based on the local density functional and ab initio pseudopotentials. The results of these calculations for the bandgap of strained SiGe alloys on Ge substrates are shown by the dashed line in FIGURE 3. Eberl et al [12] calculated the splitting and the shift of the energy bands in strained SiGe alloys by using deformation potential parameters, which were determined by linear interpolation between the respective values of Si and Ge. For deformation potential parameters for which no experimental data exist, i.e. for the X minimum in Ge or for the L minimum in Si, the values calculated by Van de Walle et al [11] were used. Using the bandgaps determined by Weber and Alonso [2] for the unstrained alloy, Eberl et al [12] obtained the results shown by the dashed-dotted lines in FIGURE 3.

C DIRECT ENERGY GAPS

The following direct energy gaps were determined for unstrained SiGe alloys, using either ellipsometric techniques [13,14] or electroreflectance [15]: E_0 (Γ_{25}'-Γ_2'), the spin-orbit split energy gap $E_0 + \Delta_0$, E_0' (Γ_{25}'-Γ_{15}), E_1 (Λ_3-Λ_1), the spin-orbit split energy gap $E_1 + \Delta_1$, and the E_2 (close to the X-point). The data from [14,15] are shown in FIGURE 4. The most precise data for the higher gaps at the end points Si and Ge were recently obtained by Etchegoin et al [16] using a piezomodulation technique.

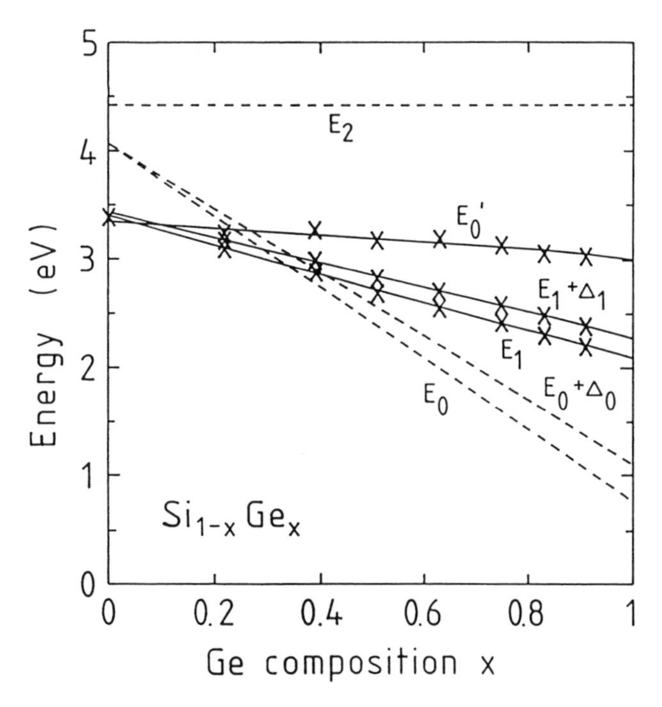

FIGURE 4 Composition dependence of the direct energy gaps of bulk $Si_{1-x}Ge_x$ alloys. The data for the E_0, $E_0 + \Delta_0$ and the E_2 gaps (E_0', E_1 and $E_1 + \Delta_1$ gaps) are taken from [15] ([16]).

In the $Si_{1-x}Ge_x$ alloys the following functional dependences were found to fit the data for the E_1 and $E_1 + \Delta_1$ gaps:

$$E_1(x) = (3.395 - 1.440x + 0.153x^2)\ \text{eV} \tag{6}$$

according to [13] and

$$(E_1 + \Delta_1)(x) = (3.428 - 1.294x + 0.062x^2) \text{ eV} \qquad (7)$$

after [16].

Experimental data on the strain dependence of the direct energy gaps (E_1, $E_1 + \Delta_1$ and E_2) exist only for pure Si and Ge under uniaxial stress ([16] and references therein). These data are not directly applicable to pseudomorphic SiGe alloy layers, as in this case the strain is biaxial. Although uniaxial and biaxial stress are equivalent from the symmetry point of view, they differ by a hydrostatic contribution which causes energy shifts apart from the level splitting due to the uniaxial stress. Once the deformation potential parameters are known for the SiGe alloys, the splitting and the shift of the higher energy gaps due to biaxial strain can be calculated [17]. In [16], the deformation potential parameters were determined for the E_1, $E_1 + \Delta_1$ transitions for Si and Ge and compared with the literature. Neither experimental nor calculated data exist so far on the strain dependence of these direct gaps in SiGe alloys.

D CONCLUSION

In this Datareview the experimental findings on the indirect minimum gap of $Si_{1-x}Ge_x$ alloys for unstrained bulk crystals as well as for strained epilayers grown on Si substrates were summarised and compared with theoretical results. Calculated data on the minimum indirect gap for pseudomorphic SiGe epilayers grown on either (001) oriented $Si_{0.5}Ge_{0.5}$ or Ge substrates were given as well. Furthermore the composition dependence of several direct energy gaps (E_0, $E_0 + \Delta_0$, E_0', E_1, $E_1 + \Delta_1$) of bulk $Si_{1-x}Ge_x$ alloys was presented.

REFERENCES

[1] R. Braunstein, A.R. Moor, F. Herman [*Phys. Rev. (USA)* vol.109 (1958) p.695]
[2] J. Weber, M.I. Alonso [*Phys. Rev. B (USA)* vol.40 (1989) p.5683]
[3] S. Krishnamurthy, A. Sher, A.-B. Chen [*Appl. Phys. Lett. (USA)* vol.47 (1985) p.160]; S. Krishnamurthy, A. Sher [*Phys. Rev. B (USA)* vol.33 (1986) p.1026]
[4] M.M. Rieger, P. Vogl [*Phys. Rev. B (USA)* vol.48 (1993) p.14276]
[5] F. Schäffler, E. Kasper [*Semicond. Semimet. (USA)* vol.33 (1991) p.223]
[6] D.V. Lang, R. People, J.C. Bean, A.M. Sergent [*Appl. Phys. Lett. (USA)* vol.47 (1985) p.1333]
[7] D.J. Robbins, L.T. Canham, S.J. Barnett, A.D. Pitt, P. Calcott [*J. Appl. Phys. (USA)* vol.71 (1992) p.1407]
[8] D. Dutartre, G. Brémond, A. Souifi, T. Benyattou [*Phys. Rev. B (USA)* vol.44 (1991) p.115252]
[9] N.O. Lipari, A. Baldereschi [*Phys. Rev. B (USA)* vol.3 (1971) p.2497]
[10] R. People [*Phys. Rev. B (USA)* vol.32 (1985) p.1405]; R. People [*Phys. Rev. B (USA)* vol.34 (1986) p.2508]
[11] C.G. Van de Walle, R.M. Martin [*Phys. Rev. B (USA)* vol.34 (1986) p.5621]
[12] K. Eberl, W. Wegscheider [*Handbook of Semiconductors* vol.3 (North-Holland, Amsterdam) in press]
[13] J. Humlicek, M. Garriga, M.I. Alonso, M. Cardona [*J. Appl. Phys. (USA)* vol.65 (1989) p.2827]

[14] C. Pickering et al [*J. Appl. Phys. (USA)* vol.73 (1993) p.239]
[15] J.S. Kline, F.H. Pollak, M. Cardona [*Helv. Phys. Acta (Switzerland)* vol.41 (1968) p.968]
[16] P. Etchegoin, J. Kircher, M. Cardona, C. Grein [*Phys. Rev. B (USA)* vol.45 (1992) p.11721]; P. Etchegoin, J. Kircher, M. Cardona [*Phys. Rev. B (USA)* vol.47 (1993) p.10292]
[17] I. Balslev [*Phys. Rev. (USA)* vol.143 (1966) p.636]; G.L. Bir, G.E. Pikus [*Symmetry and Strain-Induced Effects in Semiconductors* (Halsted Press, New York, 1974)]

4.2 Strain effects on the valence-band structure of SiGe

C.G. Van de Walle

December 1993

A INTRODUCTION AND NOTATION

Strain has two main effects on the band structure of a semiconductor: hydrostatic strain shifts the energetic position of a band, and uniaxial or biaxial strain splits degenerate bands. The effect is schematically illustrated in FIGURE 1.

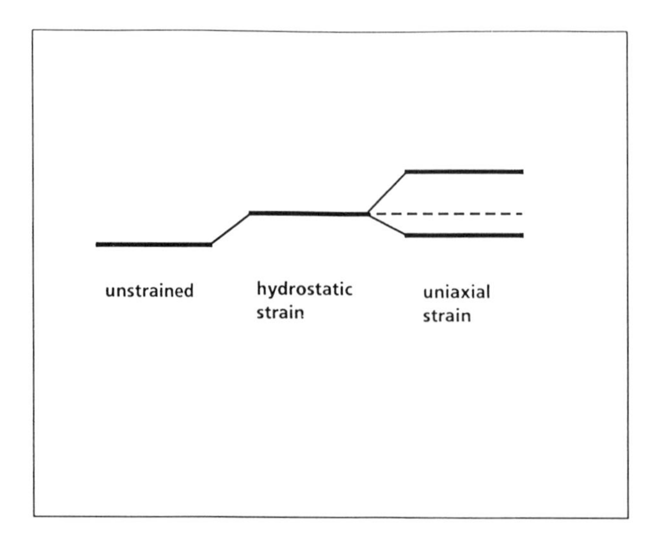

FIGURE 1 Schematic representation of the effect of strain on a triply degenerate band. Hydrostatic strain shifts the absolute energy position of the band. Uniaxial strain splits the degeneracy; illustrated is a case where the threefold degenerate band is split into a set of twofold degenerate bands at lower energy plus a singly degenerate band at higher energy. Note that the average over the three bands (dotted line) is unaffected by uniaxial strain.

A1 Strain

The strain state of the semiconductor (in the bulk, or in an epitaxial layer) can be expressed by the strain tensor $\overset{\leftrightarrow}{\varepsilon}$. We will assume here that the strains are sufficiently small to be in the linear regime; this assumption is likely to be satisfied in pseudomorphically strained layers. Note that strains large enough to lead to nonlinear effects normally do not occur under pseudomorphic conditions.

The hydrostatic strain, corresponding to the fractional volume change, is given by the trace of the strain tensor:

$$\frac{\Delta V}{V} = \mathrm{Tr}(\overset{\leftrightarrow}{\varepsilon}) \tag{1}$$

Pseudomorphic growth of an overlayer on a substrate with a different lattice constant introduces uniaxial strain components (we use the terms 'uniaxial' and 'biaxial' interchangeably; the strain arises because the in-plane lattice constant has to conform to that of the substrate, causing strain along the axes parallel to the interface, but simultaneously introducing strain in the perpendicular direction). Here we will assume that the magnitude of the strain components is known, and we will use the following notation, which applies to any growth direction ([001], [111], [110], ...): we place the z axis along the growth direction, i.e. perpendicular to the interface between substrate and overlayer, and designate this direction with the subscript '⊥' (for 'perpendicular'). The strain component along this direction is then $\varepsilon_\perp$. The x and y axes lie in the plane of the interface, and the strain along each of these is $e_\parallel$ (subscript '||' for 'parallel'). The strain directions that we will explicitly consider are those for which the z axis ('perpendicular' direction) is oriented along the [001], the [111] or the [110] crystallographic direction.

A2 Spin-Orbit Splitting

The topmost valence band at the zone centre in tetrahedral semiconductors with the diamond structure such as Si and Ge would be threefold degenerate, in the absence of spin-orbit splitting the strain. In unstrained material, the spin-orbit interaction lifts the degeneracy and splits and bands by an amount Δ_0 (in unstrained material); with respect to the average band position, which we denote $E_{v,av}$, two bands are shifted up by an amount $\Delta_0/3$ (the light and heavy hole bands, which we denote $E_{v,1}$ and $E_{v,2}$, respectively), while one band is shifted down by an amount $2\Delta_0/3$ (the spin-orbit split-off band, which we denote $E_{v,3}$). The magnitude of the spin-orbit splitting in Si and Ge is listed in TABLE 1. It is a good approximation to assume that the spin-orbit splitting in $Si_{1-x}Ge_x$ alloys is given by linear interpolation.

TABLE 1 Theoretical and experimental values for valence-band parameters in Si and Ge. Δ_0 is the spin-orbit splitting; a_v is the absolute deformation potential; b and d are the uniaxial deformation potentials for tetragonal and trigonal strains, respectively. All values are in eV.

	Si		Ge	
	Theoretical	Experimental	Theoretical	Experimental
Δ_0		0.04[a]		0.30[a]
a_v	2.46[b]	1.80[c]	1.24[b]	-
b	-2.35[d]	-2.10 ± 0.10[e]	-2.55[d]	-2.86 ± 0.15[f]
d	-5.32[d]	-4.85 ± 0.15[e]	-5.50[d]	-5.28 ± 0.50[f]

[a] [1].
[b] [2].
[c] Obtained from $a_v = a_c - a$, where a is the deformation potential for the gap. Values for a and a_c are given in Datareview 4.3, Table 1.
[d] [3].
[e] [4].
[f] [5].

A3 Deformation Potentials

In the following sections, we will discuss how the effect of hydrostatic and uniaxial strains on the valence-band structure is expressed via deformation potentials. Deformation potentials for hydrostatic strains will be denoted by the symbol a; deformation potentials for uniaxial deformations are denoted by b and d. These deformation potentials have been determined experimentally and theoretically for Si and for Ge; to our knowledge, no direct determinations have been performed for $Si_{1-x}Ge_x$ alloys. However, linear interpolation tends to be a good approximation for obtaining these band structure related quantities in alloys. Since the values for Si and Ge are usually quite similar to begin with, a linear interpolation procedure is expected to give good results.

B HYDROSTATIC STRAIN

Hydrostatic strain shifts the average position of the valence band, $E_{v,av}$, according to the formula

$$\Delta E_{v,av} = a_v \frac{\Delta V}{V} \tag{2}$$

where

$$\frac{\Delta V}{V} = Tr(\overleftrightarrow{\varepsilon}) = 2\varepsilon_{\parallel} + \varepsilon_{\perp}$$

and a_v is the hydrostatic deformation potential for the valence band.

Eqn (2) describes how the position of the valence band (on the absolute energy scale) changes when a hydrostatic strain is applied. Note that this quantity is not straightforward to obtain experimentally; the quantities usually observed in experiments are energy differences between bands, and how they change when strain is applied. For instance, the hydrostatic deformation potential for the bandgap (energy difference between conduction band and valence band) can be directly measured. The so-called absolute deformation potential of the valence band is not directly experimentally observable, although it has an effect on various physical quantities, such as the mobility. Indeed, the shifts in energy bands due to local deformations caused by acoustic phonons constitute an important scattering mechanism for carriers, as discussed by Bardeen and Shockley [6]. More directly relevant to the present Datareview, the absolute deformation potentials enter into a theoretical description of the band offsets at a heterojunction in which one (or both) of the materials is strained (see Datareview 4.5).

A proper formulation of the absolute deformation potential problem actually requires careful attention [7,8]; for instance, the absolute deformation potentials may depend on the orientation of the deformation. Here we will not explore these intricacies, and assume (as supported by first-principles calculations [7]) that the hydrostatic, orientation-independent terms in the deformation potential dominate, and that an 'absolute' deformation potential can be associated with the valence band and with the conduction band.

In principle it is possible to extract the absolute deformation potentials from experimental information; however, usually the determination is indirect, requiring complicated analysis or various assumptions. A discussion can be found in [2]. In TABLE 1 we therefore prefer to list theoretical values which have been obtained within a consistent first-principles framework [2,7]. In Datareview 4.3 we will list an experimental result for the absolute deformation potential for the conduction band in Si; the value for the valence band listed in TABLE 1 has been obtained by subtracting the bandgap deformation potential from the conduction-band value (a_c).

The values for absolute deformation potentials listed in TABLE 1 were obtained using the 'model-solid theory' [2], which is based on first-principles pseudopotential density-functional calculations. These values agree well with those obtained using more elaborate first-principles calculations [7,8].

C UNIAXIAL STRAIN

Uniaxial strain leads to a splitting of the valence bands (above and beyond the splitting introduced by the spin-orbit interaction). These splittings can also be expressed in terms of deformation potentials.

For perpendicular strains along [001] or [111], the following expressions [9] describe the shift of the bands with respect to the average $E_{v,av}$:

$$\Delta E_{v,1} = -\frac{1}{6}\Delta_0 + \frac{1}{4}\delta E + \frac{1}{2}\left[\Delta_0^2 + \Delta_0\delta E + \frac{9}{4}(\delta E)^2\right]^{\frac{1}{2}} \tag{3}$$

$$\Delta E_{v,2} = \frac{1}{3}\Delta_0 - \frac{1}{2}\delta E \tag{4}$$

$$\Delta E_{v,3} = -\frac{1}{6}\Delta_0 + \frac{1}{4}\delta E - \frac{1}{2}\left[\Delta_0^2 + \Delta_0\delta E + \frac{9}{4}(\delta E)^2\right]^{\frac{1}{2}} \tag{5}$$

In these equations, δE is given by

$$\text{for [001] strain: } \delta E_{001} = 2b(\varepsilon_\perp - \varepsilon_{||}) \tag{6}$$

$$\text{for [111] strain: } \delta E_{111} = 2\frac{\sqrt{3}}{3}d(\varepsilon_\perp - \varepsilon_{||}) \tag{7}$$

Eqns (3), (4) and (5) include the effects of uniaxial strain as well as spin-orbit coupling; in the absence of strain they reduce to the spin-orbit shifts. We note that the band v_2 (heavy hole band) is a pure $|\frac{3}{2}, \frac{3}{2}>$ state, while v_1 and v_3 are mixtures of $|\frac{3}{2}, \frac{1}{2}>$ and $|\frac{1}{2}, \frac{1}{2}>$.

The quantity b is the deformation potential for a strain of tetragonal symmetry; it is negative for Si and Ge. The quantity d is the deformation potential for a strain of trigonal symmetry; it is also negative. Theoretical [3] and experimental values for b and d are listed in TABLE 1. We consider the experimental values obtained by Pollak et al [4,5,9] to be most reliable; they were obtained by wavelength derivative transmission (for Si) or piezoreflectance (for Ge).

The case of uniaxial strain along [110] is somewhat more complicated. Analytical expressions can not easily be written down for the shifts of the valence bands; the strain splitting is a consequence of a mixture of the deformation potentials b and d. The shifts can be obtained as the eigenvalues of a matrix, as discussed in [2] and [9]. However, since only the parameters b and d enter, the values listed in TABLE 1 suffice to obtain the solution.

D CONCLUSION

TABLE 1 gives values for all parameters required to evaluate effects of strain on the valence bands in Si and Ge. Values for $Si_{1-x}Ge_x$ alloys can be obtained by linear interpolation. Where available, the experimental values are likely to be more reliable than the theoretical numbers, except for the absolute deformation potential a_v, for which the experimental uncertainty is at least as large as the theoretical error bar.

Note that the strain-induced splitting of the valence bands leads to preferential occupation of either the light-hole or heavy-hole bands. Under compressive biaxial strain (growth on a substrate with smaller lattice constant), the heavy-hole band will be at the band edge; under tensile strain (growth on a substrate with larger lattice constant), the light-hole band will be preferentially occupied. The different effective masses of the light and heavy hole bands lead to different behaviour in terms of mobility of carriers or confinement shifts in quantum wells. One can therefore use strain to tailor the band structure according to the needs for a particular application.

Examples of the practical use of the deformation potentials are given in Datareview 4.5.

REFERENCES

[1] M. Cardona, G. Harbeke, O. Madelung, U. Rösler [in *Landolt-Börnstein: Numerical Data and Functional Relationships in Science and Technology* vol.III/17a, Ed. O. Madelung (Springer Verlag, New York, 1982)]
[2] C.G. Van de Walle [*Phys. Rev. B (USA)* vol.39 (1989) p.1871]
[3] C.G. Van de Walle, R.M. Martin [*Phys. Rev. B (USA)* vol.34 (1986) p.5621]
[4] L.D. Laude, F.H. Pollak, M. Cardona [*Phys. Rev. B (USA)* vol.3 (1971) p.2623]
[5] M. Chandrasekhar, F.H. Pollak [*Phys. Rev. B (USA)* vol.15 (1977) p.2127]
[6] J. Bardeen, W. Shockley [*Phys. Rev. (USA)* vol.80 (1950) p.72]
[7] C.G. Van de Walle, R.M. Martin [*Phys. Rev. Lett. (USA)* vol.62 (1989) p.2028]
[8] R. Resta, L. Colombo, S. Baroni [*Phys. Rev. B (USA)* vol.41 (1990) p.12358; vol.43 (1991) p.14273]
[9] F.H. Pollak, M. Cardona [*Phys. Rev. (USA)* vol.172 (1968) p.816]

4.3 Strain effects on the conduction-band structure of SiGe

C.G. Van de Walle

December 1993

A INTRODUCTION AND NOTATION

Strain effects on the conduction band are similar to those discussed for the valence band in Datareview 4.2: hydrostatic strain shifts the overall energetic position of the band, and uniaxial or biaxial strain introduces a splitting of degenerate bands. The degeneracy of the conduction bands in Si and Ge is different in nature from that of the valence bands. In the case of the valence-band maximum at Γ there exists an orbital degeneracy; these bands contain different mixtures of orbitals with p_x, p_y and p_z character, which explains why application of uniaxial strain affects the relative energies of the bands. For the conduction bands, the degeneracy is spatial in nature: in Si, the conduction-band minima are found along the <100> directions (along the line from Γ to X; also called Δ) in reciprocal space; this implies there are six minima, which occur at the same energy in unstrained material. Application of a uniaxial strain along [001] will affect the minimum oriented along [001] differently from those oriented along [100] and [010]. In Ge, the conduction-band minima occur at the L point, i.e. at the zone boundary along the <111> directions, leading to eight equivalent minima. The relevant expressions for splittings are well documented [1-3] and will be discussed below.

As discussed in Datareview 4.2, values for $Si_{1-x}Ge_x$ can in principle be obtained by linear interpolation. A complication arises because the character of the lowest conduction band changes (from Δ to L) at about 85% Ge content. In order to perform the interpolation properly, one should determine values for the Δ minimum by interpolating the Δ-related values for the endpoints (Si and Ge), and values for the L minimum by interpolating between L-related values for Si and Ge. Experimental values for the deformation potentials have only been determined for the lowest conduction-band minimum, so that values for the L-point in Si and for the Δ minimum in Ge are not available. Theoretical values, however, have been calculated for both types of conduction-band valley. We will see that the values in Si and Ge are quite similar.

B HYDROSTATIC STRAIN

Hydrostatic strain shifts the average position of the conduction band, $E_{c,av}$, according to the formula

$$\Delta E_{c,av} = a_c \frac{\Delta V}{V} \qquad (1)$$

where

$$\frac{\Delta V}{V} = \mathrm{Tr}(\overleftrightarrow{\varepsilon}) = 2\varepsilon_{\parallel} + \varepsilon_{\perp}$$

and a_c is the hydrostatic deformation potential for the conduction band. Eqn (1) describes how the position of the conduction band (on an absolute energy scale) changes when a hydrostatic strain is applied. See Datareview 4.2 for a discussion about absolute deformation potentials.

The values for a_c and a_v defined here and in Datareview 4.2 should of course be consistent with the deformation potential for the bandgap, a_g. Since $E_g = E_c - E_v$, we have $a_g = a_c - a_v$. Values for a_g are listed in TABLE 1. Note that the value of a_c^{Δ} is positive, i.e. the bandgap (at Δ) in Si increases when the volume increases, or, alternatively, the gap decreases under pressure. In contrast, a_c^L has the opposite sign, indicating that the gap at L will increase under pressure. Still, the bandgap of Ge eventually decreases under large pressure, because the conduction-band minimum at X drops down in energy below the L minimum when pressure is applied.

TABLE 1 Theoretical and experimental values for conduction-band parameters in Si and Ge. a_c is the absolute deformation potential for the conduction-band minimum (Δ in Si, L in Ge); a_g is the deformation potential for the bandgap (i.e. $a_g = a_c - a_v$); Ξ_u is the deformation potential for uniaxial strains. The superscripts refer to the type of conduction-band valley: Δ (along Γ to X), which is lower in Si, or L, which is lower in Ge. All values are in eV.

	Si		Ge	
	Theoretical	Experimental	Theoretical	Experimental
a_c^{Δ}	4.18[a]	3.3[b]	2.55[a,c]	-
a_c^L	-0.66[a,c]	-	-1.54[a]	-
a_g^{Δ}	1.72[a]	1.50 ± 0.30[d]	1.31[c]	-
a_g^L	-3.12[c]	-	-2.78[a]	-2.0 ± 0.5[e]
Ξ_u^{Δ}	9.16[c]	8.6 ± 0.4[d]	9.42[c]	-
Ξ_u^L	16.14[c]	-	15.13[c]	16.2 ± 0.4[e]

[a] [1]; [b] [4]; [c] [5]; [d] [6]; [e] [3].

TABLE 1 lists theoretical values for a_c which have been obtained using the 'model-solid theory' [1], which is based on first-principles pseudopotential density-functional calculations. The experimental value for the absolute deformation potential for the conduction band in Si was obtained from a measurement of the effect of heavy doping on the lattice constant [4].

C UNIAXIAL STRAIN

Uniaxial strain leads to a splitting of the conduction bands that are degenerate in the absence of strain. These splittings are expressed with respect to the average band position, which is unaffected by the uniaxial components, and is shifted only by the hydrostatic component of the strain. We use the notation of Herring and Vogt [2] for the deformation potentials for the conduction band.

The general expression for the energy shift of conduction-band valley i, for a deformation described by the strain tensor ($\overleftrightarrow{\varepsilon}$), is

$$\Delta E_c^i = \left(\Xi_d \overset{\leftrightarrow}{1} + \Xi_u \left\{ \hat{a}_i \, \hat{a}_i \right\} \right) : \overset{\leftrightarrow}{\varepsilon} \tag{2}$$

where $\overset{\leftrightarrow}{1}$ is the unit tensor, $\hat{a}_i$ is a unit vector parallel to the **k** vector of valley i, and { } denotes a dyadic product. The shift of the average energy of the conduction-band extrema is

$$\Delta E_{c,av} = \left(\Xi_d + \frac{1}{3} \Xi_u \right) \overset{\leftrightarrow}{1} : \overset{\leftrightarrow}{\varepsilon} = \left(\Xi_d + \frac{1}{3} \Xi_u \right) \mathrm{Tr}(\overset{\leftrightarrow}{\varepsilon}) \tag{3}$$

Note that in the notation of Herring and Vogt, the hydrostatic deformation potential for the conduction band is given by $a_c = \Xi_d + \frac{1}{3} \Xi_u$. This quantity is also often denoted by E_1 [7].

We proceed to give specific expressions for strains along [001], [110] and [111] (i.e. uniaxial strains caused by pseudomorphic growth along these directions); the general case is obtained by using Eqns (2) and (3), and involves the same deformation potentials.

Let us first consider conduction-band minima along Δ (i.e. along the Γ-X direction). Uniaxial strain along [111] has no effect on these minima. Under uniaxial strain along [001], the bands along [100] and [010] split off from the one along [001]. The energy shifts of the bands are given by:

$$\Delta E_c^{001} = \frac{2}{3} \Xi_u^{\Delta} (\varepsilon_{\perp} - \varepsilon_{||}) \tag{4}$$

$$\Delta E_c^{100,010} = -\frac{1}{3} \Xi_u^{\Delta} (\varepsilon_{\perp} - \varepsilon_{||}) \tag{5}$$

The superscript Δ on Ξ_u indicates which type of conduction-band valley (at Δ or at L) we are considering, while the superscripts on ΔE_c refer to the direction of the particular conduction-band minimum. Ξ_u is often denoted as E_2 (see [7]). Under uniaxial strain along [110], the energy shifts of the bands are given by

$$\Delta E_c^{001} = -\frac{1}{3} \Xi_u^{\Delta} (\varepsilon_{\perp} - \varepsilon_{||}) \tag{6}$$

$$\Delta E_c^{100,010} = \frac{1}{6} \Xi_u^{\Delta} (\varepsilon_{\perp} - \varepsilon_{||}) \tag{7}$$

Next we consider conduction bands at L. These are unaffected by strain along [001]. Strain along [111] leads to

$$\Delta E_c^{111} = \frac{2}{3} \Xi_u^{L} (\varepsilon_{\perp} - \varepsilon_{||}) \tag{8}$$

$$\Delta E_c^{\bar{1}11,1\bar{1}1,11\bar{1}} = -\frac{2}{9} \Xi_u^{L} (\varepsilon_{\perp} - \varepsilon_{||}) \tag{9}$$

Finally, strain along [110] yields

$$\Delta E_c^{111,11\bar{1}} = \frac{1}{3} \Xi_u^{L} (\varepsilon_{\perp} - \varepsilon_{||}) \tag{10}$$

$$\Delta E_c^{\bar{1}11,1\bar{1}1} = -\frac{1}{3}\,\Xi_u^L\,(\varepsilon_\perp - \varepsilon_\parallel) \qquad (11)$$

Theoretical [5] and experimental values for Ξ_u^Δ and Ξ_u^L are listed in TABLE 1. The experimental values were obtained by wavelength derivative transmission (for Si) [6], and by piezoabsorption (for Ge) [3].

D CONCLUSION

TABLE 1 gives values for all parameters required to evaluate effects of strain on the conduction bands in Si and Ge. Values for SiGe alloys can be obtained by linear interpolation. Where available, the experimental values are likely to be more reliable than the theoretical numbers, except for the absolute deformation potential a_c, for which the experimental uncertainty is at least as large as the theoretical error bar.

Note that the strain-induced splitting of the degenerate conduction-band minima offers the particular advantage that, depending on whether the strain is compressive or tensile, different conduction-band valleys become lower in energy, allowing one to pick carriers with a particular (longitudinal or transverse) effective mass. This provides flexibility in the design of devices where the transport properties along a particular direction are important; it also affects the magnitude of confinement shifts in quantum wells.

Examples of the practical use of the deformation potentials are given in Datareview 4.5.

REFERENCES

[1] C.G. Van de Walle [*Phys. Rev. B (USA)* vol.39 (1989) p.1871]
[2] C. Herring, E. Vogt [*Phys. Rev. (USA)* vol.101 (1956) p.944]
[3] I. Balslev [*Phys. Rev. (USA)* vol.143 (1966) p.636]
[4] G.S. Cargill III, J. Angilello, K.L. Kavanagh [*Phys. Rev. Lett. (USA)* vol.61 (1988) p.1748]
[5] C.G. Van de Walle, R.M. Martin [*Phys. Rev. B (USA)* vol.34 (1986) p.5621]
[6] L.D. Laude, F.H. Pollak, M. Cardona [*Phys. Rev. B (USA)* vol.3 (1971) p.2623]
[7] E.O. Kane [*Phys. Rev. (USA)* vol.178 (1969) p.1368]

4.4 Effective masses in SiGe

J.F. Nützel, C.M. Engelhardt and G. Abstreiter

July 1994

A INTRODUCTION

For a long time the determination of band parameters in SiGe alloys and compound systems was aggravated by extreme experimental difficulties in the preparation of high quality bulk and layered structures. Within the last few years substantial progress has been made in the preparation and measurement of 2D carrier systems in SiGe heterostructures.

B CONDUCTION BAND

B1 Bulk Electron Masses of Unstrained SiGe

The conduction band in unstrained pure Si consists of six degenerate minima close to the X-points, i.e. at 0.85 π/a_0 in the [100] direction. The energy can be written as

$$E = \frac{\hbar^2}{2}\frac{1}{m_e}\left(\frac{k_t^2}{m_t} + \frac{k_l^2}{m_l}\right) \tag{1}$$

The measured values for the transverse mass m_t and longitudinal mass m_l are listed in TABLE 1.

In pure unstrained germanium the lowest conduction band minima are four-fold degenerate in the 111-directions at the L-point with a longitudinal mass in the 111-direction and a transverse mass perpendicular to it. For the measured values see TABLE 1.

To our knowledge there are no experimental data available on the electron masses of SiGe alloys. However theoretical calculations (non-local pseudopotential calculations [26]) predict that the conduction band remains Si-like up to approximately 85% Ge, where a crossover of the lowest lying minima takes place from the X valleys to the L valleys, which are expected to be Ge-like. This is in good agreement with experimental results on photoluminescence of bulk SiGe alloys [4,35]. There are no significant changes expected in the effective masses as can be seen from FIGURE 1. This expectation is strongly supported from 2D data presented below.

Measured effective masses of pure Si and Ge show that the variation with temperature due to the influence of non-parabolicity is rather small [24].

TABLE 1 Effective mass-relevant conduction and valence band parameters for bulk Si and Ge.

Conduction band:	Si	(Temperature) Ref	Ge	(Temperature) Ref
m_l (m_e)	0.1905(1)	(1.26 K) [13]	0.082	(1.4 K) [9]
m_t (m_e)	0.9163(4)	(1.26 K) [13]	1.58	(1.4 K) [9]

Valence band:	Si	(Temperature) Ref	Ge	(Temperature) Ref
Δ_0 (meV)	44	[5]	290	[16]
b (eV)	-2.1(1)	(77 K) [17]	-2.86(15)	(77 K) [7]
d (eV)	-4.85(15)	(77 K) [17]	-5.28(50)	(77 K) [7]
L	-6.64	(1.26 K) [2]	-31.34	(1.2 K) [14]
M	-4.61	(1.26 K) [2]	-5.9	(1.2 K) [14]
N	-8.68	(1.26 K) [2]	-34.14	(1.2 K) [14]

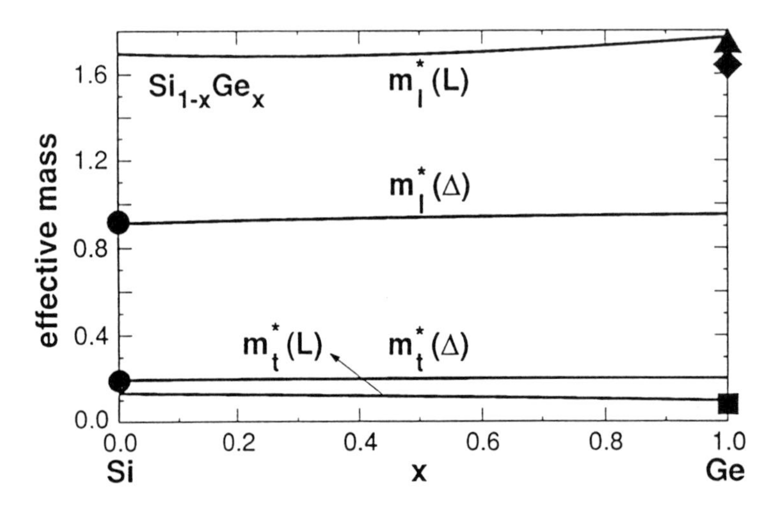

FIGURE 1 Effective conduction band masses of SiGe alloys: dependence on Ge content (from Rieger and Vogl [26] with permission).

B2 Influence of Strain on the Bulk Electron Masses

The degeneracy of the conduction band minima is lifted by uniaxial strain, resulting in a genuine heavy mass in the direction of compression ($m_l = 0.92\ m_e$) and genuine light mass ($m_t = 0.19\ m_e$) perpendicular to the compression in the (100) case and vice versa.

There is no experimental data on the strain dependence of the bulk electron masses available. Theoretical calculations [26] show only slight changes of the mass parameters due to strain and Ge content.

B3 Two-Dimensional Data

Cyclotron resonance measurements on high mobility 2D electron gases in strained Si on a relaxed $Si_{0.7}Ge_{0.3}$ substrate [27,28,31] show an in-plane mass of 0.195 m_e in good agreement with the expectations shown above. Variations of the carrier density in similar 2D channels result only in minor changes of this effective mass [21]. These findings are in agreement with earlier experiments on the effect of uniaxial stress on the cyclotron resonance in inversion layers on Si [30].

Experimental verifications of other structures are expected to be difficult due to small band offsets (see Datareview 4.5).

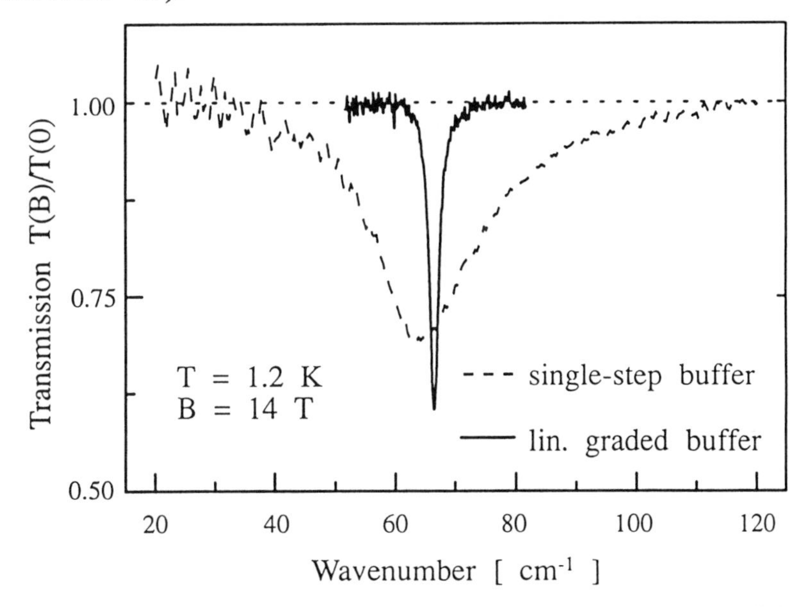

FIGURE 2 Cyclotron resonance measurements on 2D electron gases in strained Si on SiGe. The resonance position corresponds to an effective mass of 0.195 m_e (from Többen et al [31] with permission).

C VALENCE BAND

C1 Valence Band Hamiltonian

The valence band structure is more complex even in the unstrained pure Si case. Ignoring interaction with the conduction band and lower lying valence bands it can be described by the **k•p**-theory with a 6x6 Hamiltonian [3,9,13]. The addition of the spin-orbit split off Hamiltonian lifts the degeneracy at $k = 0$ of the third valence band from the still degenerate heavy- and light-hole band. This degeneracy leads to a strong anisotropy of the heavy- and light-hole bands as one can see in FIGURE 3.

The small value of the splitting energy of only 44 meV in Si leads to strong non-parabolicity of the band in all practical cases for energies not negligible compared to the splitting energy. The application of strain can be included within the linear deformation potential theory by the addition of a strain Hamiltonian

$$H = H_{kp} + H_{so} + H_{strain} \tag{2}$$

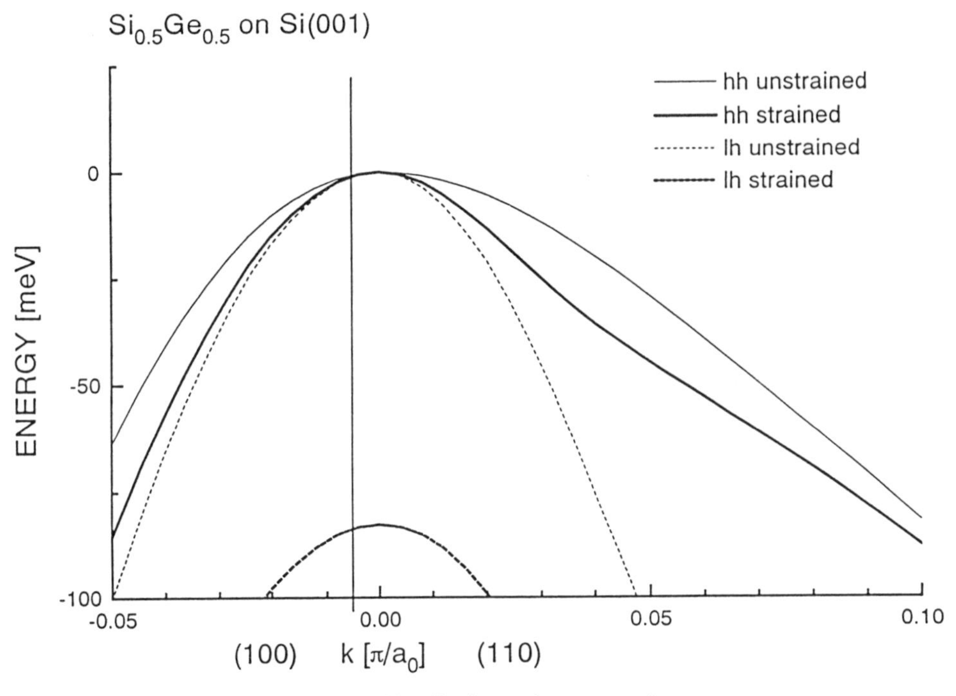

FIGURE 3 hh-, lh-dispersion vs. strain.

Practical implementations can be found for example in [15,19,37].

C2 Experimental Values for Pure Si and Ge

The valence band parameters used in Eqn (2) were determined by cyclotron resonance measurements with systematic strain variations performed for the first time by Dresselhaus et al [9] and more precisely by Hensel and Feher in 1963 (with a small correction in interpretation by Balslev [2]) for Si and by Hensel and Suzuki [14] for Ge. The values are given in TABLE 1.

C3 Interpolation of Band Parameters for Unstrained SiGe Alloys

Experimental data on SiGe bulk alloys are from Braunstein [5], who investigated SiGe alloys by studies of free-hole absorption spectra. The spin-orbit splitting was determined to vary linearly with the Ge content within experimental accuracy. He also measured the ratio of the effective masses of the different bands in the Ge rich region which was in disagreement with interpolation formulae suggested by him and used by several other groups [33]. This approach assumes linearly varying parameters F, G, H, H1 which is based on matrix element considerations (for the conversion of the different parameter sets see [1]). Recent calculations by Rieger and Vogl [26] suggest a non-linear interpolation of the L, M, N parameters. However, this calculation does not reproduce the experimental values for germanium. A non-linear interpolation of the band parameters based on the scheme of Lawaetz [18] qualitatively reproduces the x-dependency as it results from the calculation additionally being consistent with the values for both pure Si and Ge [20] (FIGURE 4). For the definition of band parameters see [1].

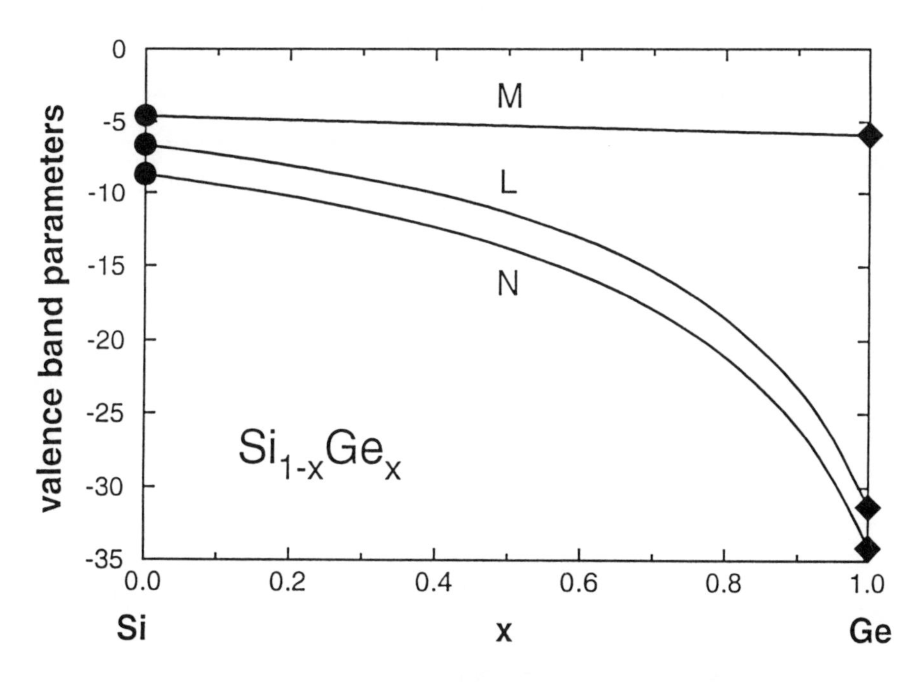

FIGURE 4 Non-linear interpolation of valence band parameters following the scheme of Lawaetz [18] (from Merkler [20] with permission).

C4 Influence of Strain

Application of strain lifts the degeneracy of the heavy and light hole bands at the Γ-point. Tensile strain in the z-direction (pseudomorphic SiGe on Si, or strained Ge on relaxed SiGe) shifts the heavy hole band upward, but modifies its dispersion to be light-hole-like near the Γ-point in the plane perpendicular to the strain direction. Compressive strain shifts the lh band up, but modifies its dispersion to be heavy-hole-like in the perpendicular direction (FIGURE 3).

C5 Experimental In-Plane Mass

Experimental data based on cyclotron resonance are available from measurements of 2D samples for compressively strained Ge channels on SiGe [10] and pseudomorphic SiGe channels on Si [8,22].

The Ge channel data show clearly both the reduced in-plane mass of the top-most heavy hole band and a strong increase of the CR mass with increasing carrier density due to the strong band non-parabolicity (TABLE 2). Even lower hole effective masses in the range of 0.04 - 0.10 m_e have been reported from temperature dependent SdH measurements but no dependence of the masses on strain or carrier density has been given [36].

The CR data for pseudomorphic SiGe channels on Si also show a reduction of the in-plane heavy hole mass compared to the values expected for the unstrained case. Temperature dependent SdH measurements [25,33,34] are in good agreement with the model described above compared with the difficulties of the band structure. Systematic quantitative studies however are still lacking.

TABLE 2 Measured cyclotron resonance masses for SiGe systems. * Due to partial relaxation of the channel the strain in the system corresponds to an effective Ge content of about 0.9. ** Derived from the temperature dependence of the Shubnikov-de Haas oscillations. *** At 13 tesla.

Ge-content substrate	Ge-content carrier system	N_s (10^{11} cm^{-2})	m_{CR} (m_e)	T (K)	Ref
0.0	0.2	3.5	0.30(2)**	1.8 - 4.2	[25]
0.0	0.15	12	0.44(3)**	1.4 - 4.2	[33]
0.0	0.37	23	0.29(2)	3.0	[8]
0.0	0.13	2.2	0.23(2)** 0.26	0.3 - 2	[34] [22]
0.8*	1.0	10.6(5) 11.0(5) 17.8(5)	0.142(5) 0.151(5) 0.20(1)	1.5	[10]
0.75	1.0	10.5(9)	0.120(1)	1.5	[23]
1.0	1.0	5.0(5)***	0.274(3)	0.4	[23]

C6 Quantisation Mass for Pseudomorphic SiGe on Si

For negligible $k_{\|}$ and small k_z Eqn (1) reduces to

$$\frac{1}{m_z(x)} = A(x) - B(x) = M(x) + 1 \tag{3}$$

for the z-quantisation mass. Photoluminescence studies show systematic energy shifts for different well widths in excellent agreement with Eqn (3) [6,32].

Systematic experimental investigations of valence intersubband transition energies for various Ge contents and well widths [11,12] are in good agreement with calculations based on the quantisation masses (Eqn (3)) using the band parameters cited above. However because of the cosensitivity to other parameters (band offsets, finite $k_{\|}$ etc.) these results can not yet be used for a determination of band parameters better than 10%. For the moment these measurements nevertheless represent the most advanced verification of the parameters used in the model discussed above.

Resonant magnetotunnelling experiments also give valuable insight into the complexity of the valence band structure [27-29,38,39]. Quantitative analysis of band parameters due to such experiments might be possible in the future.

REFERENCES

[1] Landolt-Börnstein [*Numerical Data and Functional Relationships in Science and Technology* vol.III/17a (Springer Verlag, Berlin, 1982)]

[2] I. Balslev, P. Lawaetz [*Phys. Lett. (Netherlands)* vol.19 (1965) p.6]

[3] G.L. Bir, G.E. Pikus [*Symmetry and Strain-Induced Effects in Semiconductors* (J. Wiley & Sons, New York, 1974)]

[4] R. Braunstein, A.R. Moore, F. Herman [*Phys. Rev. (USA)* vol.109 (1958) p.695]

[5] R. Braunstein [*Phys. Rev. (USA)* vol.130 (1963) p.869]

[6] J. Brunner, J.F. Nützel, M. Gail, U. Menczigar, G. Abstreiter [*J. Vac. Sci. Technol. B (USA)* vol.11 (1993) p.1079]
[7] M. Chandrasekhar, F. Pollak [*Phys. Rev. B (USA)* vol.15 (1977) p.2127]
[8] J.-P. Cheng, V.P. Kesan, D.A. Grutzmacher, T.O. Sedgwick, J.A. Ott [*Appl. Phys. Lett. (USA)* vol.62 (1993) p.1522]
[9] G. Dresselhaus, A.F. Kip, C. Kittel [*Phys. Rev. (USA)* vol.98 (1955) p.368]
[10] C.M. Engelhardt, D. Többen, M. Aschauer, F. Schäffler, G. Abstreiter, E. Gornik [*Solid-State Electron. (UK)* vol.37 (1994) p.949]; C.M. Engelhardt et al [to be published]
[11] T. Fromherz, E. Koppensteiner, M. Helm, G. Bauer, J.F. Nützel, G. Abstreiter [*Extended Abstracts of the 1993 Int. Conf. on Solid State Devices and Materials*, Makuhari (1993) p.410]; T. Fromherz, M. Helm, G. Bauer, J.F. Nützel, G. Abstreiter [*Solid-State Electron. (UK)* vol.37 (1994) p.941]
[12] T. Fromherz, E. Koppensteiner, M. Helm, G. Bauer, J.F. Nützel, G. Abstreiter [*Jpn. J. Appl. Phys. (Japan)* (in press)]
[13] J.C. Hensel, H. Hasegawa, M. Nakayama [*Phys. Rev. (USA)* vol.138 (1965) p.A225]
[14] J.C. Hensel, K. Suzuki [*Phys. Rev. B (USA)* vol.9 (1974) p.4219]
[15] J.M. Hinckley, J. Singh [*Phys. Rev. B (USA)* vol.41 (1990) p.2912]
[16] E.O. Kane [*J. Phys. Chem. Solids (UK)* vol.1 (1956) p.82]
[17] L. Laude, F. Pollak, M. Cardona [*Phys. Rev. B (USA)* vol.3 (1971) p.2623]
[18] P. Lawaetz [*Phys. Rev. B (USA)* vol.4 (1971) p.3460]
[19] T. Manku, A. Nathan [*Phys. Rev. B (USA)* vol.43 (1991) p.12634]
[20] M. Merkler [Diploma thesis, University of Regensburg, Germany (1994)]
[21] S.Q. Murphy, Z. Schlesinger, S.F. Nelson, J.O. Chu, B.S. Meyerson [*Appl. Phys. Lett. (USA)* vol.63 (1993) p.222]
[22] R.J. Nicholas [unpublished, cited in [34]]
[23] J.F. Nützel et al [to be published]
[24] J.C. Ousset, J. Leotin, S. Askenazy, M.S. Skolnick, R.A. Stradling [*J. Phys. C (UK)* vol.9 (1976) p.2803]
[25] R. People et al [*Appl. Phys. Lett. (USA)* vol.45 (1984) p.1231]
[26] M. Rieger, P. Vogl [*Phys. Rev. B (USA)* vol.48 (1993) p.14276]
[27] G. Schuberth, G. Abstreiter, E. Gornik, F. Schäffler, J.F. Luy [*Phys. Rev. B (USA)* vol.43 (1991) p.2280]
[28] G. Schuberth, F. Schäffler, M. Besson, G. Abstreiter, E. Gornik [*Appl. Phys. Lett. (USA)* vol.59 (1991) p.3318]
[29] G.D. Shen, D.X. Xu, M. Willander, G.V. Hansson [*Superlattices Microstruct. (UK)* vol.12 (1992) p.481]
[30] P. Stallhofer, J.P. Kotthaus, G. Abstreiter [*Solid State Commun. (USA)* vol.32 (1979) p.655]
[31] D. Többen et al [*J. Cryst. Growth (Netherlands)* vol.127 (1993) p.421]
[32] M. Wachter, K. Thonke, R. Sauer, F. Schäffler, H.-J. Herzog, E. Kasper [*Thin Solid Films (Switzerland)* vol.222 (1992) p.10]
[33] P.J. Wang, F.F. Fang, B.S. Meyerson, J. Nocera, B. Parker [*Appl. Phys. Lett. (USA)* vol.54 (1989) p.2701]
[34] T.E. Whall et al [*Appl. Phys. Lett. (USA)* vol.64 (1994) p.357]
[35] J. Weber, M.I. Alonso [*Phys. Rev. B (USA)* vol.40 (1989) p.5683]
[36] Y.H. Xie, D. Monroe, E.A. Fitzgerald, P.J. Silverman, F.A. Thiel, G.P. Watson [*Appl. Phys. Lett. (USA)* vol.63 (1993) p.2263]
[37] E.T. Yu, J.O. McCaldin, T.C. McGill [*Solid State Phys. (USA)* vol.46 (1992) p.1]
[38] A. Zaslavsky, D.A. Grützmacher, S.Y. Lin, T.P. Smith III, R.A. Kiehl, T.O. Sedgwick [*Phys. Rev. B (USA)* vol.47 (1993) p.16036]
[39] A. Zaslavsky, T.P. Smith III, D.A. Grützmacher, S.Y. Lin, T.O. Sedgwick [*Phys. Rev. B (USA)* vol.48 (1993) p.15112]

4.5 SiGe heterojunctions and band offsets

C.G. Van de Walle

January 1994

A INTRODUCTION AND NOTATION

When two semiconductors are joined at a heterojunction, discontinuities occur in the valence bands and in the conduction bands. For an atomically abrupt interface, these discontinuities are sharp on an atomic length scale (i.e. on the order of a few atomic distances). This is in contrast with band-bending effects, which are associated with depletion layers, and which occur on much larger length scales (hundreds of Å, up to several μm). The band discontinuities actually enter as boundary conditions in the solution of Poisson's equation, which would produce the band bending.

In the absence of strain, i.e. for a lattice-matched interface, the band-lineup problem simply consists of determining how the band structures of the two materials line up at the interface; the lineup then produces values for the valence-band discontinuity, ΔE_v, and the conduction-band discontinuity, ΔE_c. When the materials are strained, the strains will produce additional shifts (due to hydrostatic strain) and splittings (due to uniaxial strain) of the bands; these changes in the positions of valence and conduction bands will, of course, affect the band discontinuities.

If the heterostructure exhibits a lattice mismatch (such as is the case for Si/Ge, or interfaces with $Si_{1-x}Ge_x$ alloys), the band discontinuities are well defined only if the interface is pseudomorphic; i.e. one or both of the materials has to be appropriately strained so that the in-plane lattice constant is continuous across the interface. The effect of strain on the band structure is discussed in Datareviews 4.2 and 4.3. Since strain has significant effects on the band structure, the problem of band offsets will be intimately coupled with the strain effects.

The treatment of strains in Datareviews 4.2 and 4.3 actually allows for a straightforward separation between the 'lineup' problem and the 'strain' problem. The effect of strain on the bands is a 'bulk' issue, i.e. once the strain tensor is known, the individual shifts of the valence band and the conduction band can be determined, as well as the splitting of these bands. In Datareviews 4.2 and 4.3 these shifts and splittings were expressed with reference to the 'average' band positions, $E_{v,av}$ and $E_{c,av}$. The 'strain' problem can therefore be handled completely on the basis of Datareviews 4.2 and 4.3. The 'lineup' problem then consists of determining the proper offset between $E_{v,av}$ (or $E_{c,av}$) on either side of the interface, which really amounts to determining the lineup between the unstrained materials.

A note on notation: in Datareviews 4.2 and 4.3, the notation Δ was used to denote the shift in the band position due to strain; e.g. $\Delta E_{v,av}$ was the shift in the average valence-band position, within a specific material, due to hydrostatic strain. In the present Datareview, the notation Δ is used to denote band offsets, i.e. energy differences between two bands in different materials, across an interface; e.g. $\Delta E_{v,av}$ would refer to the offset between the average valence bands in the two materials that make up the heterojunction. The difference between these two uses of the same symbol should be clear from the context, but to avoid any possible confusion, we only use Δ in this Datareview to refer to a band offset.

This approach to the band-offset problem is the most rigorous, as well as elegant, from the theoretical point of view. Theoretical values for the band offset $\Delta E_{v,av}$ will be given in Section B. From an experimental point of view, the problem is more complicated. Experimental measurements usually provide information only about the discontinuity in the highest-lying valence band in each material (or similarly, the lowest-lying conduction band). A single measurement does not provide information about how to separate the offset into a 'lineup' and a 'strain' part. Fortunately, virtually all experimental results have turned out to be consistent with the theoretical predictions, so that theory can provide a reliable framework. Experiments will be discussed in Section C.

B THEORY

The most direct way to predict values for $E_{v,av}$ in Si and Ge is provided by the 'model solid theory' [1]. This approach treats the band offsets as linear quantities, which can be obtained as differences between reference values which have been calculated once and for all for each semiconductor. The reference potential for each material is determined for a 'model solid', which consists of a superposition of neutral atomic charge densities. The nature of these neutral-atom building blocks renders the electrostatic potential independent of the details of the surface (or interface) structure, producing a reference level which can be used for a lineup procedure. Details are given in [1]. The values are $E_{v,av}$ = -7.03 eV in Si, and $E_{v,av}$ = -6.35 eV in Ge. Note that the absolute values do not carry physical meaning; only differences between values are relevant. These values lead to $\Delta E^0_{v,av}$ = 0.68 eV between Si and Ge, where the superscript '0' refers to an offset between unstrained materials.

Let us illustrate how this value, combined with the strain treatment of Datareviews 4.2 and 4.3, produces values for the actual band lineups. Theoretical values for the deformation potentials will be used; the use of experimental values would produce only minor changes. As an example we choose the lineup between a pseudomorphically strained Ge overlayer on a Si (001) substrate. The procedure is illustrated in FIGURE 1. On the Si side, there is no strain. However, we still have to keep in mind that the position of the highest-lying valence band (which is degenerate at Γ, consisting of light and heavy hole bands) differs from $E_{v,av}$ because of spin-orbit splitting. In Si, we have:

$$E_{v,1,2} = E_{v,av} + \frac{1}{3}\Delta_0 = -7.03 + \frac{1}{3}0.04 = -7.02 \qquad (1)$$

Δ_0 is the spin-orbit splitting, and the subscripts 1 and 2 on E_v refer to the light and heavy hole bands; this notation is discussed in Datareview 4.2.

The Ge side of the junction is pseudomorphically strained. Using the notation for strains established in Datareview 4.2, and elastic theory, we obtain $\varepsilon_{\parallel}$ = -0.039 and $\varepsilon_{\perp}$ = 0.029 (we use lattice constants and elastic constants as listed in [1]). We then obtain

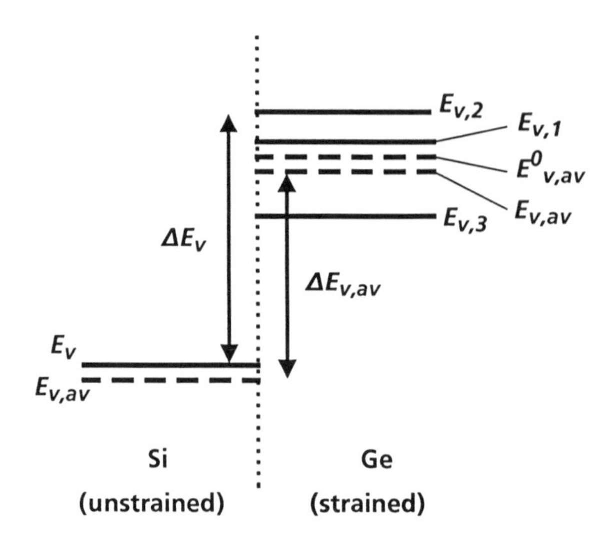

FIGURE 1 Theoretical valence-band lineups at an interface between unstrained Si and strained Ge (not to scale). The discontinuity in the average valence bands, $\Delta E_{v,av}$, is obtained from the model-solid theory. Strain shifts and splittings are obtained as described in the text, based upon the expressions in Datareviews 4.2 and 4.3.

$$E_{v,av} = E^{0}_{v,av} + a_v \frac{\Delta V}{V} = -6.35 + 1.24(2 \times (-0.039) + 0.029) = -6.41 \text{ eV} \qquad (2)$$

where $E^{0}_{v,av}$ refers to the value in unstrained material. The offset between the average valence bands in Si and strained Ge is therefore: $\Delta E_{v,av}$ = -6.41 - (-7.03) = 0.62 eV. Note that this offset does not contain the effects of uniaxial strain or spin-orbit coupling.

Finally, the splitting of the Ge valence bands has to be determined, based on Equations (5) to (7) in Datareview 4.2; the shifts in $E_{v,1}$, $E_{v,2}$ and $E_{v,3}$ are, respectively, 0.12 eV, 0.27 eV and -0.39 eV. Note that these three values add up to zero; indeed, these values should only express the splitting of the bands with respect to the average $E_{v,av}$, and not introduce any shift of the average.

The highest-lying valence band on the Ge side is therefore at $E_{v,av}$ + 0.27 = -6.14 eV, resulting in a valence-band discontinuity: ΔE_v = -6.14 - (-7.02) = 0.88 eV. This result may be compared with values of 0.84 eV [2], 0.86 eV [3] and 0.74 eV [4], obtained from full first-principles calculations, which do not make the approximations involved in the model-solid approach.

Similar arguments for a Si/Ge structure grown on Ge(001) lead to $\Delta E_{v,av}$ = -6.35 - (-6.91) = 0.56 eV, and ΔE_v = 0.32 eV. This value may be compared with the following values obtained from full first-principles interface calculations: 0.31 eV [2], 0.20 eV [3] and 0.20 eV [4].

When determining offsets for systems involving $Si_{1-x}Ge_x$ alloys, the strain effects in the individual materials can be assessed using linear interpolation, as described in Datareviews 4.2 and 4.3. It is also a good approximation to apply linear interpolation to obtain the band offset between average valence bands, based upon the band lineups between pure Si and pure Ge. Linear interpolation would, of course, not apply to the conduction bands, since it is known that the bandgap in the alloys exhibits substantial bowing (see Datareview 4.1). Once the valence-band lineup is known, however, one can use experimental information about the bandgap in unstrained alloys, combined with deformation-potential theory, to obtain the conduction-band lineup.

Finally, it has been shown that interface interdiffusion has a negligible influence on the band offset: Hybertsen [5] modelled the effect of a non-abrupt interface by including an interface layer of $Si_{0.5}Ge_{0.5}$, and found a change in the valence-band offset of less than 0.01 eV. This result is consistent with the notion that exchange of atoms across this interface between isovalent semiconductors does not generate any additional dipoles, as can be concluded from linear response theory [6].

C EXPERIMENT

As pointed out above, experiments can yield a specific offset value for a specific structure, but this offset value is influenced by a number of parameters (the unstrained lineup, and the hydrostatic and uniaxial deformation potentials). To our knowledge, no experiments have been carried out aimed at systematically determining these various components. Our approach in this section will therefore be to show that the theoretical 'model solid' values provide a reliable and accurate description of the band lineups in this system, by comparing the theoretical predictions with a number of experimental values for specific systems.

We only discuss experiments on pseudomorphic interfaces, in which the strain state of the heterostructure was accurately known.

Schwartz et al [7] carried out core-level photoemission measurements on Si/Ge interfaces on Si(100) and on Ge(100) substrates. Their data, combined with calculated core-valence-band deformation potentials and uniaxial deformation potentials, lead to $\Delta E_v = 0.74 \pm 0.13$ eV in the case of the Si substrate, and $\Delta E_v = 0.17 \pm 0.13$ eV in the case of the Ge substrate. The model solid values, calculated above, were 0.88 eV and 0.32 eV.

Yu et al [8] used X-ray photoelectron spectroscopy not only to obtain the core level lineups across the heterojunction, but also to measure the strain effect on the energy difference between core level and valence band. They obtained $\Delta E_v = 0.83 \pm 0.11$ eV for strained Ge on Si(100), and $\Delta E_v = 0.22 \pm 0.13$ eV for strained Si on Ge(100).

Ni and Hansson [9,10] also used X-ray photoelectron spectroscopy to obtain core level lineups, as well as to measure the band positions in strained materials. Their results are all in good agreement with the model-solid theory; for instance, for a $Si_{0.52}Ge_{0.48}$/Si interface on a Si(100) substrate they found $\Delta E_v = 0.36 \pm 0.6$ eV; the model-solid value is 0.42 eV.

Vescan et al [11] have studied $Si_{0.7}Ge_{0.3}$/Si single and multiple quantum wells grown by low-pressure chemical vapour deposition on a Si(100) substrate; they obtained $\Delta E_v = 0.27$ eV. The same group [12] has also performed DLTS measurements to obtain the valence-band offset at a $Si_{0.7}Ge_{0.3}$/Si heterojunction grown on a Si substrate, leading to $\Delta E_v = 0.22 \pm 0.02$ eV. The model-solid value is 0.25 eV for this structure.

Nauka et al [13] have used admittance spectroscopy to measure the valence-band offset in $Si/Si_{1-x}Ge_x$ heterojunctions ($0 < x < 0.45$) grown by chemical vapour deposition on Si(100). Within the model-solid approach it is possible to derive an analytical expression for the valence-band offset for such structures as a function of the composition x, keeping only terms linear in x. For $x < 0.5$, the model solid prediction is $\Delta E_v = 0.85x$. The values reported in [13] agree with this prediction to within 0.01 eV, as long as $x < 0.4$. As pointed out in [13] higher values of x lead to the breakdown of the assumption of a pseudomorphic interface.

Rodrigues et al [14] have performed photoreflectance measurements on $Ge_{0.7}Si_{0.3}/Ge$ superlattices grown on Ge(001). Their analysis leads to $\Delta E_{v,av} = 0.14 \pm 0.03$ eV. This value still includes the contribution from hydrostatic strain. To compare with the model solid, we therefore need to proceed as follows: the unstrained model-solid lineup for this structure would be $\Delta E^0_{v,av} = -6.35 - (-7.03 \times 0.3 - 6.35 \times 0.7) = -6.35 - (-6.55) = 0.20$ eV. Linear elastic theory [1] predicts a hydrostatic deformation in the $Ge_{0.7}Si_{0.3}$ layer of $\Delta V/V = 0.015$. The linearly interpolated absolute deformation potential for the valence band (see Datareview 4.2) is $a_v = 1.61$ eV, leading to an average valence-band position in the alloy of $E_{v,av} = -6.55 + 0.02 = -6.53$ eV. This leads to a final lineup of $\Delta E_{v,av} = -6.35 - (-6.53) = 0.18$ eV, which is consistent with the experimental value.

Similar arguments for a Si/Ge structure grown on Ge(001) would lead to $\Delta E_{v,av} = -6.35 - (-6.91) = 0.56$ eV. Comparing this value with the value of $\Delta E_{v,av} = 0.62$ eV which we previously obtained for a Si/Ge structure grown on Si(001), we note that the offsets between the average valence bands are very similar, and relatively unaffected by the specific strain situation. This point was already made in [2] based on full first-principles calculations for these interfaces.

Finally, Morar et al [15] have derived an unstrained valence-band offset between pure Si and Ge, based on spatially resolved electron-energy-loss spectroscopy in SiGe alloys, and the assumption that core level positions in the alloy can serve as references for determining the band lineup. They find $\Delta E^0_v = 0.78$ eV, where we use the superscript 'zero' to denote the absence of strain. The model-solid value would be 0.77 eV (note that, even though strain is absent, spin-orbit splitting has to be taken into account in determining the position of the top of the valence band).

D CONCLUSION

The band offsets at interfaces between Si, Ge and/or SiGe alloys are strongly influenced by strain effects, which depend on the alloy compositions of the overlayer and the substrate. Rather than list individual values for specific heterostructures, we have provided a comprehensive formalism that allows the determination of the band offsets for any heterojunction. The theoretical approach for the band lineup was based on the 'model solid theory' [1], which yields a band-offset between the average valence bands in unstrained Si and Ge of $\Delta E^0_{v,av} = 0.68$ eV.

We have shown that the model-solid predictions agree with more thorough first-principles calculations, as well as with experiment, to better than 0.15 eV (for the extreme of pure Si/Ge interfaces; for alloys, the error bar can be scaled with the alloy composition). Since the error bar on model-solid theory predictions was estimated to be $\pm$0.2 eV [1], the agreement is satisfactory. It is easy to check, however, that the model-solid values have a tendency to overestimate the offset, by about 0.1 eV for the pure materials. Even better agreement could therefore be obtained by reducing the unstrained valence-band offset by this amount; i.e. rather than using the model-solid value of $\Delta E^0_{v,av}$ = 0.68 eV (from [1]), one may use $\Delta E^0_{v,av}$ = 0.58 eV. However, we want to caution that it only makes sense to use this adjusted value if all other parameters (i.e. the hydrostatic and uniaxial deformation potentials) remain fixed as specified above.

REFERENCES

[1] C.G. Van de Walle [*Phys. Rev. B (USA)* vol.39 (1989) p.1871]

[2] C.G. Van de Walle, R.M. Martin [*Phys. Rev. B (USA)* vol.34 (1986) p.5621]

[3] M. Ikeda, K. Terakura, T. Oguchi [*Phys. Rev. B (USA)* vol.48 (1993) p.1571]

[4] L. Colombo, R. Resta, S. Baroni [*Phys. Rev. B (USA)* vol.44 (1991) p.5572]

[5] M.S. Hybertsen [in *Chemistry and Defects in Semiconductor Heterostructures* Eds M. Kawabe, T.D. Sands, E.R. Weber, R.S. Williams (Materials Research Society, Pittsburgh, 1989) vol.148 p.329]

[6] S. Baroni, R. Resta, A. Baldereschi, M. Peressi [in *Spectroscopy of Semiconductor Microstructures* Eds G. Fasol, A. Fasolino, P. Lugli (Plenum, London, 1989) p.251]

[7] G.P. Schwartz et al [*Phys. Rev. B (USA)* vol.39 (1989) p.1235]

[8] E.T. Yu, E.T. Croke, T.C. McGill, R.H. Miles [*Appl. Phys. Lett. (USA)* vol.56 (1990) p.569]

[9] W.-X. Ni, J. Knall, G.V. Hansson [*Phys. Rev. B (USA)* vol.36 (1987) p.7744]

[10] W.-X. Ni, G.V. Hansson [*Phys. Rev. B (USA)* vol.42 (1990) p.3030]

[11] L. Vescan et al [*Appl. Phys. Lett. (USA)* vol.60 (1992) p.2183]

[12] L. Vescan, R. Apetz, H. Lüth [*J. Appl. Phys. (USA)* vol.73 (1993) p.7427]

[13] K. Nauka et al [*Appl. Phys. Lett. (USA)* vol.60 (1992) p.195]

[14] P.A.M. Rodrigues, F. Cerdeira, J.C. Bean [*Phys. Rev. B (USA)* vol.46 (1992) p.15263]

[15] J.F. Morar, P.E. Batson, J. Tersoff [*Phys. Rev. B (USA)* vol.47 (1993) p.4107]

4.6 Optical spectroscopy of SiGe

J. Humlíček

September 1993

A INTRODUCTION

Optical spectroscopy contributes substantially to our understanding of the fundamental physical properties of solids, and enables an efficient contactless characterisation of materials and structures. The $Si_{1-x}Ge_x$ alloys provide suitable degrees of freedom as the composition x can be chosen and structures prepared with built-in strain; their optical properties have already been treated in a recent review by Pearsall [1]. The relevant review literature also includes discussions of the results for both constituents by Edwards [2] (Si) and Potter [3] (Ge), and for $Si_{1-x}Ge_x$ alloy [4].

We discuss here selected results of absorption (Section B), reflectance (Section C), ellipsometry (Section D), modulation spectroscopy (Section E) and photoluminescence (Section F) studies of $Si_{1-x}Ge_x$ alloys. We include plots of the spectral and compositional dependences of the refractive index n and absorption coefficient K. These quantities are useful to describe the wave propagation in terms of reflections at interfaces and attenuation in absorbing media; in particular, 1/K measures the penetration depth of light [5].

B ABSORPTION

The light transmitted through a free-standing (or supported by a transparent substrate) slab of material loses its intensity by the back reflections at interfaces and by absorption. For $Si_{1-x}Ge_x$ alloys, the reflection losses are fairly high even in the far-infrared region of weak lattice absorption and near-infrared region of weak absorption via indirect electronic transitions. The reason is the fairly high values of the refractive index n shown in FIGURE 1. The normal incidence reflectance at the interface with air is $R = (n - 1)^2/(n + 1)^2$; the transmission of a non-absorbing slab thick compared to the coherence length of light should be $T_0 = (1 - R)/(1 + R)$. The lower values observed occasionally on $Si_{1-x}Ge_x$ alloys are very likely to be due to the scattering of the probing beam on inhomogeneities of the composition (and, consequently, refractive index), which directs a part of the transmitted light to pass the detector [4].

B1 Lattice Vibrations

Braunstein [6] measured the lattice absorption in the whole range of compositions; however, his data showed a strong free carrier background. A more recent study [7], limited to the Si- and Ge-rich alloys, identifies several single-phonon processes in addition to the two-phonon bands analogous to those of pure Si and Ge. The study of Shen and Cardona [8] assigns local and quasilocal modes of Si in the Ge-rich alloys. The absorption of the $x \sim 0.5$ alloy reported in [4] peaks near 300, 400, and 500 wavenumbers, which are the energies of the 'Ge-Ge', 'Ge-Si' and 'Si-Si' vibrations [9], respectively.

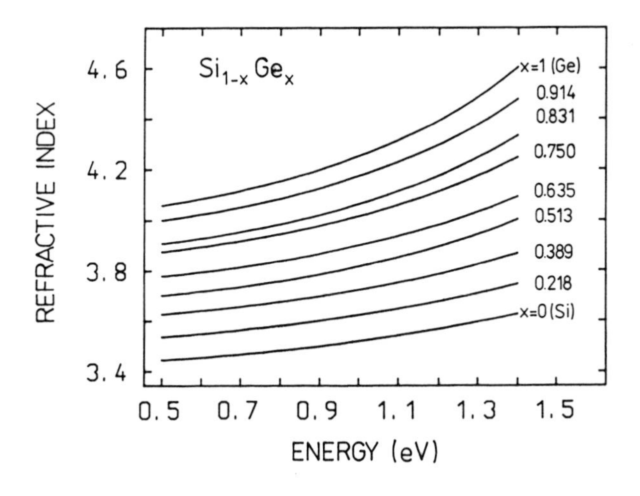

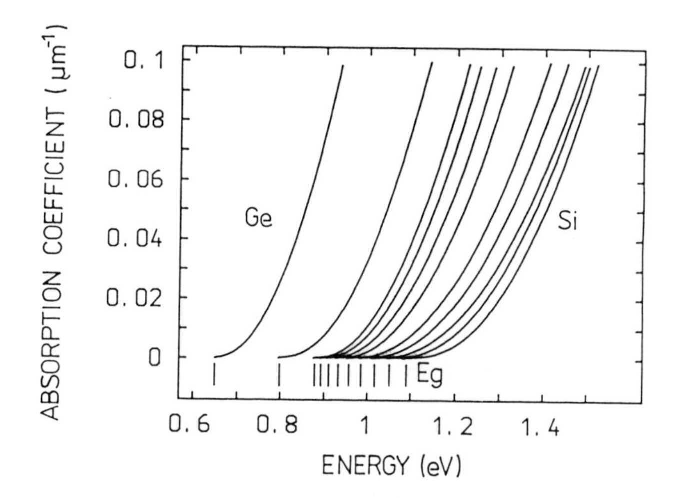

FIGURE 1 Refractive index for the listed compositions, obtained from the Kramers-Kronig analysis of the extinction index measured ellipsometrically above 1.66 eV [11]. Absorption coefficient at room temperature with the parameters of the Macfarlane-Roberts expression interpolated for x = 1 (Ge), 0.9, 0.8, ..., 0 (Si) using absorption spectra [11,4]; the vertical lines at the bottom show the corresponding gap energy.

B2 Indirect Absorption Edge

The near infrared absorption was studied in detail by Braunstein et al [11]. They fitted the absorption coefficient K to the Macfarlane-Roberts expression and plotted its parameters, which can be used in an interpolation scheme to obtain both the composition and temperature dependence [4]. The results are summarised in FIGURE 1. The spectra of absorption coefficients show the penetration depth of a few mm at E_g, decreasing to ~10 μm about 0.3 eV above E_g, still very large for the absorption to be seen in the filmed structures of nm scale.

C REFLECTANCE

The normal-incidence reflectance measured in the wide interval 1 - 13 eV by Schmidt [12] was analysed using Kramers-Kronig relations to obtain the complex response functions and their temperature changes [13]. With proper extrapolations above and below the measurement range, these data differ from the more recent ellipsometric results mainly due to thicker surface overlayers [4]. The technique is fairly sensitive to the strong direct interband transitions; the contributions of critical points of the joint density of states can be enhanced by numerical differentiation [13].

D ELLIPSOMETRY

The advantageous feature of the spectroscopic ellipsometry pioneered by Aspnes [14] is the way of obtaining the complex optical functions on a wavelength-by-wavelength basis. The work of Aspnes and Studna [15] on a number of carefully prepared semiconductor surfaces including Si and Ge has provided extremely accurate results. In the case of Si, they seem to be slightly improved by recent ellipsometric measurements on very clean epitaxial surfaces [16]. The ellipsometric technique has been used to measure the $Si_{1-x}Ge_x$ alloys throughout the whole composition range [17]. The results shown in FIGURE 2 display a fairly high optical contrast of the $Si_{1-x}Ge_x$ alloys of different compositions in the region of E_1 transitions. The critical point energy and composition are related by [17]

$$E_1(x) = 3.395 - 1.287x - 0.153x(1-x), \; x = 4.707 - \sqrt{(6.538\,E_1 - 0.0397)} \quad (1)$$

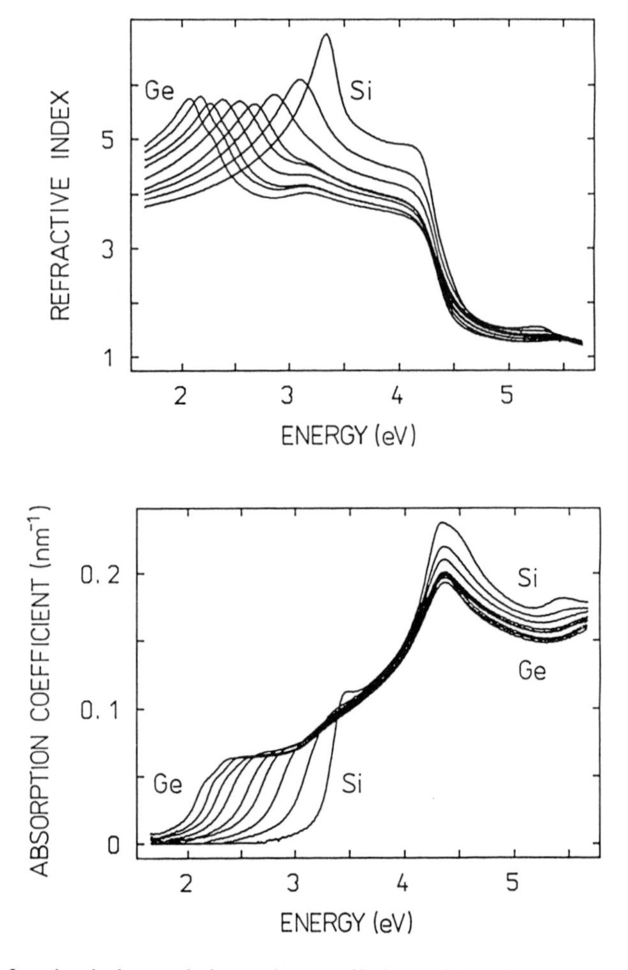

FIGURE 2 Refractive index and absorption coefficient of $Si_{1-x}Ge_x$ alloys with x = 1 (Ge), 0.915, 0.831, 0.750, 0.635, 0.513, 0.389, 0.218, 0 (Si) obtained from ellipsometric data [17].

The penetration depth of ~10 - 20 nm at E_1 increases to ~100 nm at about 0.3 eV below it for any composition. The lowest penetration depth of 4 nm is reached at 4.3 eV for Si. Both relaxed and strained Si-rich alloys on Si have been measured ellipsometrically by Pickering et al [18]. The technique is a very sensitive tool to characterise thin-film heterostructures using the interference patterns due to multiple reflections in the regions of high penetration depth [18,19].

E MODULATION SPECTROSCOPY

Modulation techniques enhance the spectral structures due to the critical points of the density of states [20]. For $Si_{1-x}Ge_x$ alloys, electroreflectance was used to obtain transition energies [21] and their shifts with hydrostatic pressure [22]. Thermoreflectance measurements provided the temperature coefficients of the transition energies from -0.2 to -0.5 meV/K and broadening energies from 0.1 to 0.2 meV/K [13]. Note that the numerical differentiation of either reflectance or ellipsometric spectra produces similar enhancement of the critical point structures.

F PHOTOLUMINESCENCE

A comprehensive photoluminescence study of bulk and relaxed epitaxial $Si_{1-x}Ge_x$ alloys has been done by Weber and Alonso [23]. They identified the emission of free and bound excitons, identified lines due to no-phonon radiative recombination and their phonon replicas, and extracted the compositional dependence of the X and L conduction band minima:

$$E_g^X(x) = 1.155 - 0.43x + 0.206x^2,\ E_g^L(x) = 2.010 - 1.270x \qquad (2)$$

Both the gaps are in eV and have been obtained at low temperatures (4.2 K); the L minimum is lower in energy for the Ge-rich alloys with the crossover at $x = 0.843$. The low-temperature extrapolation of the gap energies deduced from the absorption data using the Macfarlane-Roberts expression [10] gives lower values than Eqn (2), with the maximum deviation of about 40 meV. This is probably due to the oversimplification of the model absorption lineshapes with a single phonon energy [23].

A systematic study of pseudomorphic (fully strained) Si-rich $Si_{1-x}Ge_x$ layers has been reported by Robbins et al [24]. They observed the energy of no-phonon lines to decrease with the slope $dE_g/dx = 0.874$ eV at $x = 0$, twice the value given by Eqn (2) for the relaxed alloy. Since only the narrow interval $0.12 < x < 0.24$ has been covered, the linear and quadratic coefficients of the analytical expression given for the excitonic energy gap [24] can be rather unreliable.

G CONCLUSION

We have discussed selected results of various optical spectroscopies on $Si_{1-x}Ge_x$ alloys. The data for unstrained alloys are reasonably complete, except for the infrared spectra of lattice vibrations.

REFERENCES

[1] T.P. Pearsall [*CRC Crit. Rev. Solid State Mater. Sci. (USA)* vol.15 (1989) p.551]

[2] D.F. Edwards [in *Handbook of Optical Constants of Solids* Ed. E.D. Palik (Academic Press, New York, 1985) p.547]

[3] R.F. Potter [in *Handbook of Optical Constants of Solids* Ed. E.D. Palik (Academic Press, New York, 1985) p.465]

[4] J. Humlíček, F. Lukeš, E. Schmidt [in *Handbook of Optical Constants of Solids II* Ed. E.D. Palik (Academic Press, New York, 1991) p.607]

[5] J. Humlíček [Datareview in this book: 4.7 Optical functions of the relaxed SiGe alloy: influence of strain]

[6] R. Braunstein [*Phys. Rev. (USA)* vol.130 (1958) p.879]

[7] A.E. Cosand, W.G. Spitzer [*J. Appl. Phys. (USA)* vol.42 (1971) p.5241]

[8] S.C. Shen, M. Cardona [*Solid State Commun. (USA)* vol.36 (1980) p.327]

[9] G.M. Zinger, I.P. Ipatova, A.V. Subashiev [*Sov. Phys.-Semicond. (USA)* vol.11 (1977) p.383]

[10] J. Humlíček, A. Roeseler, T. Zettler, M.G. Kekoua, E. Khoutsishvili [*Appl. Opt. (USA)* vol.31 (1992) p.90]

[11] R. Braunstein, A.R. Moore, F. Herman [*Phys. Rev. (USA)* vol.109 (1958) p.695]

[12] E. Schmidt [*Phys. Status Solidi (Germany)* vol.27 (1968) p.57]

[13] J. Humlíček, F. Lukeš, E. Schmidt, M.G. Kekoua, E. Khoutsishvili [*Phys. Rev. B (USA)* vol.33 (1986) p.1092]

[14] D.E. Aspnes [in *Handbook of Optical Constants of Solids* Ed. E.D. Palik (Academic Press, New York, 1985) p.89]

[15] D.E. Aspnes, A.A. Studna [*Phys. Rev. B (USA)* vol.27 (1983) p.985]

[16] V. Nayar, W.Y. Leong, C. Pickering, A.J. Pidduck, R.R. Carline, D.J. Robbins [*Appl. Phys. Lett. (USA)* vol.61 (1992) p.1304]

[17] J. Humlíček, M. Garriga, M.I. Alonso, M. Cardona [*J. Appl. Phys. (USA)* vol.65 (1989) p.2827]

[18] C. Pickering et al [*J. Appl. Phys. (USA)* vol.73 (1993) p.239]

[19] U. Schmid et al [*Phys. Rev. B (USA)* vol.45 (1992) p.6793]

[20] M. Cardona [*Modulation Spectroscopy* (Academic Press, New York, 1969)]

[21] J.S. Kline, F.H. Pollak, M. Cardona [*Helv. Phys. Acta. (Switzerland)* vol.41 (1968) p.968]

[22] E. Schmidt, K. Vedam [*Solid State Commun. (USA)* vol.9 (1971) p.1187]

[23] J. Weber, M.I. Alonso [*Phys. Rev. B (USA)* vol.40 (1989) p.5683]

[24] D.J. Robbins, L.T. Canham, S.J. Barnett, A.D. Pitt, P. Calcott [*J. Appl. Phys. (USA)* vol.71 (1992) p.1407]

4.7 Optical functions of the relaxed SiGe alloy and influence of strain

J. Humlíček

September 1993

A INTRODUCTION

The optical response of an isotropic material (e.g. relaxed $Si_{1-x}Ge_x$ alloy) is conveniently described by several sets of complex scalar functions; these functions become tensorial due to the strain-induced anisotropy. The optical functions of both constituents have been discussed and tabulated by Edwards [1] (Si) and Potter [2] (Ge); those of unstrained bulk $Si_{1-x}Ge_x$ alloy have already been reviewed and tabulated [3] for three compositions x ~ 0.20, 0.50 and 0.75. We comment here on the use of different optical functions (Section B), discuss the optical properties of $Si_{1-x}Ge_x$ alloys in different spectral regions (Section C), provide tables for several alloy compositions (Section D), and discuss the effects of strain (Section E).

B OPTICAL FUNCTIONS

The most common optical functions include, for any photon energy $\hbar\omega$ or wavelength λ, the complex dielectric function $\varepsilon = \varepsilon_1 + i\varepsilon_2$ and the complex refractive index $N = n + ik = \sqrt{\varepsilon}$. The dielectric function links the macroscopic electric displacement D and field E via $D = \varepsilon E = E + 4\pi P$, where P is the induced dipole-moment density. Since the polarisation (measured by P) due to different physical mechanisms is approximately additive, ε is the appropriate function to deal with an additional polarisability superposed on the intrinsic background (such as plasma of free carriers in doped materials or impurities). The complex refractive index is useful to describe the wave propagation, including the reflections at interfaces of layered structures. The intensity of a wave traversing a distance d in a homogeneous medium of extinction coefficient k is attenuated by the factor exp(-Kd), where $K = 4\pi k/\lambda$ is the absorption coefficient. Thus, 1/K is a convenient measure of the penetration depth.

C OPTICAL FUNCTIONS OF RELAXED SiGe ALLOY FROM IR TO UV

C1 IR, Lattice Vibrations

The lattice absorption has been measured for Si- and Ge-rich $Si_{1-x}Ge_x$ alloys in the far-IR region 12 - 87 meV [4]; the spectra are rather complex due to a number of two- and single-phonon bands. The absorption is weak for the whole composition range, with the peak values of $k \sim 10^{-3}$ [3]. This is related to weak variations of a smooth spectral dependence of the real refractive index of $Si_{1-x}Ge_x$ alloy as a function of x and the photon energy $\hbar\omega$ in eV, $\hbar\omega \leq 0.5$ eV [5],

$$n(\hbar\omega) = n_0 + n_1 (\hbar\omega)^2,\ n_0 = 3.42 + 0.37x + 0.22x^2,$$
$$n_1 = 0.094 + 0.033x + 0.089x^2 \qquad (1)$$

C2 Near IR, Indirect Bandgap

A detailed study of the indirect absorption edge has been performed by Braunstein et al [6]; the resulting spectra of absorption coefficient can be represented efficiently by means of the compositionally dependent Macfarlane-Roberts expression [6,3]. The strength of the indirect absorption up to a few tenths eV above the gap energy is comparable to the strength of the lattice vibrations. Consequently, there is no pronounced structure in the refractive index, which can be obtained from ellipsometric data in vis-UV [5].

C3 Vis-UV, Direct Interband Transitions

A comprehensive set of optical data above the fundamental absorption edge was reported by Schmidt [7] who used the normal incidence reflectance and Kramers-Kronig analysis. Extensive ellipsometric measurements have been performed on bulk and LPE samples in the whole composition range [8] and on Si-rich MBE layers [9]. The strong absorption due to the E_1 interband transitions shifts about 1.3 eV downwards going from Si to Ge. Since the E_1 energy can be determined fairly accurately (to ~4 meV), the optical functions in this region provide a convenient way of determining the alloy composition (to ~0.3%) [8]. Temperature effects on the direct interband absorption, and the related temperature dependence of the infrared refractive index, have been recently studied using ellipsometry [10].

D TABLES OF OPTICAL FUNCTIONS OF RELAXED SiGe ALLOY FOR SEVERAL COMPOSITIONS

Values of the complex dielectric function $\varepsilon_1 + i\varepsilon_2$, complex refractive index $n + ik$, and absorption coefficient K, interpolated from various sources, are listed in TABLE 1. The horizontal lines divide the segments of

- indirect absorption edge; K and k are from the parametrisation [3] of the absorption data [6], n is from the Kramers-Kronig transform of the vis-UV absorption [5],
- 2 - 5.6 eV, ellipsometric data [8],
- 6 - 12 eV, reflectance data [7].

TABLE 1 Optical functions of relaxed SiGe.

$Si_{1-x}Ge_x$, x = 0.00

Energy (eV)	Wavelength (nm)	ε_1	ε_2	n	k	K (cm^{-1})
1.00	1239.8	12.39	0.0001	3.52	0.00001	1.01
1.05	1180.8	12.46	0.0001	3.53	0.00001	1.06
1.10	1127.1	12.53	0.0001	3.54	0.00002	2.23
1.15	1078.1	12.60	0.0007	3.55	0.00010	1.17 x 10
1.20	1033.2	12.74	0.0027	3.57	0.00038	4.62 x 10
1.25	991.9	12.82	0.0063	3.58	0.00088	1.11×10^2
1.30	953.7	12.96	0.0113	3.60	0.00157	2.07×10^2
1.35	918.4	13.03	0.0176	3.61	0.00244	3.34×10^2
1.40	885.6	13.18	0.0251	3.63	0.00346	4.91×10^2
2.0	619.9	15.52	0.08	3.94	0.01	2.03×10^3
2.1	590.4	16.08	0.08	4.01	0.01	2.13×10^3
2.2	563.6	16.65	0.08	4.08	0.01	2.23×10^3
2.3	539.1	17.39	0.08	4.17	0.01	2.33×10^3
2.4	516.6	18.15	0.09	4.26	0.01	2.43×10^3
2.5	495.9	19.10	0.17	4.37	0.02	5.07×10^3
2.6	476.9	20.25	0.18	4.50	0.02	5.27×10^3
2.7	459.2	21.62	0.19	4.65	0.02	5.47×10^3
2.8	442.8	23.22	1.06	4.82	0.11	3.12×10^4
2.9	427.5	25.37	1.71	5.04	0.17	5.00×10^4
3.0	413.3	28.13	2.65	5.31	0.25	7.60×10^4
3.1	400.0	32.01	4.20	5.67	0.37	1.16×10^5
3.2	387.5	37.77	8.03	6.18	0.65	2.11×10^5
3.3	375.7	44.88	19.59	6.85	1.43	4.78×10^5
3.4	364.7	34.06	38.27	6.53	2.93	1.01×10^6
3.5	354.2	20.86	35.25	5.56	3.17	1.12×10^6
3.6	344.4	17.77	32.87	5.25	3.13	1.14×10^6
3.8	326.3	14.01	33.37	5.01	3.33	1.28×10^6
4.0	310.0	10.12	36.73	4.91	3.74	1.52×10^6
4.2	295.2	-1.04	44.83	4.68	4.79	2.04×10^6
4.4	281.8	-19.35	30.84	2.92	5.28	2.35×10^6
4.6	269.5	-17.35	17.18	1.88	4.57	2.13×10^6
4.8	258.3	-12.73	12.51	1.60	3.91	1.90×10^6
5.0	248.0	-9.98	10.74	1.53	3.51	1.78×10^6
5.2	238.4	-8.55	10.26	1.55	3.31	1.74×10^6
5.4	229.6	-8.98	9.50	1.43	3.32	1.82×10^6
5.6	221.4	-8.38	7.81	1.24	3.15	1.79×10^6
6	206.6	-6.57	5.82	1.05	2.77	1.68×10^6
7	177.1	-4.39	3.09	0.70	2.21	1.57×10^6
8	155.0	-2.70	2.10	0.60	1.75	1.42×10^6
9	137.8	-1.80	1.60	0.55	1.45	1.32×10^6
10	124.0	-1.23	1.30	0.53	1.23	1.25×10^6
11	112.7	-0.84	1.07	0.51	1.05	1.17×10^6
12	103.3	-0.56	0.90	0.50	0.90	1.09×10^6

TABLE 1 continued

$Si_{1-x}Ge_x$, x = 0.10

Energy (eV)	Wavelength (nm)	ε_1	ε_2	n	k	K (cm^{-1})
1.00	1239.8	12.72	0.0001	3.57	0.00001	1.01
1.05	1180.8	12.82	0.0002	3.58	0.00003	3.02
1.10	1127.1	12.89	0.0008	3.59	0.00011	1.19 x 10
1.15	1078.1	12.96	0.0023	3.60	0.00032	3.68 x 10
1.20	1033.2	13.11	0.0051	3.62	0.00071	8.64 x 10
1.25	991.9	13.18	0.0094	3.63	0.00130	1.64 x 10^2
1.30	953.7	13.33	0.0150	3.65	0.00205	2.70 x 10^2
1.35	918.4	13.43	0.0217	3.67	0.00296	4.05 x 10^2
1.40	885.6	13.58	0.0295	3.69	0.00400	5.67 x 10^2
2.0	619.9	16.14	0.08	4.02	0.01	2.03 x 10^3
2.1	590.4	16.75	0.08	4.09	0.01	2.13 x 10^3
2.2	563.6	17.44	0.08	4.18	0.01	2.23 x 10^3
2.3	539.1	18.24	0.12	4.27	0.01	3.40 x 10^3
2.4	516.6	19.14	0.41	4.37	0.05	1.14 x 10^4
2.5	495.9	20.23	0.68	4.50	0.08	1.90 x 10^4
2.6	476.9	21.58	1.08	4.65	0.12	3.07 x 10^4
2.7	459.2	23.20	1.52	4.82	0.16	4.31 x 10^4
2.8	442.8	25.15	2.63	5.02	0.26	7.42 x 10^4
2.9	427.5	27.76	4.17	5.28	0.39	1.16 x 10^5
3.0	413.3	30.86	6.95	5.59	0.62	1.89 x 10^5
3.1	400.0	33.49	11.67	5.87	0.99	3.12 x 10^5
3.2	387.5	34.44	17.62	6.05	1.46	4.73 x 10^5
3.3	375.7	34.32	25.69	6.21	2.07	6.92 x 10^5
3.4	364.7	25.81	34.50	5.87	2.94	1.01 x 10^6
3.5	354.2	17.89	32.40	5.24	3.09	1.10 x 10^6
3.6	344.4	15.52	31.05	5.01	3.10	1.13 x 10^6
3.8	326.3	12.08	31.86	4.80	3.32	1.28 x 10^6
4.0	310.0	8.22	34.90	4.69	3.72	1.51 x 10^6
4.2	295.2	-2.13	41.28	4.43	4.66	1.98 x 10^6
4.4	281.8	-18.17	28.69	2.81	5.11	2.28 x 10^6
4.6	269.5	-16.09	16.61	1.88	4.43	2.06 x 10^6
4.8	258.3	-12.12	12.26	1.60	3.83	1.86 x 10^6
5.0	248.0	-9.64	10.48	1.52	3.45	1.75 x 10^6
5.2	238.4	-8.26	9.87	1.52	3.25	1.71 x 10^6
5.4	229.6	-8.41	9.25	1.43	3.23	1.77 x 10^6
5.6	221.4	-8.08	7.82	1.26	3.11	1.76 x 10^6
6	206.6	-6.26	6.11	1.12	2.74	1.67 x 10^6
7	177.1	-4.28	3.23	0.74	2.19	1.56 x 10^6
8	155.0	-2.56	2.17	0.63	1.72	1.39 x 10^6
9	137.8	-1.70	1.64	0.58	1.42	1.30 x 10^6
10	124.0	-1.11	1.35	0.57	1.19	1.21 x 10^6
11	112.7	-0.71	1.15	0.57	1.01	1.13 x 10^6
12	103.3	-0.45	1.00	0.57	0.88	1.07 x 10^6

TABLE 1 continued

$Si_{1-x}Ge_x$, x = 0.20

Energy (eV)	Wavelength (nm)	ε_1	ε_2	n	k	K (cm^{-1})
1.00	1239.8	13.04	0.0001	3.61	0.00001	1.01
1.05	1180.8	13.18	0.0003	3.63	0.00005	4.97
1.10	1127.1	13.26	0.0014	3.64	0.00019	2.17 x 10
1.15	1078.1	13.33	0.0039	3.65	0.00053	6.19 x 10
1.20	1033.2	13.48	0.0076	3.67	0.00104	1.27 x 10^2
1.25	991.9	13.55	0.0126	3.68	0.00171	2.17 x 10^2
1.30	953.7	13.70	0.0188	3.70	0.00253	3.34 x 10^2
1.35	918.4	13.84	0.0259	3.72	0.00348	4.76 x 10^2
1.40	885.6	13.99	0.0339	3.74	0.00453	6.43 x 10^2
2.0	619.9	16.78	0.08	4.10	0.01	2.03 x 10^3
2.1	590.4	17.43	0.08	4.18	0.01	2.13 x 10^3
2.2	563.6	18.26	0.09	4.27	0.01	2.23 x 10^3
2.3	539.1	19.11	0.17	4.37	0.02	4.47 x 10^3
2.4	516.6	20.15	0.75	4.49	0.08	2.03 x 10^4
2.5	495.9	21.39	1.20	4.63	0.13	3.30 x 10^4
2.6	476.9	22.93	2.04	4.79	0.21	5.60 x 10^4
2.7	459.2	24.81	2.95	4.99	0.30	8.08 x 10^4
2.8	442.8	27.12	4.31	5.22	0.41	1.17 x 10^5
2.9	427.5	30.16	6.85	5.53	0.62	1.82 x 10^5
3.0	413.3	33.47	11.66	5.87	0.99	3.02 x 10^5
3.1	400.0	34.27	19.65	6.07	1.62	5.08 x 10^5
3.2	387.5	29.85	26.79	5.91	2.26	7.34 x 10^5
3.3	375.7	23.76	30.16	5.57	2.71	9.05 x 10^5
3.4	364.7	18.44	30.72	5.21	2.95	1.02 x 10^6
3.5	354.2	15.10	29.64	4.92	3.01	1.07 x 10^6
3.6	344.4	13.38	29.27	4.77	3.07	1.12 x 10^6
3.8	326.3	10.23	30.36	4.60	3.30	1.27 x 10^6
4.0	310.0	6.41	33.09	4.48	3.69	1.50 x 10^6
4.2	295.2	-3.12	37.86	4.18	4.53	1.93 x 10^6
4.4	281.8	-17.03	26.63	2.70	4.93	2.20 x 10^6
4.6	269.5	-14.87	16.04	1.87	4.29	2.00 x 10^6
4.8	258.3	-11.53	12.01	1.60	3.75	1.83 x 10^6
5.0	248.0	-9.30	10.22	1.50	3.40	1.72 x 10^6
5.2	238.4	-7.97	9.48	1.49	3.19	1.68 x 10^6
5.4	229.6	-7.85	9.00	1.43	3.15	1.72 x 10^6
5.6	221.4	-7.78	7.83	1.28	3.07	1.74 x 10^6
6	206.6	-5.95	6.40	1.18	2.71	1.65 x 10^6
7	177.1	-4.16	3.36	0.77	2.18	1.55 x 10^6
8	155.0	-2.42	2.23	0.66	1.69	1.37 x 10^6
9	137.8	-1.60	1.68	0.60	1.40	1.28 x 10^6
10	124.0	-0.98	1.39	0.60	1.16	1.18 x 10^6
11	112.7	-0.57	1.22	0.62	0.98	1.09 x 10^6
12	103.3	-0.33	1.10	0.64	0.86	1.05 x 10^6

TABLE 1 continued

$Si_{1-x}Ge_x$, x = 0.30

Energy (eV)	Wavelength (nm)	ε_1	ε_2	n	k	K (cm^{-1})
1.00	1239.8	13.45	0.0002	3.67	0.00003	3.44
1.05	1180.8	13.60	0.0010	3.69	0.00014	1.50 x 10
1.10	1127.1	13.67	0.0031	3.70	0.00042	4.64 x 10
1.15	1078.1	13.78	0.0066	3.71	0.00089	1.04 x 10^2
1.20	1033.2	13.93	0.0115	3.73	0.00154	1.87 x 10^2
1.25	991.9	14.04	0.0175	3.75	0.00234	2.96 x 10^2
1.30	953.7	14.19	0.0247	3.77	0.00328	4.32 x 10^2
1.35	918.4	14.35	0.0328	3.79	0.00433	5.93 x 10^2
1.40	885.6	14.50	0.0419	3.81	0.00550	7.80 x 10^2
2.0	619.9	17.53	0.08	4.19	0.01	2.03 x 10^3
2.1	590.4	18.29	0.17	4.28	0.02	4.17 x 10^3
2.2	563.6	19.19	0.38	4.38	0.04	9.71 x 10^3
2.3	539.1	20.24	0.83	4.50	0.09	2.14 x 10^4
2.4	516.6	21.50	1.50	4.64	0.16	3.94 x 10^4
2.5	495.9	23.15	2.32	4.82	0.24	6.10 x 10^4
2.6	476.9	24.90	3.55	5.00	0.35	9.35 x 10^4
2.7	459.2	27.20	5.41	5.24	0.52	1.41 x 10^5
2.8	442.8	29.62	8.69	5.50	0.79	2.24 x 10^5
2.9	427.5	30.74	13.75	5.68	1.21	3.56 x 10^5
3.0	431.3	29.98	18.87	5.72	1.65	5.02 x 10^5
3.1	400.0	27.71	23.85	5.67	2.10	6.61 x 10^5
3.2	387.5	23.39	27.49	5.45	2.52	8.17 x 10^5
3.3	275.7	18.56	28.79	5.14	2.80	9.37 x 10^5
3.4	364.7	15.19	28.33	4.86	2.91	1.00 x 10^6
3.5	354.2	13.10	27.83	4.68	2.97	1.05 x 10^6
3.6	344.4	11.65	27.80	4.57	3.04	1.11 x 10^6
3.8	326.3	8.81	28.97	4.42	3.28	1.26 x 10^6
4.0	310.0	5.10	31.57	4.31	3.67	1.49 x 10^6
4.2	295.2	-3.97	35.37	3.98	4.45	1.89 x 10^6
4.4	281.8	-16.24	24.92	2.60	4.79	2.14 x 10^6
4.6	269.5	-14.09	15.35	1.84	4.18	1.95 x 10^6
4.8	258.3	-11.11	11.62	1.58	3.69	1.79 x 10^6
5.0	248.0	-9.04	9.85	1.47	3.35	1.70 x 10^6
5.2	238.4	-7.78	9.09	1.45	3.14	1.66 x 10^6
5.4	229.6	-7.48	8.68	1.41	3.08	1.68 x 10^6
5.6	221.4	-7.55	7.76	1.28	3.03	1.72 x 10^6
6	206.6	-5.89	6.47	1.20	2.71	1.65 x 10^6
7	177.1	-4.04	3.41	0.79	2.16	1.53 x 10^6
8	155.0	-2.28	2.33	0.70	1.67	1.35 x 10^6
9	137.8	-1.50	1.80	0.65	1.39	1.27 x 10^6
10	124.0	-0.93	1.48	0.64	1.16	1.17 x 10^6
11	112.7	-0.53	1.31	0.66	0.98	1.10 x 10^6
12	103.3	-0.28	1.19	0.68	0.87	1.06 x 10^6

TABLE 1 continued

$Si_{1-x}Ge_x$, x = 0.50

Energy (eV)	Wavelength (nm)	ε_1	ε_2	n	k	K (cm^{-1})
0.95	1305.1	14.37	0.0003	3.79	0.00005	4.41
1.00	1239.8	14.51	0.0018	3.81	0.00023	2.33 x 10
1.05	1180.8	14.67	0.0054	3.83	0.00071	7.51 x 10
1.10	1127.1	14.74	0.0111	3.84	0.00145	1.61×10^2
1.15	1078.1	14.90	0.0186	3.86	0.00241	2.81×10^2
1.20	1033.2	15.12	0.0279	3.89	0.00359	4.36×10^2
1.25	991.9	15.28	0.0386	3.91	0.00494	6.25×10^2
1.30	953.7	15.50	0.0507	3.94	0.00644	8.49×10^2
2.0	619.9	19.21	1.42	4.39	0.16	3.29×10^4
2.1	590.4	20.25	1.72	4.50	0.19	4.07×10^4
2.2	563.6	21.56	2.41	4.65	0.26	5.78×10^4
2.3	539.1	23.18	3.20	4.83	0.33	7.72×10^4
2.4	516.6	25.09	4.67	5.03	0.46	1.13×10^5
2.5	495.9	27.20	7.17	5.26	0.68	1.73×10^5
2.6	476.9	29.38	11.13	5.51	1.01	2.66×10^5
2.7	459.2	29.18	17.57	5.62	1.56	4.28×10^5
2.8	442.8	24.76	22.14	5.38	2.06	5.83×10^5
2.9	427.5	20.27	23.50	5.06	2.32	6.82×10^5
3.0	431.3	17.44	23.20	4.82	2.41	7.32×10^5
3.1	400.0	15.91	23.12	4.69	2.47	7.75×10^5
3.2	387.5	14.47	23.69	4.60	2.58	8.36×10^5
3.3	375.7	12.67	24.12	4.47	2.70	9.03×10^5
3.4	364.7	11.03	24.30	4.34	2.80	9.64×10^5
3.5	354.2	9.62	24.34	4.23	2.88	1.02×10^6
3.6	344.4	8.52	24.47	4.15	2.95	1.08×10^6
3.8	326.3	6.18	25.53	4.03	3.17	1.22×10^6
4.0	310.0	2.73	27.56	3.90	3.53	1.43×10^6
4.2	295.2	-4.68	29.65	3.56	4.17	1.77×10^6
4.4	281.8	-13.72	21.32	2.41	4.42	1.97×10^6
4.6	269.5	-12.01	13.81	1.77	3.89	1.81×10^6
4.8	258.3	-9.72	10.72	1.54	3.48	1.69×10^6
5.0	248.0	-8.03	9.09	1.43	3.17	1.61×10^6
5.2	238.4	-6.89	8.34	1.40	2.97	1.57×10^6
5.4	229.6	-6.45	8.05	1.39	2.89	1.58×10^6
5.6	221.4	-6.58	7.59	1.32	2.88	1.64×10^6
6	206.6	-5.68	6.58	1.23	2.68	1.63×10^6
7	177.1	-3.77	3.63	0.85	2.12	1.51×10^6
8	155.0	-2.07	2.52	0.77	1.63	1.32×10^6
9	137.8	-1.33	2.01	0.73	1.37	1.25×10^6
10	124.0	-0.80	1.64	0.72	1.15	1.16×10^6
11	112.7	-0.42	1.48	0.75	0.99	1.10×10^6
12	103.3	-0.19	1.35	0.77	0.88	1.07×10^6

TABLE 1 continued

$Si_{1-x}Ge_x$, x = 0.75

Energy (eV)	Wavelength (nm)	ε_1	ε_2	n	k	K (cm^{-1})
0.90	1377.6	15.84	0.0003	3.98	0.00004	3.65
0.95	1305.1	16.00	0.0023	4.00	0.00029	2.79 x 10
1.00	1239.8	16.16	0.0076	4.02	0.00094	9.53 x 10
1.05	1180.8	16.32	0.0156	4.04	0.00193	2.05 x 10^2
1.10	1127.1	16.48	0.0261	4.06	0.00322	3.59 x 10^2
1.15	1078.1	16.73	0.0391	4.09	0.00478	5.57 x 10^2
1.20	1033.2	16.89	0.0540	4.11	0.00657	7.99 x 10^2
2.0	619.9	24.22	2.86	4.93	0.29	5.88 x 10^4
2.1	590.4	25.93	4.39	5.11	0.43	9.15 x 10^4
2.2	563.6	28.00	6.82	5.33	0.64	1.43 x 10^5
2.3	539.1	30.40	10.98	5.60	0.98	2.28 x 10^5
2.4	516.6	30.25	17.96	5.72	1.57	3.82 x 10^5
2.5	495.9	25.25	23.00	5.45	2.11	5.35 x 10^5
2.6	476.9	19.99	24.23	5.07	2.39	6.30 x 10^5
2.7	459.2	16.28	23.13	4.72	2.45	6.70 x 10^5
2.8	442.8	14.39	21.78	4.50	2.42	6.87 x 10^5
2.9	427.5	13.65	21.03	4.40	2.39	7.02 x 10^5
3.0	413.3	13.35	20.75	4.36	2.38	7.24 x 10^5
3.1	400.0	13.01	21.36	4.36	2.45	7.70 x 10^5
3.2	387.5	12.23	22.31	4.34	2.57	8.33 x 10^5
3.3	375.7	10.94	23.06	4.27	2.70	9.03 x 10^5
3.4	364.7	9.63	23.41	4.18	2.80	9.65 x 10^5
3.5	354.2	8.38	23.64	4.09	2.89	1.03 x 10^6
3.6	344.4	7.36	24.02	4.03	2.98	1.09 x 10^6
3.8	326.3	5.00	25.24	3.92	3.22	1.24 x 10^6
4.0	310.0	1.62	27.21	3.80	3.58	1.45 x 10^6
4.2	295.2	-5.75	29.13	3.46	4.21	1.79 x 10^6
4.4	281.8	-14.60	20.52	2.30	4.46	1.99 x 10^6
4.6	269.5	-12.34	12.99	1.67	3.89	1.81 x 10^6
4.8	258.3	-9.94	10.06	1.45	3.47	1.69 x 10^6
5.0	248.0	-8.23	8.56	1.35	3.17	1.61 x 10^6
5.2	238.4	-6.99	7.87	1.33	2.96	1.56 x 10^6
5.4	229.6	-6.44	7.69	1.34	2.87	1.57 x 10^6
5.6	221.4	-6.71	7.40	1.28	2.89	1.64 x 10^6
6	206.6	-5.07	6.55	1.27	2.58	1.57 x 10^6
7	177.1	-3.29	4.26	1.02	2.08	1.48 x 10^6
8	155.0	-2.00	2.71	0.83	1.64	1.33 x 10^6
9	137.8	-1.21	2.17	0.80	1.36	1.24 x 10^6
10	124.0	-0.64	1.80	0.80	1.13	1.15 x 10^6
11	112.7	-0.28	1.66	0.84	0.99	1.10 x 10^6
12	103.3	-0.09	1.52	0.85	0.90	1.09 x 10^6

TABLE 1 continued

$Si_{1-x}Ge_x$, x = 1.00

Energy (eV)	Wavelength (nm)	ε_1	ε_2	n	k	K (cm^{-1})
0.65	1907.5	16.81	0.0002	4.10	0.00002	1.32
0.70	1771.2	16.97	0.0032	4.12	0.00039	2.77 x 10
0.75	1653.1	17.06	0.0124	4.13	0.00150	1.14×10^2
0.80	1549.8	17.22	0.0270	4.15	0.00325	2.64×10^2
0.85	1458.6	17.47	0.0463	4.18	0.00554	4.77×10^2
0.90	1377.6	17.64	0.0694	4.20	0.00826	7.53×10^2
2.0	619.9	30.48	11.80	5.62	1.05	2.13×10^5
2.1	590.4	29.51	20.10	5.71	1.76	3.75×10^5
2.2	563.6	22.93	23.19	5.27	2.20	4.91×10^5
2.3	539.1	18.85	24.55	4.99	2.46	5.73×10^5
2.4	516.6	13.90	23.52	4.54	2.59	6.30×10^5
2.5	495.9	11.81	21.25	4.25	2.50	6.33×10^5
2.6	476.9	10.87	19.80	4.09	2.42	6.38×10^5
2.7	459.2	10.40	18.75	3.99	2.35	6.43×10^5
2.8	442.8	10.28	18.05	3.94	2.29	6.50×10^5
2.9	427.5	10.46	17.73	3.94	2.25	6.61×10^5
3.0	413.3	10.73	17.99	3.98	2.26	6.87×10^5
3.1	400.0	10.68	18.81	4.02	2.34	7.35×10^5
3.2	387.5	10.06	19.86	4.02	2.47	8.01×10^5
3.3	375.7	9.00	20.64	3.97	2.60	8.70×10^5
3.4	364.7	7.92	21.06	3.90	2.70	9.30×10^5
3.5	354.2	7.02	21.35	3.84	2.78	9.86×10^5
3.6	344.4	6.05	21.70	3.78	2.87	1.05×10^6
3.8	326.3	4.14	22.87	3.70	3.09	1.19×10^6
4.0	310.0	1.27	24.76	3.61	3.43	1.39×10^6
4.2	295.2	-5.17	26.99	3.34	4.04	1.72×10^6
4.4	281.8	-14.96	18.92	2.14	4.42	1.97×10^6
4.6	269.5	-11.53	11.31	1.52	3.72	1.73×10^6
4.8	258.3	-9.11	9.00	1.36	3.31	1.61×10^6
5.0	248.0	-7.46	7.79	1.29	3.02	1.53×10^6
5.2	238.4	-6.31	7.22	1.28	2.82	1.49×10^6
5.4	229.6	-5.76	7.10	1.30	2.73	1.49×10^6
5.6	221.4	-5.79	7.04	1.29	2.73	1.55×10^6
6	206.6	-5.04	7.21	1.37	2.63	1.60×10^6
7	177.1	-3.58	4.28	1.00	2.14	1.52×10^6
8	155.0	-2.30	2.87	0.83	1.73	1.40×10^6
9	137.8	-1.49	2.26	0.78	1.45	1.32×10^6
10	124.0	-1.00	1.88	0.75	1.25	1.27×10^6
11	112.7	-0.62	1.60	0.74	1.08	1.20×10^6
12	103.3	-0.37	1.42	0.74	0.96	1.17×10^6

E INFLUENCE OF STRAIN

Except for the E_0' and E_2 transitions, the interband separation in $Si_{1-x}Ge_x$ is fairly sensitive to the alloy composition [11]. The sensitivity to interatomic spacing can be expected to be seen also in the pressure dependence of the transition energies. The influence of hydrostatic pressure p on the electroreflectance spectra has been reported [12], with the pressure coefficients of the E_1 and E_0' transitions

$$dE_1/dp = 6.2 + 1.6x, \quad dE_0'/dp = 1.0 + 0.4x \text{ meV/kbar} \qquad (2)$$

The strain in thin layers grown pseudomorphically on substrates with different lattice constants has both a hydrostatic and a uniaxial component. The $Si_{1-x}Ge_x$ alloy films on Si are compressed along the substrate plane and under tensile stress perpendicular to it, which induces optical anisotropy. At normal incidence, the light sees the in-plane components of the response tensor. The same is approximately valid for the oblique incidence, because of the strong attenuation of the field component normal to the surface for the typical optical functions of semiconductors [13]. The strain effects in Si-rich alloy films have been studied by Pickering et al using ellipsometry; they observed shifts of the critical point energies in agreement with the predictions of deformation potentials and elastic constants interpolated between Si and Ge [9].

The strain effects on the absorption at the fundamental indirect edge are of considerable interest. The strain applied by the Si substrate to the $Si_{1-x}Ge_x$ alloy lowers the bandgap as calculated by People [14]. Since the absorption coefficient is of the order of 10 cm^{-1} near the gap energy, the shift of the optical absorption is difficult to observe on the very thin unrelaxed films. To our knowledge, there has been no relevant experimental result.

F CONCLUSION

We have discussed the optical functions of relaxed $Si_{1-x}Ge_x$ alloys and provided tables for their spectral and compositional dependences. Data on the influence of strain available at present just start to cover the three-dimensional (photon energy, composition, strain) space of the optical response of this alloy system.

REFERENCES

[1] D.F. Edwards [in *Handbook of Optical Constants of Solids* Ed. E.D. Palik (Academic Press, New York, 1985) p.547]

[2] R.F. Potter [in *Handbook of Optical Constants of Solids* Ed. E.D. Palik (Academic Press, New York, 1985) p.465]

[3] J. Humlíček, F. Lukeš, E. Schmidt [in *Handbook of Optical Constants of Solids II* Ed. E.D. Palik (Academic Press, New York, 1991) p.607]

[4] A.E. Cosand, W.G. Spitzer [*J. Appl. Phys. (USA)* vol.42 (1971) p.5241]

[5] J. Humlíček, A. Roeseler, T. Zettler, M.G. Kekoua, E. Khoutsishvili [*Appl. Opt. (USA)* vol.31 (1992) p.90]

[6] R. Braunstein, A.R. Moore, F. Herman [*Phys. Rev. (USA)* vol.109 (1958) p.695]

[7] E. Schmidt [*Phys. Status Solidi (Germany)* vol.27 (1968) p.57]

[8] J. Humlíček, M. Garriga, M.I. Alonso, M. Cardona [*J. Appl. Phys. (USA)* vol.65 (1989) p.2827]
[9] C. Pickering et al [*J. Appl. Phys. (USA)* vol.73 (1993) p.239]
[10] J. Humlíček, M. Garriga [*Appl. Phys. A (Germany)* vol.56 (1993) p.259]
[11] J.S. Kline, F.H. Pollak, M. Cardona [*Helv. Phys. Acta (Switzerland)* vol.41 (1968) p.968]
[12] E. Schmidt, K. Vedam [*Solid State Commun. (USA)* vol.9 (1971) p.1187]
[13] D.E. Aspnes [*J. Opt. Soc. Am. (USA)* vol.70 (1980) p.1275]
[14] R. People [*Phys. Rev. B (USA)* vol.32 (1985) p.1405]

CHAPTER 5

TRANSPORT PROPERTIES

5.1 Electron and hole mobilities in the SiGe/Si system

F. Schäffler

March 1994

A INTRODUCTION

The carrier mobility μ is the most important transport parameter of a semiconducting material. It describes the linear relation between an electric field $\vec{E}$ and the carrier drift velocity $\vec{v}_d$, which is valid at low electric (and in the absence of magnetic) fields in homogeneous, isothermal semiconductors:

$$\vec{v}_d = \mu \vec{E} \quad (1)$$

μ can be expressed as a function of the electron charge e, the transport effective mass m* and the transport scattering time τ_t

$$\mu = \frac{e}{m^*} \tau_t \quad (2)$$

τ_t, and thus μ, is a fundamental material parameter that represents all scattering mechanisms a carrier experiences when moving through a semiconductor in the presence of an electric field. Within the limits of the wave-vector-independent relaxation time approximation $\frac{1}{\tau_t}$ is the sum of all reciprocal scattering times associated with the respective scattering mechanism (Mathiessen's rule) [1]:

$$\frac{1}{\tau_t} = \sum_i \frac{1}{\tau_i} \quad (3)$$

Hence the mobility in a semiconductor is limited by the scattering mechanism with the smallest relaxation time.

B SCATTERING MECHANISMS AND MODELLING

Because of their relevance for both basic understanding and for device applications, there has always been a strong interest in accurate model descriptions of the mobility as a function of temperature and dopant concentration. For obvious reasons, the main emphasis was put on the elemental semiconductors Ge and Si, whereas SiGe alloys only became important in recent years in connection with the development of modern epitaxy techniques that allowed the fabrication of high quality SiGe layers. The main scattering mechanisms to be considered in the non-polar elemental semiconductors Si and Ge are [2] (i) lattice scattering (scattering at acoustic and non-polar optical phonons) [3], (ii) ionised impurity scattering [4], and (iii) neutral impurity scattering [5]. In $Si_{1-x}Ge_x$ crystals alloy scattering [6] (iv) becomes important as a fourth independent mechanism. Also, the strain distribution in the lattice mismatched SiGe layer will affect the relative importance of intra- and inter-valley scattering due to strain-induced changes in the conduction and valence bands [7].

Despite obvious improvements over the years, the agreement between model predictions and experimental data is not entirely satisfactory. This is in part due to the unfortunate fact that most simulations refer to the aforementioned drift mobility μ_d, whereas experiments usually provide the Hall mobility μ_H. Since the presence of a magnetic field during the Hall measurements affects the scattering mechanisms, the two quantities are not identical:

$$\mu_H = r\,\mu \tag{4}$$

r is called the Hall scattering factor, which is usually on the order of 1, but can in certain situations reach values close to 2, e.g. in the case of the room-temperature hole mobility in pure Ge [8]. Because the scattering mechanisms listed above are affected by a magnetic field in different ways, r is a complex function of temperature and doping concentration [9]. Hence, it was recognised that the correct prediction of the Hall scattering factor r is a more stringent test for the validity range of a model description than the agreement with mobility data [10].

B1 Bulk Si and Ge

The available theoretical models work best for electrons in unstrained n-type bulk material at reasonably high temperatures (some above 100 K) and relatively low doping concentrations (below 10^{17} cm^{-3}). Under these conditions intra- and inter-valley lattice scattering dominate. With the freezing-out of phonons at cryogenic temperatures ionised impurity scattering becomes the most important mechanism, whereas the influence of neutral dopant scattering remains moderate. With increasing doping concentration, the calculated mobilities are generally higher than experimentally observed, which has partly been attributed to an inadequate treatment of electron-electron scattering [11].

Modelling of hole mobilities even in the elemental semiconductors is much more demanding because of the complicated valence band structure consisting of three strongly interacting bands for heavy, light and spin-orbit-split holes. Moreover, significant non-parabolicities and the warping of the hole bands have been shown to be essential for a correct model description [12,13]. Hence, satisfactory agreement to within $\pm$10% between theoretical results and experiments are only available over limited temperature and doping concentration ranges.

The remaining discrepancies between theoretical and experimental mobilities, and the requirement for simple analytical formulae to be used in device simulation programs, have led to the development of phenomenological expressions that are based on a combination of physical models and parameter fits to experimental data [14,15]. The most concise of such model descriptions was recently published by Klaassen [16], who treated electron and hole mobilities in unstrained bulk Si over very wide ranges of doping concentrations and temperatures. He also accounted for the quite significant differences between majority and minority carriers.

As a reference for the following treatment of $Si_{1-x}Ge_x$ alloys, the mobility values of intrinsic Si (x = 0) and Ge (x = 1) are listed in TABLE 1. Because of the relevance for mobility, the effective masses of the two materials are also listed. For a complete set of experimental and theoretical data concerning the influence of temperature and doping concentration on the electron and hole mobilities, the reader is referred to [9].

TABLE 1 Electron and hole drift mobilities μ_e and μ_h of intrinsic bulk Si and Ge at room temperature. Also, the transverse (m_t*) and longitudinal (m_l*) effective masses of electrons and the heavy- (m_{hh}*), light- (m_{lh}*) and spin-orbit-split- (m_{so}*) hole effective masses are tabulated for the two materials. Note that the hole masses are band-edge masses, which are relevant for low-doped structures at low electric fields. Higher doping levels and/or electric fields can lead to a significant (up to about a factor of two) increase of these values due to the pronounced non-parabolicity of the valence bands. (Data from [9].)

	μ_e (cm^2/V s)	μ_h (cm^2/V s)	m_t (m_0)	m_l (m_0)	m_{hh} (m_0)	m_{lh} (m_0)	m_{so} (m_0)
Si	1450	505	0.191	0.916	0.537	0.153	0.234
Ge	3900	1800	0.082	1.59	0.284	0.044	0.095

B2 Bulk $Si_{1-x}Ge_x$

While the efforts to find accurate mobility models for Si (and earlier for Ge) were triggered by the technical importance of this material, and by the large amount of experimental data, the treatment of $Si_{1-x}Ge_x$ alloys is still in a quite rudimentary state. This is mainly due to the problems with the fabrication of $Si_{1-x}Ge_x$ bulk crystals with homogeneous Ge content x and controlled doping concentration over the complete range of x ($0 \leq x \leq 1$) that led to a significant scatter of experimental data [17,18]. The experimental situation improved with the availability of modern epitaxy techniques, which allow the growth of well-defined $Si_{1-x}Ge_x$ layers deposited either on Si or Ge substrates. However, the relatively large lattice mismatch of 4.17% between Si and Ge limits the thicknesses for pseudomorphic growth of such epilayers, while simultaneously leading to a tetragonal distortion of the lattice. Thicker layers exceeding the critical thickness for pseudomorphic growth, which can then relax to the intrinsic cubic lattice constant at their given composition x, are possible, but until very recently such layers were heavily dislocated throughout. New epitaxy strategies have overcome this problem, but no systematic mobility investigation on relaxed bulk-like $Si_{1-x}Ge_x$ epilayers has been performed to date. Instead, recent experimental data are mainly concentrated on two-dimensional electron and hole gases in Si/SiGe and SiGe/Ge heterostructures, which are treated in some detail below.

Lacking adequate experimental data, most of the investigations concerning carrier drift mobilities in $Si_{1-x}Ge_x$ are based on theoretical model calculations. It is obvious from the above discussion of scattering mechanisms that not only is alloy scattering added as an additional mechanism, but also that the other mechanisms and the effective masses are modified through the changes of the band structure and the phonon spectra. The most prominent example, which is at least qualitatively confirmed by experimental data, is the drop of the electron mobility at a Ge content of about 85%, which corresponds to the transition from a Ge-like conduction band with eight equivalent L-point minima to a Si-like conduction band with six minima near the X-points [19]. The main effect on mobility is in this case associated with the increase of the effective electron mass in the Si-like conduction band.

As in the elemental semiconductors, the behaviour of holes in $Si_{1-x}Ge_x$ alloys is more complex because of the three types of interacting holes [20]. Again, of major influence on the mobility is the lowering of the effective hole mass [21] concomitant with the lifting of the heavy-hole/light-hole degeneracy [22]. These effects become more pronounced in strained $Si_{1-x}Ge_x$ layers pseudomorphically grown on a Si (100) substrate, which is now the most commonly employed epitaxial condition. This situation was recently treated in a Monte Carlo approach by Hinckley and Singh [23]. Besides the x-dependence of the valence band structure, and strain effects, both Si-Si and Ge-Ge type optical phonons, which differ significantly in energy, were considered. Based on this calculation they predict a monotonic increase of the hole mobility in strained, bulk-like $Si_{1-x}Ge_x$ layers with increasing Ge content, reaching the bulk-Ge values already at an x value of about 40%. It was argued that the reduced effective mass due to both the change in the valence band via x and the compressive in-plane strain increases the hole mobility more that it is reduced by alloy scattering. A similar but somewhat smaller trend was also predicted by Manku and Nathan [24] for strained $Si_{1-x}Ge_x$ layers, whereas for unstrained alloys they reproduced the U-shaped mobility vs. x behaviour, which was - despite discrepancies in the absolute values - observed in all early experiments on bulk material [16,17,25].

It is presently not clear if the predicted monotonic increase of the hole mobility with increasing x is realistic. Manku et al [26] evaluated published sheet resistances of the strained $Si_{1-x}Ge_x$ layers of heterobipolar transistors (HBT) in terms of doping-dependent hole mobilities. They found a drastic (a factor of 10 between x = 0 and x = 10%) decrease of μ_d for increasing x under all conditions investigated (10^{16} cm^{-3} $\leq N_A$ (boron) $\leq 10^{19}$ cm^{-3}). This contradicts a recent more precisely controlled experimental investigation of a series of Si_xGe_{1-x} layers with x ranging from 0 to 20%, which were boron doped at 2 x 10^{19} cm^{-3} [3,27]. These samples showed a 20 to 40% increase of the in-plane drift mobility with increasing Ge content. Since the mobility values were derived from simple sheet resistance measurements, and error bars are large due to uncertainties in geometrical and doping parameters, further experiments over a wider range of composition and doping concentration are required to clarify this topic.

Most of the theoretical treatments and experimental data discussed so far refer to the carrier mobilities at low electric fields, as defined in Eqn (1). For an increasing number of device applications, however, down-scaling of critical device dimensions can lead to relatively high electric fields within the active semiconducting layers (e.g. in the SiGe base layer of a heterobipolar transistor). For such situations electric-field dependent mobility data are important. Although experimental results are not yet available, some field-dependent Monte-Carlo-type calculations have been reported in the literature [7]. The general trend of these calculations is a reduction of the predicted, strain-induced mobility enhancement at higher electric fields.

C TWO-DIMENSIONAL CARRIERS IN Si/SiGe AND Ge/SiGe HETEROSTRUCTURES

As mentioned, the technical relevance of the $Si_{1-x}Ge_x$-alloys lies in their use for Si-based heterostructures, not in their potential applications as a bulk semiconductor. Hence most available experimental data refer to the electronic properties of $Si/Si_{1-x}Ge_x$ and $Si_{1-x}Ge_x/Ge$ heterostructures. Some experimental bulk data based on Si/SiGe HBTs with a highly p-doped $Si_{1-x}Ge_x$ base layer were presented in Section B2. Here, modulation doped structures, which utilise a heterobarrier for the separation of doping atoms and mobile carriers, are treated. In such structures the mobile carriers are usually confined to a narrow layer (channel) consisting of the material with the energetically favourable band edge, while the ionised impurities are located beyond the heterointerface. A two-dimensional carrier gas (2DCG) develops within the channel with free movement in, and momentum quantisation perpendicular to the plane of this layer. The principal advantage of modulation-doped heterostructures is the mobility enhancement due to the spatial separation of the ionised impurities. On the other hand, interface roughness scattering at the heterointerface(s) appears as an additional scattering mechanism. Also, the other scattering mechanisms listed above are modified due to the reduced dimensionality of the carrier gas [28].

The strain-induced type-II band offset in the SiGe/Si heterosystem (see Chapter 4 of this volume) requires two distinct layer sequences for modulation doped n- and p-type structures. Without external potentials electron confinement is only possible in strained Si channels (tensile in-plane strain) with the barrier consisting of SiGe-layers. Hole confinement, on the other hand, occurs always in the smaller bandgap material $Si_{1-x}Ge_x$, which can be grown pseudomorphically at moderate Ge contents, but requires, like the n-channel situation, relaxed buffer layers at higher x, especially for the most interesting case x = 100%.

C1 Electron Mobility in Strained Si Channels

The electron mobility in the strained Si channels of n-type modulation doped quantum well structures (MODQW) has reached a high level in the last years, mainly due to improvements in the quality of the strain-adjusting $Si_{1-x}Ge_x$ buffer layers employed [29]. FIGURE 1 shows the temperature dependence of the electron Hall mobility of the best samples published so far [28,30,31]. The three upper curves represent the highest low-temperature mobilities (maximum value about 175000 cm^2/V s at 0.5 K), but underestimates the 2DCG mobility at around room temperature. This is due to parasitic parallel channels of low mobility and unknown carrier concentration in these samples, which freeze out at low temperatures, but lead to a reduced average value of the Hall mobility at higher temperatures. By carefully designing the doping concentration in a series of samples, Nelson et al [32] could separate the contribution of the 2DCG at room temperature; their data points are included in FIGURE 1.

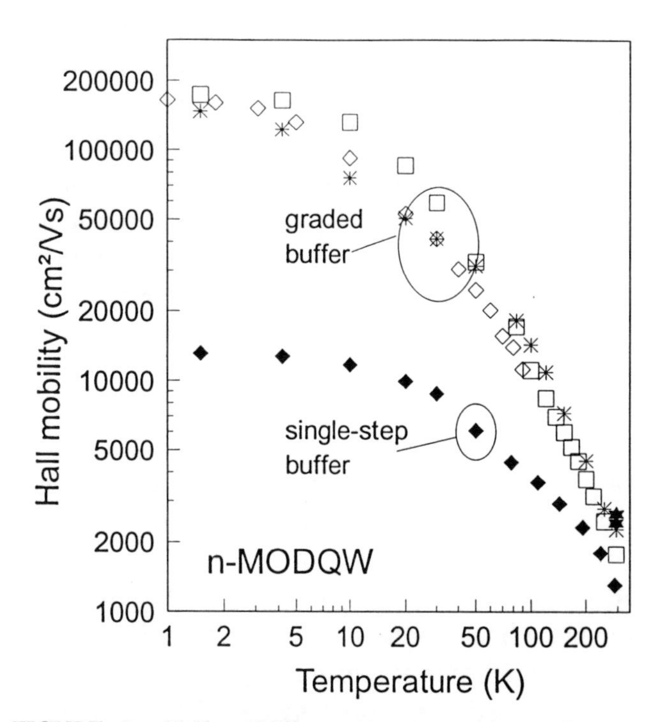

FIGURE 1 Hall mobility versus temperature of n-type modulation doped Si/SiGe heterostructures with strained Si channels. The three upper curves are from samples grown on high-quality strain adjusting SiGe buffer layers, whereas the lowest curve is from a sample grown on a highly dislocated SiGe buffer. Data are from [29] (unfilled diamond), [30] (unfilled square, filled diamond), [31] (*) and [32] (filled star).

The available data demonstrate the order of improvement achievable with Si/SiGe MODQWs as compared to MOSFETs (metal oxide silicon field effect transistor), which are the standard devices in large scale integrated circuits and also depend on current transport via 2DCGs. The best 4.2 K electron mobility ever reported of a MOSFET (41000 cm^2/V s) [33] is more than a factor of four smaller than the low-temperature mobilities in FIGURE 1. But even at room temperature, where intrinsic effects are reduced because of the dominance of lattice scattering, MODQWs with reported values exceeding 2500 cm^2/V s significantly improve MOSFET mobilities which are limited to about 1000 cm^2/V s.

The basic mechanisms that favour the n-MODQW over the MOSFET with respect to electron mobility are well understood [34]. The strain induced band ordering in the Si/SiGe MODQW allows only a population of the two electron valleys with effective mass $m_t = 0.19m_0$ along the channel and $m_l = 0.92m_0$ perpendicular to the channel [35], whereas in a MOSFET at room temperature most carriers are in the four valleys with inverse assignment of the effective masses [36]. The significantly lower transport mass and, somewhat less important, reduced intervalley-scattering, are the main reasons why Si/SiGe MODQWs do not only surpass the room temperature values of MOSFETs, but also the (3D) mobility of intrinsic bulk Si, where all six valleys are equally occupied. The heavy effective electron mass perpendicular to the channel helps the MODQW structure to keep interface roughness scattering small, since the electron wavefunction penetrates only little into the heterobarrier. This effect is enhanced by the small discontinuity of the dielectric constant at the Si/SiGe interface, which is much smaller than at the Si/SiO_2 interface. Also, the structural perfection of the Si/SiGe interface can be better than between (crystalline) Si and (amorphous) SiO_2. Finally, in the MODQW structure ionised impurity scattering can be significantly reduced by the spatial separation of the donors from the 2D channel. Hence, the mobility limiting scattering mechanism at low temperatures is in a well designed MODFET remote impurity scattering and scattering at the unintentional background doping in the channel, whereas the MOSFET is limited by interface roughness and interface charge scattering.

In order to achieve this limit, it is essential for the Si/SiGe MODQW to keep the number of threading dislocations low, which are launched by the strain-adjusting, relaxed SiGe buffer layer [29]. It could be shown that a threading dislocation density on the order of 10^9 cm^{-2}, which is typical for the simple SiGe buffer layers employed in early studies, reduces the electron mobility by a factor of ten or more (lowest curve in FIGURE 1).

Assuming a background doping on the order of 10^{15} cm^{-3}, the best low temperature mobilities in FIGURE 1 are in reasonable agreement with the calculations of Stern and Laux [37], who also determined the ratio between the transport scattering time τ_t and the single-particle relaxation time τ_s. This ratio is a measure of the amount of small angle scattering, and thus is indicative of the dominant scattering mechanism. Többen et al found for high-quality Si/SiGe MODQWs transport times τ_t exceeding τ_s by more than an order of magnitude [38], in good agreement with the theoretical predictions. This result confirms the above assignment of long-range impurity scattering being the limiting scattering mechanism in MODQWs.

The model calculations of Stern and Laux predict over the range investigated ($0 < l_s < 20$ nm) a monotonic increase of the low-temperature mobility with increasing spacer width l_s. No systematic experimental investigation of mobility versus spacer width is presently available for high quality samples comparable to those in FIGURE 1. However, for samples with somewhat lower mobilities (around 100000 cm^2/V s) the beneficial influence of an increasing spacer width has been demonstrated [39]. The authors also investigated the influence of scattering at remote ionised impurities, which leads to a decrease of the mobility with decreasing carrier concentration in the channel (the spacer was kept constant in these experiments). This effect is caused by the reduced screening properties of the 2DCG, and is in agreement with the predictions of Gold [40]. Hence, in order to further increase the low-temperature electron mobilities in Si/SiGe MODQW structures, an increase of the undoped spacer thickness is required, but simultaneously the carrier concentration in the channel should be kept as high as possible. Because the charge transfer from the doped layer into the channel is controlled by the spacer width and the band offset [41], an increase of the band offset by increasing the tensile in-plane strain in the Si channel is desirable. Also, symmetric doping at either side of the Si channel will be helpful, because it will double the number of transferable carriers.

C2 Hole Mobility

2D hole gases in Si/SiGe heterostructures were for a long time based on $Si_{1-x}Ge_x$ channels with relatively low Ge contents (up to typically 30%) grown pseudomorphically on Si substrates [40,42-44]. The advantage of such MODQW structures lies in the relative ease of epitaxial growth, because no strain adjusting buffer layer is required. In addition, the aforementioned theoretical considerations concerning the hole mobilities in strained $Si_{1-x}Ge_x$ predicted an enhancement of the hole mobility already at low Ge contents. In recent years layer sequences with pure Ge channels became available, which were initially grown on Ge substrates [45], but a few groups started with relaxed $Si_{1-x}Ge_x$ (x up to 80%) buffer layers deposited on Si substrates [46,47]. FIGURE 2 shows a selection of Hall data for both types of channel over the same temperature range as in FIGURE 1.

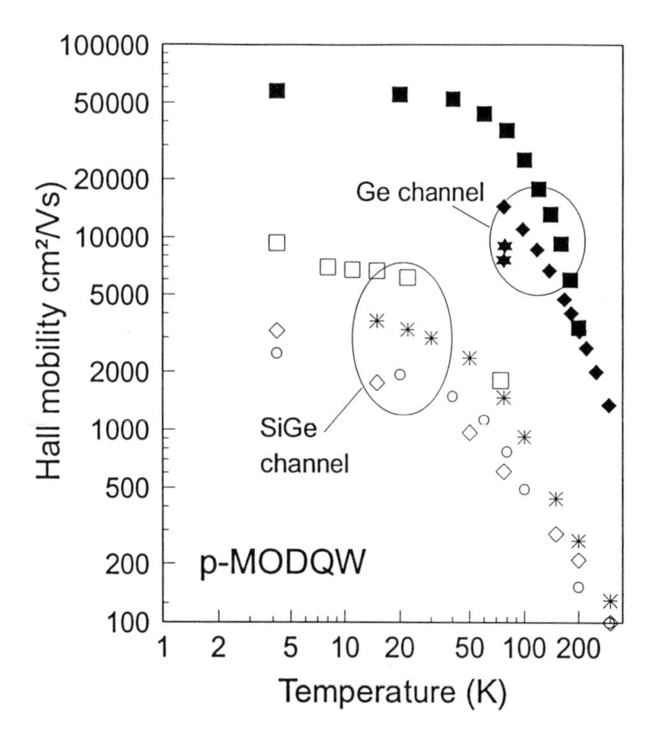

FIGURE 2 Hall mobility versus temperature of p-type modulation doped SiGe/Si and Ge/SiGe heterostructures with pseudomorphic, strained $Si_{1-x}Ge_x$ channels (x > 30%, lower set of data) and strained Ge-channels on relaxed SiGe buffer layers (upper data set). SiGe channel data are from [41] (unfilled diamond), [42] (*), [43] (unfilled circle) and [44] (unfilled square); Ge-channel data from [45] (filled star), [46] (filled diamond) and [47] (filled square).

It is obvious from FIGURE 2 that the theoretically predicted, strain-induced mobility enhancement at low Ge contents is far from being confirmed experimentally. Despite gradual improvements of the low temperature mobility in such samples, which can be attributed to progress in growth techniques, the device-relevant room temperature mobilities remained significantly below the corresponding hole mobilities in undoped bulk silicon. As can be judged from the behaviour of 2D electron gases (see Section C1), such a comparison between 2D and 3D mobilities is meaningful despite the differences in screening and scattering properties. The discrepancies between theoretical predictions and experiments are presently unresolved.

In contrast to the pseudomorphic $Si_{1-x}Ge_x$ channel MODQWs, the advantage of a pure Ge channel becomes evident from the experimental data in FIGURE 2. The best room temperature Hall mobility reported so far amounts to more than 1300 cm^2/V s, which was determined on a sample with a significant parallel channel [45]. Thus, as mentioned above, the intrinsic mobility of the 2DCG in the Ge channel is expected to be higher, and might come close to the bulk mobility of intrinsic Ge (1800 cm^2/V s). On the other hand, the material quality of the Ge-channel samples is not yet ideal, as can be seen in the temperature variation at low temperatures. All data published so far show saturation behaviour below about 100 K, which is indicative of unintended scattering mechanisms. It is known that especially relaxed $Si_{1-x}Ge_x$ buffer layers of high Ge content impose severe growth problems due to the tendency towards a 3D growth mode, which leads to wavy interfaces. Such behaviour has indeed been reported for the sample with the presently highest hole mobility of 55000 cm^2/V s, which clearly revealed a corrugated interface between the Ge channel and the upper $Si_{1-x}Ge_x$ barrier [46].

Further improvements of the material quality are expected, and will lead to even higher hole mobilities. By analogy with the electron gases in the Si channels of n-MODQWs, one may speculate that Ge-channel p-MODQWs will finally reach or even surpass the mobilities of intrinsic bulk Ge. The bulk data in TABLE 1 reveal that under these conditions almost ideally matched p- and n-type MODQWs could be fabricated, which is a truly unique condition among the established semiconductors.

Despite the unexpectedly low hole mobilities in $Si_{1-x}Ge_x$ channels with $x < 30\%$, p-type MOSFETs with such channels have been fabricated by several groups [48-50]. In fact, improvements of the hole field-effect mobility as compared to all-silicon reference MOSFETs were reported, which are attributed to the SiGe channel keeping the carriers away from the SiO_2 interface and thus reducing interface scattering. However, the best room temperature field-effect mobilities reached values of only 220 cm^2/V s, which is in reasonable agreement with the best Hall mobilities observed in similar SiGe-channel p-MODQW structures, but is not sufficient to allow matched p- and n-devices as far as carrier mobility is concerned.

D CONCLUSION

The bulk properties of pure Si and Ge are well characterised experimentally and at least phenomenological, analytical expressions are available that describe the electron mobility over a large range of temperature and doping concentration. This is not the case for $Si_{1-x}Ge_x$ alloys, for which only an incomplete and to some extent questionable body of data exists. This is mainly due to the problems in pulling homogeneous $Si_{1-x}Ge_x$ crystals of arbitrary composition, and because of the limited technical relevance of $Si_{1-x}Ge_x$ bulk material. On the other hand, low temperature epitaxial regrowth techniques and the degree of freedom introduced by band structure engineering via heterointerfaces have led to increasing interest in Si/SiGe heterostructures which offer the advantage of being compatible with standard Si technologies. Modulation doped structures both for electrons and holes have been fabricated, which reached unprecedented Hall mobilities in Si-based systems. The highest values have been observed in pure Si channels for electrons and pure Ge channels for holes, whereas alloy channels did not live up to theoretical expectations. Further investigations are required to resolve this discrepancy.

REFERENCES

[1] Matthiesen's rule and its limitations are treated in most standard textbooks on solid state physics, e.g. N.W. Ashcroft, N.D. Mermin [*Solid State Physics* (Holt, Rinehart and Winston, New York, 1976)]

[2] For an overview of the relevant scattering mechanisms see e.g. S.S. Li, R. Thurber [*Solid-State Electron. (UK)* vol.20 (1977) p.609-16]; S.S. Li [*Solid-State Electron. (UK)* vol.21 (1978) p.1109-17]

[3] C. Herring, E. Vogt [*Phys. Rev. (USA)* vol.101 (1956) p.944]

[4] H. Brooks [in *Theory of the Electrical Properties of Germanium and Silicon, Advances in Electronics and Electron Physics* Ed. L. Marton (Academic Press, New York, 1955) vol.7 p.85-182]

[5] N. Sclar [*Phys. Rev. (USA)* vol.104 (1956) p.1559]

[6] e.g. L. Makowski, M. Glicksman [*J. Phys. Chem. Solids (UK)* vol.34 (1973) p.487]

[7] L.E. Kay, T.-W. Tang [*J. Appl. Phys. (USA)* vol.70 (1991) p.1483-8]

[8] F.J. Morin [*Phys. Rev. (USA)* vol.93 (1954) p.62]

[9] For a collection of experimental mobility data including values of r see e.g. Landolt-Börnstein [*Numerical Data and Functional Relationships in Science and Technology*, New Series, Group III (Springer-Verlag, Berlin) vol.17a (1982) and vol.22a (1989)]

[10] F. Szmulowicz [*Appl. Phys. Lett. (USA)* vol.43 (1983) p.485-7]

[11] S.S. Li, W.R. Thurber [*Solid-State Electron. (UK)* vol.20 (1977) p.609-16]

[12] G. Ottaviani, L. Reggiani, C. Canali, F. Nava, A. Alberigi-Quaranta [*Phys. Rev. B (USA)* vol.12 (1975) p.3318]

[13] S.S. Li [*Solid-State Electron. (UK)* vol.21 (1978) p.1109-17]

[14] N.D. Arora, J.R. Hauser, D.J. Roulston [*IEEE Trans. Electron Devices (USA)* vol.29 (1982) p.292-5]

[15] G. Masetti, M. Severi, S. Solmi [*IEEE Trans. Electron Devices (USA)* vol.30 (1983) p.764-9]

[16] D.B.M. Klaassen [*Solid-State Electron. (UK)* vol.35 (1992) p.953-9 and p.961-7]

[17] G. von Busch, O. Vogt [*Helv. Phys. Acta (Switzerland)* vol.33 (1960) p.437]

[18] E. Braunstein [*Phys. Rev. (USA)* vol.130 (1963) p.869-79]

[19] S. Krishnamurthy, A. Sher, An-Ban Chen [*Appl. Phys. Lett. (USA)* vol.47 (1985) p.160-2]

[20] T. Manku, A. Nathan [*Phys. Rev. B (USA)* vol.43 (1991) p.12634-7]

[21] T. Manku, A. Nathan [*J. Appl. Phys. (USA)* vol.69 (1991) p.8414-6]

[22] K. Takeda, A. Taguchi, M. Sakata [*J. Phys. C (UK)* vol.16 (1983) p.2237-49]

[23] J.M. Hinckley, J. Singh [*Phys. Rev. B (USA)* vol.41 (1990) p.2912]

[24] T. Manku, A. Nathan [*IEEE Electron Device Lett. (USA)* vol.12 (1991) p.704-6]

[25] A. Levitas [*Phys. Rev. (USA)* vol.99 (1955) p.1810-4]

[26] T. Manku, S.C. Jain, A. Nathan [*J. Appl. Phys. (USA)* vol.71 (1992) p.4618-9]

[27] T. Manku, J.M. McGregor, A. Nathan, D.J. Roulston, J.-P. Noel, D.C. Houghton [*IEEE Trans. Electron Devices (USA)* vol.40 (1993) p.1990-6]

[28] For an overview of scattering mechanisms in 2D carrier systems see for example T. Ando, A.B. Fowler, F. Stern [*Rev. Mod. Phys. (USA)* vol.54 (1982) p.437-672]

[29] For a recent overview of this development see e.g. E.A. Fitzgerald et al [*J. Vac. Sci. Technol. B (USA)* vol.10 (1992) p.1807-19]

[30] F. Schäffler, D. Többen, H.-J. Herzog, G. Abstreiter, B. Holländer [*Semicond. Sci. Technol. (UK)* vol.7 (1992) p.260-6]

[31] S.F. Nelson, K. Ismail, T.N. Jackson, J.J. Nocera, J.O. Chu, B.S. Meyerson [*Appl. Phys. Lett. (USA)* vol.63 (1993) p.794-6]

[32] S.F. Nelson, K. Ismail, J.O. Chu, B.S. Meyerson [*Appl. Phys. Lett. (USA)* vol.63 (1993) p.367-9]

[33] I.V. Kukushkin, V.B. Timofeev [*Sov. Phys.-JETP (USA)* vol.67 (1988) p.594-9]

[34] For an overview see e.g. D. Monroe, Y.H. Xie, E.A. Fitzgerald, P.J. Silverman, G.P. Watson [*J. Vac. Sci. Technol. B (USA)* vol.11 (1993) p.1731-7]

[35] G. Abstreiter, H. Brugger, T. Wolf, H. Jorke, H.-J. Herzog [*Phys. Rev. Lett. (USA)* vol.54 (1985) p.2441-4]

[36] F. Schäffler, F. Koch [*Solid State Commun. (USA)* vol.37 (1981) p.365-8]

[37] F. Stern, S.E. Laux [*Appl. Phys. Lett. (USA)* vol.61 (1992) p.1110-2]

[38] D. Többen, F. Schäffler, A. Zrenner, G. Abstreiter [*Phys. Rev. B (USA)* vol.46 (1992) p.4344-7]

[39] S.F. Nelson, K. Ismail, T.N. Jackson, J.J. Nocera, J.O. Chu, B.S. Meyerson [*Appl. Phys. Lett. (USA)* vol.63 (1993) p.794-6]

[40] A. Gold [*Phys. Rev. B (USA)* vol.35 (1987) p.723-33]

[41] R. People, J.C. Bean, D.V. Lang [*J. Vac. Sci. Technol. A (USA)* vol.3 (1985) p.846-50]

[42] P.J. Wang, B.S. Meyerson, F.F. Fang, J. Nocera, B. Parker [*Appl. Phys. Lett. (USA)* vol.55 (1989) p.2333-5]

[43] T. Mishima, C.W. Fredriksz, G.F.A. van de Walle, D.J. Gravesteijn, R.A. van den Heuvel, A.A. van Gorkum [*Appl. Phys. Lett. (USA)* vol.57 (1990) p.2567-9]

[44] T.E. Whall, D.W. Smith, A.D. Plews, R.A. Kubiak, P.J. Phillips, E.H.C. Parker [*Semicond. Sci. Technol. (UK)* vol.8 (1993) p.615-6]

[45] M. Miyao, E. Murakami, H. Etho, K. Nakagawa, A. Nishida [*J. Cryst. Growth (Netherlands)* vol.111 (1991) p.912-5]

[46] U. König, F. Schäffler [*IEEE Electron Device Lett. (USA)* vol.14 (1993) p.205-7]

[47] Y.H. Xie, D. Monroe, E.A. Fitzgerald, P.J. Silverman, F.A. Thiel, G.P. Watson [*Appl. Phys. Lett. (USA)* vol.63 (1993) p.2263-5]

[48] V.P. Kesan et al [*Int. Electron Devices Meet. Tech. Dig. (USA)* (1991) p.25-8]

[49] S. Verdonckt-Vandebroek et al [*IEEE Electron Device Lett. (USA)* vol.12 (1991) p.447-9]

[50] P.M. Garone, V. Venkataraman, J.C. Sturm [*IEEE Electron Device Lett. (USA)* vol.12 (1991) p.230-2]

5.2 Injection across a Si/SiGe heterojunction

H. Jorke

January 1994

A INTRODUCTION

Injection currents across heterojunctions are influenced by multiple factors such as material properties (bandgap, dielectric constant, effective carrier masses), interface properties (conduction and valence band discontinuities) and electrostatic forces that are related to the specific doping situation and the presence of an external bias.

In $Si/Si_{1-x}Ge_x$ heterojunctions material properties are largely determined by the value of the Ge fraction x ($0 < x < 1$). In contrast, interface properties (conduction and valence band discontinuities) do not solely depend on x but are also strongly or even dominantly influenced by factors such as crystal orientation (the present data are related to (100) oriented interfaces) and the presence of layer strain. The latter is an inherent property of the lattice mismatched $Si/Si_{1-x}Ge_x$ heterosystem [1]. To illustrate the specific strain influence on band discontinuities

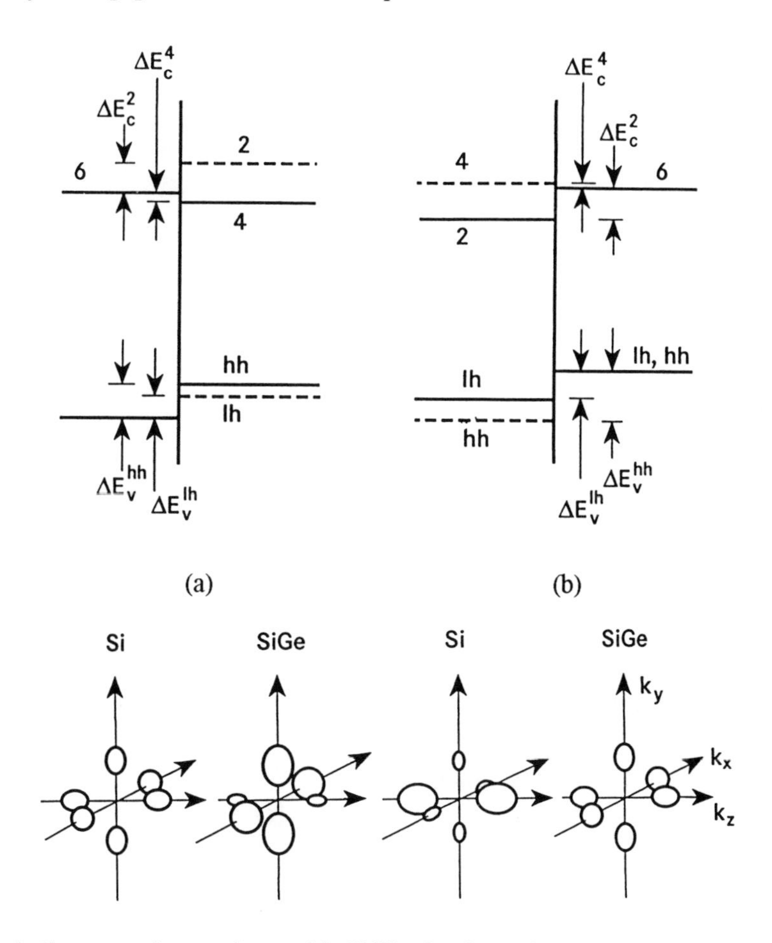

FIGURE 1 Band alignments in pseudomorphic $Si/Si_{1-x}Ge_x$ heterojunctions for (a) $Si_{1-x}Ge_x$ strained and (b) Si strained. Depending on whether strain is compressive ($Si_{1-x}Ge_x$) or tensile (Si) the transverse (fourfold degenerate) or vertical (twofold degenerate) conduction band valleys are dominantly occupied (lower part of the figure).

FIGURE 1 shows the energy band diagram for Si and $Si_{1-x}Ge_x$ isolated (i.e. without space charges) if (a) Si is unstrained ($Si_{1-x}Ge_x$ is under compressive strain) and (b) $Si_{1-x}Ge_x$ is unstrained (Si is under tensile strain).

A major influence on injection currents issues from the doping situation, especially whether the junction is isotype (n-n or p-p) or anisotype (n-p or p-n). FIGURES 2 and 3 show the corresponding energy band diagrams for isotype (FIGURE 2) and anisotype (FIGURE 3) heterojunctions under equilibrium conditions (no external bias).

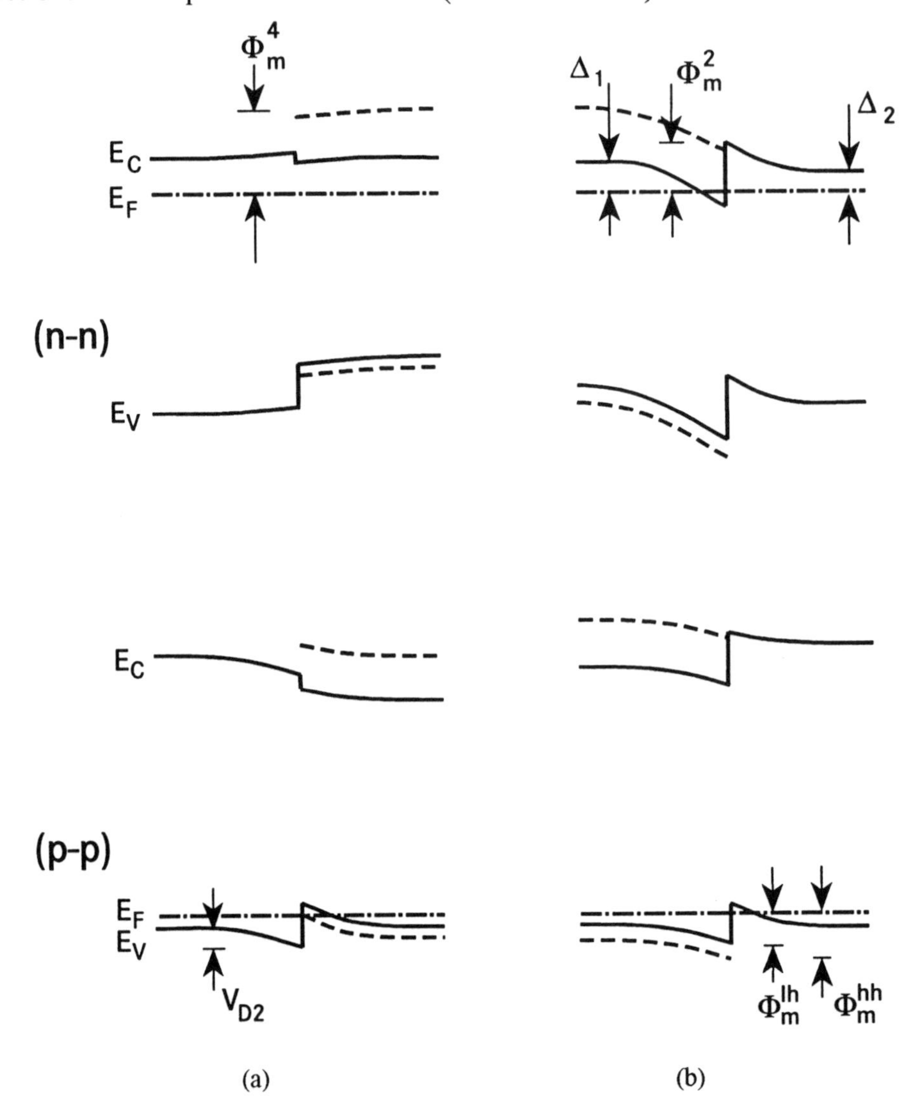

FIGURE 2 Band diagram of isotype $Si/Si_{1-x}Ge_x$ heterojunctions in equilibrium for (a) $Si_{1-x}Ge_x$ strained and (b) Si strained.

Only a minor fraction of these heterojunctions are yet of technological importance or have been at least considered for the fabrication of vertical transport devices. (p-p)&(a) is used in hole double barrier resonant tunnelling (DBRT) diodes [2,3] and in a three-terminal device to study hot hole transport [4]. Also (p-p)&(b) was considered for the fabrication of hole DBRT diodes [5]. (n-n)&(b) is the base for attempts to realise electron DBRT diodes [6]. Of great current interest is the (n-p)&(a) type junction which is the base of almost all Si/SiGe heterojunction bipolar transistor (HBT) activities so far reported.

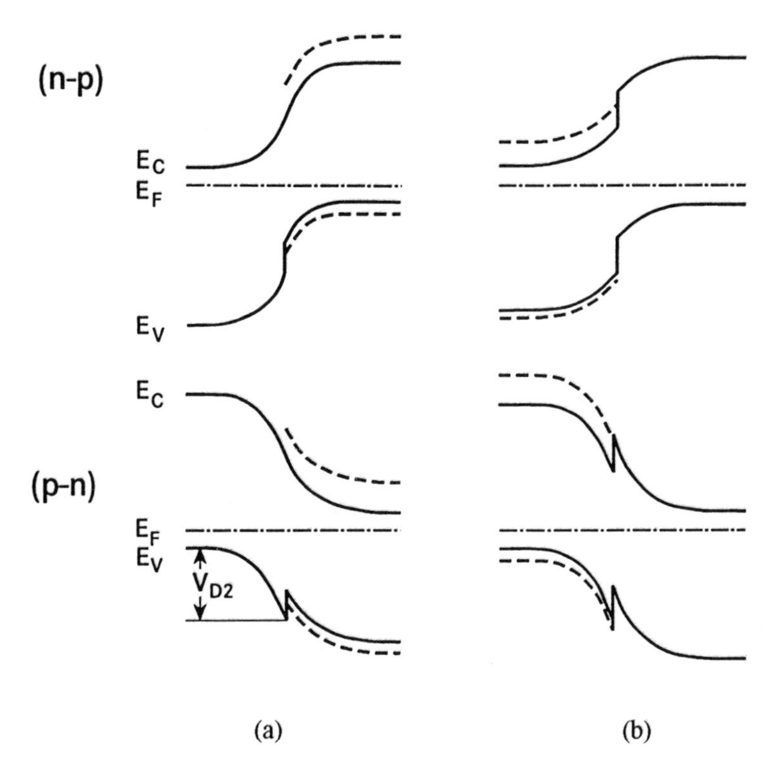

FIGURE 3 Band diagram of anisotype $Si/Si_{1-x}Ge_x$ heterojunctions in equilibrium for (a) $Si_{1-x}Ge_x$ strained and (b) Si strained.

B MODELLING

Generally, injection currents are treated either on the basis of thermionic emission theory or on the basis of diffusion theory [7]. The former applies to rapidly varying barrier potentials (where the carrier mean free path exceeds the distance in which the barrier falls by kT from its maximum value), the latter to slowly varying potentials. Isotype junctions are commonly modelled by equations of the emission type where the diffusion description is commonly used for anisotype (bipolar) heterojunctions.

B1 Isotype Injection

Setting arbitrarily the Fermi energy E_F (FIGURE 2) of the emitter layer equal to zero the equation for the individual injection currents following emission theory [8] reads

$$J_i = 4\pi\, q\, g_i\, m_i^{\parallel}\, kT\, h^{-3} \int_0^{\infty} \exp(-E_z/kT)\, T_i(E_z) dE_z \qquad i = 2, 4, \text{lh}, \text{hh} \tag{1}$$

where q is the unit charge, g_i is the valley degeneracy, $m_i^{\parallel}$ the in-plane effective mass, kT the thermal energy, h Planck's constant and T_i the barrier transmission probability.

A determination of J_i from Eqn (1) requires the knowledge of $T_i(E_z)$. If quantum mechanical tunnelling is neglected the transmission probability is simply given by

$$T_i(E_z) = \begin{cases} 0 & \text{for } E_z < \Phi_m \\ 1 & \text{for } E_z > \Phi_m \end{cases} \tag{2}$$

where Φ_m is the maximum height of the barrier potential (FIGURE 2). Insertion of Eqn (2) into Eqn (1) yields the standard current density expression for thermionic emission [7].

A more accurate estimate that also includes field emission (i.e. injection of carriers at energies below Φ_m) follows from

$$T_i(E_z) = \begin{cases} \exp\{-(4\pi/h) \int_{Z_1}^{Z_2} (2m_i^{\perp}(\Phi_i(z) - E_z))^{\frac{1}{2}}\, dz & \text{for } E_z < \Phi_m \\ 1 & \text{for } E_z > \Phi_m \end{cases} \quad (3)$$

Here $\Phi_i(z)$ denotes the barrier potential, $m_i^{\perp}$ is the vertical effective mass. In Eqn (3) tunnelling processes are accounted for in the Wentzel-Kramers-Brillouin (WKB) approximation. A method to determine (numerically) transmission probabilities that does not rely on the WKB approximation is discussed by Bhapkar and Mattauch [9]. The WKB approximation fails to predict correct transmission behaviour in more complex structures such as double barriers [10].

The barrier potential $\Phi(z)$ can be separated into the heterostructure potential Φ_h^i and the electrostatic potential $q\phi_e$

$$\Phi_i = \Phi_h^i - q\phi_e \quad (4)$$

For clarity, FIGURE 4 shows the various contributions for a single heterojunction of the (n-n)&(b) type (FIGURE 2). A refined description also includes the exchange-correlation

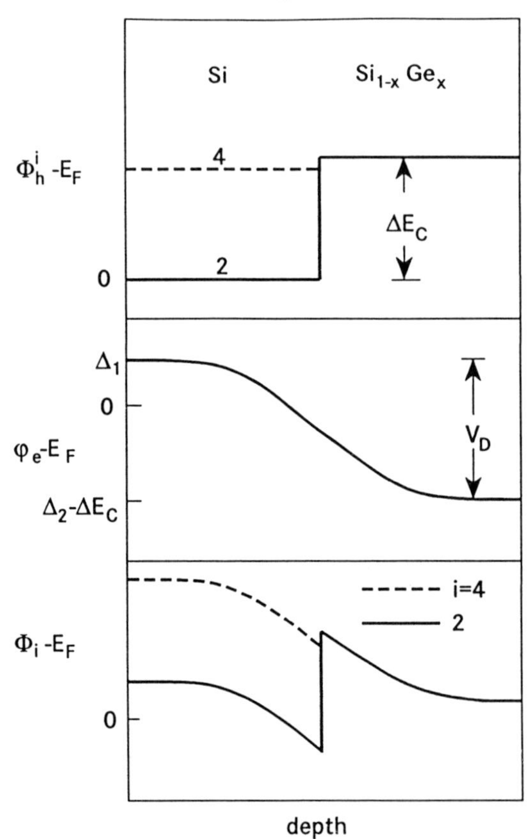

FIGURE 4 Heterostructure potential Φ_h^i, electrostatic potential ϕ_e and total potential Φ_i seen by electrons (i = 2, 4) in (n-n)&(b) type heterojunctions (FIGURE 2).

potential [11] that is not discussed here. If the electrostatic potential is also neglected, the barrier potential coincides with the (step-like) heterostructure potential that is (for the twofold degenerate valleys of the conduction band) in (n-n)&(b) type junctions given by

$$\Phi_h^2(z) = \begin{cases} 0 & \text{in the Si regions} \\ \Delta E_c^2 & \text{in the } Si_{1-x}Ge_x \text{ regions} \end{cases} \tag{5}$$

This crude approximation, however, does not account for the presence of space charges (due to free carriers and dopant atoms) and, also, does not account for an external potential qV that decisively influences the injection current.

The inclusion of the electrostatic potential $q\phi_e$ removes these insufficiencies. If carriers accumulate at the heterojunction interface forming a two-dimensional carrier gas there, $q\phi_e$ follows from an approach that solves self-consistently the Schrödinger and the Poisson equations [12]. These equations read for electrons

$$-\frac{h^2}{8\pi}\frac{d}{dz}\left(\frac{1}{m_i^{\perp}}\frac{d\Psi_{ni}}{dz}\right) + (\Phi_h^i - q\phi_e)\,\Psi_{ni} = E_{ni}\Psi_{ni} \qquad i = 2, 4 \quad n = 0, 1, 2, \ldots \tag{6}$$

and

$$\frac{d}{dz}\left(\varepsilon(z)\frac{d\phi_e}{dz}\right) = \frac{q}{\varepsilon_o}\,(n(z) - N(z)) \tag{7}$$

where $\varepsilon(z)$ is the position dependent dielectric constant and N(z) is the donor density. The electron density n(z) is commonly written as

$$n_i(z) = 4\pi g_i m_i^{\parallel} kTh^{-2}\sum_n \ln(1 + \exp((E_F - E_{ni})/kT))\,|\Psi_{ni}|^2 \qquad i = 2, 4$$

$$n(z) = n_2(z) + n_4(z) \tag{8}$$

where the sum is over all subbands.

A simplified approach with less computational effort (the above approach suffers from difficulty to reach charge neutrality [14]) uses the semiclassical Thomas-Fermi approximation

$$n(z) = N_c(z)F_{\frac{1}{2}}((E_F - \Phi_i(z))/kT) \tag{9}$$

instead of Eqn (8). Eqn (9) relates the barrier potential Φ_i - the lower of Φ_2 and Φ_4 is taken - directly to the carrier density [14]. $F_{\frac{1}{2}}$ is the Fermi integral. The effective density of states N_c in Eqn (9) is modified by band splitting. It may be written as [15]

$$N_c = 2\,(2\pi kTm_{ds,n}h^{-2})^{3/2}\,\{g_i + g_j\exp((\Phi_h^i - \Phi_h^j)/kT)\}/6 \tag{10}$$

where i and j label the downward- and upward- shifted valleys. $m_{ds,n}$ is the effective density of states mass without splitting.

For a $Si/Si_{1-x}Ge_x$ single heterostructure appropriate boundary conditions for the Poisson equation (Eqn (7)) are

$$q\phi_e(-\infty) = \Delta_1$$

$$q\phi_e(+\infty) = \Delta_2 - \Delta E_c - qV$$

where Δ_1 and Δ_2 follow from $N = N_c F_{\frac{1}{2}}(\Delta/kT)$ (N is the doping, see FIGURE 2).

The above set of equations can be used to determine injection currents as a function of the applied bias. The approach applies to single heterojunctions as well as to more complex structures such as, for example, multiple barrier structures [16]. The equations necessitate, in general, a numerical treatment even for simple structures.

For single heterojunctions older empirical treatments exist that relate the applied bias to the injection current in a closed way. For a review of such work the reader is referred to the book of Milnes and Feucht [17]. Part of this work was related to systems where interface states from misfit dislocations are present (e.g. in strain relaxed Si/Ge heterojunctions) and that are out of the scope of this review [1].

Finally, it should be mentioned that Eqn (1) presumes that the carrier gas is three-dimensional in character at least at energies where T_i is significantly different from zero. Deviations from this condition in systems where tunnelling occurs between two-dimensional carrier gases result in injection currents where momentum conservation becomes a limiting factor [18,19].

B2 Anisotype Injection

Anisotype injection across a semiconductor heterojunction was considered by Shockley [20] and Krömer [21] in their famous concept of a wide-gap emitter transistor. Modern realisation of this concept in the Si/SiGe material system (taking the (n-p)&(a) type junction shown in FIGURE 3) consists of a double heterojunction small gap base n-p-n transistor. Following Krömer [21] the electron and hole injection currents in anisotype heterojunctions read [22]

$$J_n = (qD_n n_i^2(SiGe)/L_n N_A)\,(\exp(qV/kT) - 1) \qquad (11)$$

$$J_p = (qD_p n_i^2(Si)/L_p N_D)\,(\exp(qV/kT) - 1) \qquad (12)$$

where the presence of a heterojunction is manifest in different values of the intrinsic carrier density n_i. D_n and D_p are minority diffusion constants, L_n and L_p are diffusion lengths for electrons and holes, respectively. N_D and N_A denote the dopant concentrations. Eqns (11) and (12) refer to n- and p-type layers with (i) thickness large compared to L_n and L_p and that are (ii) homogeneously doped. More general formulae given by Krömer later on [23] refer to structural dimension $<<L_n$, L_p and account for inhomogeneities (e.g. gradings) both in doping and Ge content x

$$J_n = \frac{q\exp(qV/kT)}{\int_{\text{p-type region}} p/D_n n_i^2(SiGe)dz} \qquad (13)$$

$$J_p = \frac{q \exp(qV/kT)}{\int_{\text{n-type region}} n/D_p n_i^2(\text{Si}) dz} \tag{14}$$

p (= N_A) and n (= N_D) are the majority carrier densities in the p- and n-type regions. The average transit time of an electron through the p-type region comes out to be

$$\tau = \int_0^W \left[(n_i^2/p) \int_z^W (p/D_n n_i^2)\, dy \right] dz \tag{15}$$

where W is the width of the neutral zone. For a homogeneous p-type range Eqn (15) simplifies to

$$\tau = W^2/2D_n \tag{16}$$

which describes motion through a neutral region [24].

In practice, caused e.g. by dopant outdiffusion, the p-n junction may be displaced from the Si/SiGe heterojunction. This shift gives rise to the formation of parasitic barriers. Using Eqns (13) and (14) Slotboom et al [24a] have calculated electron injection currents in SiGe HBTs including such parasitic barriers.

In the early 1960s Anderson [25] developed a theory that was first used to model injection currents in anisotype Ge/GaAs heterojunctions. The energy band diagram of this system (Ge p-type) is similar to (p-n)&(a) in FIGURE 3. Due to the large bandgap difference hole injection dominates over electron injection and the total injection current is largely determined by the former

$$J = J_o \exp(-qV_{D2}) \exp(qV/mkT) \tag{17}$$

with

$$J_o = XqN_A(D_p/\tau_p)^{1/2} \tag{18}$$

and

$$m = 1 + N_A\varepsilon_{Si}/N_D\varepsilon_{SiGe} \tag{19}$$

D_P and τ_p are the diffusion constant and the lifetime, respectively, for holes in the smaller gap material. X is the probability that an energetic hole crosses the junction energy barrier.

Depending on the forward bias the energy barrier spike occurring in the valence band in (p-n)&(a) is either above or below the valence band level in the neutral small gap layer region. In contrast to the Krömer model the Anderson model takes into account the corresponding change in the nature of the barrier. This change induces a switch in the slope of forward-bias current-voltage characteristics [26].

C MATERIAL PROPERTIES

After the outline of approaches to model injection currents given in the previous section the following section will specify the material properties relevant to injection current modelling in the $Si/Si_{1-x}Ge_x$ heterojunction system.

C1 Electron Injection in Isotype $Si/Si_{1-x}Ge_x$ Junctions

Relevant parameters are:

(a) dielectric constant ε
(b) effective masses $m_j^{\parallel}$ and $m_j^{\perp}$ (j = 2, 4)
(c) effective density of states mass $m_{ds,n}$
(d) conduction band discontinuities $\Delta E_c^{\,j}$ (j = 2, 4).

The dielectric constant is commonly assumed to change linearly between Si and Ge

$$\varepsilon(x) = 11.9 + 4.1x \qquad (20)$$

Effective masses may depend on whether electrons are injected from Si or $Si_{1-x}Ge_x$. Actually, cyclotron resonance and magnetoresistance measurements [27] revealed longitudinal and transverse effective masses in bulk $Si_{1-x}Ge_x$ layers ($0 < x < 0.8$) that do not significantly differ from Si values [28] (T = 1.26 K)

$$m_l = 0.9163$$
$$m_t = 0.1905$$

where the relative precision of these measurements was about 0.2. According to the phenomenological deformation potential theory [29] the effective masses of the conduction band, in a first order, do not depend on strain. Thus, use of the above Si values (i) for $Si_{1-x}Ge_x$ layers also ($x < 0.8$) and (ii) irrespective of the strain situation appears to be reasonable.

The effective density of states mass $m_{ds,n}$ from cyclotron resonance data (T = 300 K) is found to be [28]

$$m_{ds,n} = 1.18$$

m_l, m_t and $m_{ds,n}$ values are in units of m_0 (free electron mass).

The parallel and perpendicular effective masses appearing in Eqns (1), (3), (6) and (8) are related to m_t and m_l as follows:

$$m_2^{\parallel} = m_t$$
$$m_2^{\perp} = m_l$$
$$m_4^{\parallel} = (m_t m_l)^{1/2}$$
$$m_4^{\perp} = m_t$$

The conduction band discontinuities ΔE_c^2 and ΔE_c^4 (FIGURE 1) depend on the Ge fraction as shown in FIGURE 5(a) ($Si/Si_{1-x}Ge_x$ on Si) and FIGURE 5(b) ($Si/Si_{1-x}Ge_x$ on $Si_{1-x}Ge_x$).

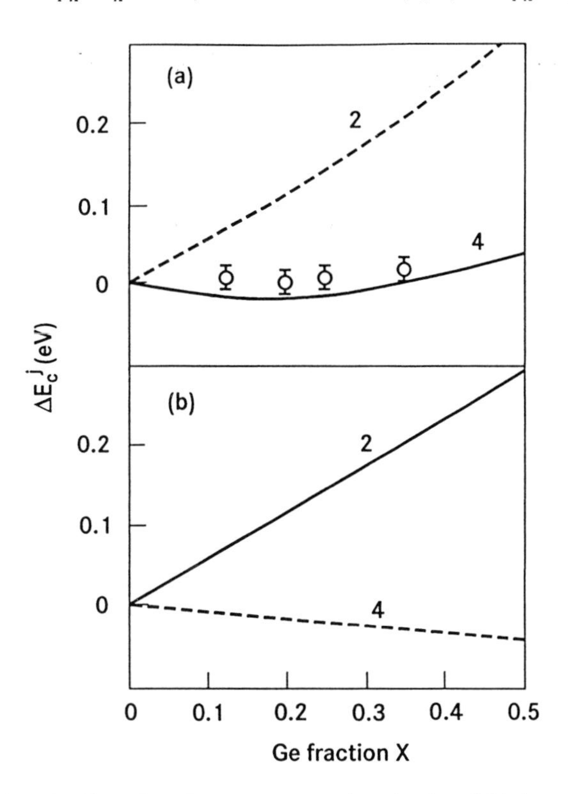

FIGURE 5 Conduction band offsets for electrons occupying the fourfold degenerate valleys and the twofold degenerate valleys at the $Si/Si_{1-x}Ge_x$ heterointerface for (a) $Si_{1-x}Ge_x$ strained and (b) Si strained (for sign see FIGURE 1). Lines were obtained from the approach suggested by People [38]. Open symbols are from Nauka et al [39].

C2 Hole Injection in Isotype $Si/Si_{1-x}Ge_x$ Junctions

The situation for hole injection is different from that for electron injection, related to the different valence band dispersion relations

$$E(k) = (h^2/8\pi^2 m_0)\,(Ak^2 \pm (B^2k^4 + C^2(k_x^2k_y^2 + k_x^2k_z^2 + k_y^2k_z^2))^{1/2}) \qquad (21)$$

with

$$k^2 = k_x^2 + k_y^2 + k_z^2 \qquad (22)$$

that describe 'warped' energy surfaces. In Eqn (21) the plus sign corresponds to light holes (lh), the minus sign to heavy holes (hh). A, B and C are valence band parameters. Values are [28] (T = 1.26 K)

$A = -4.27$
$B = -0.63$
$|C| = 4.93$

for silicon and (T = 4.2 K)

A = -13.3
B = -8.57
|C| = 12.78

for germanium.

Taking, for simplicity, the approximate injection current expression given by Eqn (1) [30], anisotropy of effective hole masses has to be taken into account [3]. The respective masses are for hole injection from unstrained Si (in units of m_0):

$m_{lh}^{\parallel} = 0.157$
$m_{lh}^{\perp} = 0.204$
$m_{hh}^{\parallel} = 0.463$
$m_{hh}^{\perp} = 0.275$

The effective hole masses for injection from strained $Si_{1-x}Ge_x$ deviate from Si values for two reasons. First, the Ge fraction changes the band parameters A, B and C and, in this way, the effective masses. Secondly, layer strain itself - in contrast to the situation in the conduction band - significantly alters the effective hole masses. For $Si_{0.77}Ge_{0.23}$ on Si substrate values are (in units of m_0) [3]:

$m_{lh}^{\parallel} = 0.195$
$m_{lh}^{\perp} = 0.219$
$m_{hh}^{\parallel} = 0.182$
$m_{hh}^{\perp} = 0.261$

The strain effect on effective hole masses has also to be considered for strained Si layers.

For a more detailed discussion of strain effects on effective masses the reader is referred to Chapter 4 of this book.

A more rigorous theoretical approach to determine transmission coefficients is used by Wessel and Altarelli [31]. A characteristic result that is beyond the decoupled description discussed in the previous section (heavy and light hole contributions are calculated separately) is the appearance of lh hh mixing in double barrier structures.

At T = 300 K the effective density of states mass in unstrained Si is given by [28]

$m_{ds,p} = 0.81$

Theoretical results for strained $Si_{1-x}Ge_x$ indicate a significant decrease in $m_{ds,p}$ compared to pure Si [32].

The valence band discontinuities ΔE_v^{lh} and ΔE_v^{hh} (FIGURE 1) depend on the Ge fraction as shown in FIGURE 6(a) ($Si/Si_{1-x}Ge_x$ on Si) and FIGURE 6(b) ($Si/Si_{1-x}Ge_x$ on $Si_{1-x}Ge_x$).

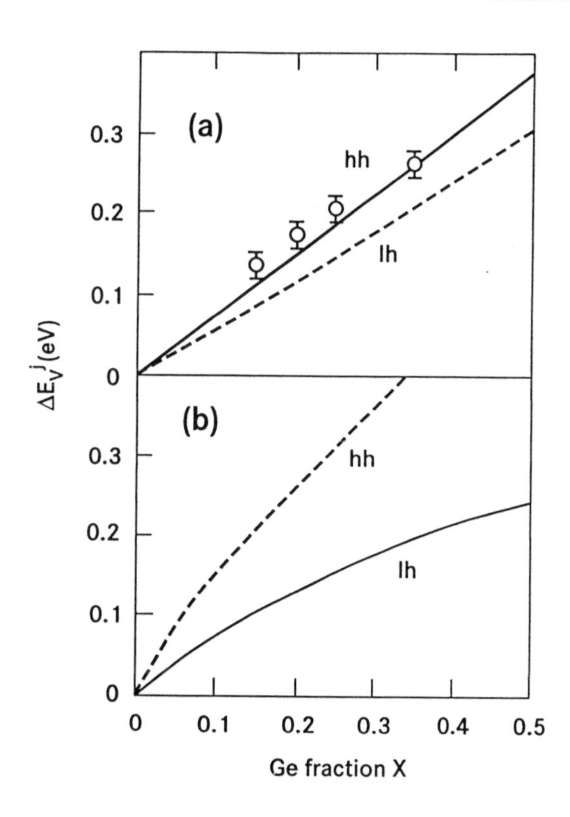

FIGURE 6 Valence band offsets for heavy and light holes at the $Si_{1-x}Ge_x$ heterointerface for (a) $Si_{1-x}Ge_x$ strained and (b) Si strained (for sign see FIGURE 1). Lines were obtained from the approach suggested by People [38]. Open symbols are from [39].

C3 Injection Currents in Anisotype $Si/Si_{1-x}Ge_x$ Heterojunctions

A look at the injection current formulae given by Eqns (13) and (14) reveals the following material properties relevant to anisotype injection:

(a) diffusion constants of minority carriers D_n and D_p
(b) intrinsic carrier densities n_i(Si) and n_i(SiGe).

In contrast to the isotype case, the heights of the individual band discontinuities ΔE_c and ΔE_v are not important for anisotype injection. Only the bandgap difference $\Delta E_g = \Delta E_c + \Delta E_v$ influences the injection currents via the corresponding change in n_i. This remarkable situation, however, applies for types of junctions where potential spikes are absent (as e.g. in (n-p)&(a) type junctions, see FIGURE 3). The presence of potential spikes as e.g. in (p-n)&(a) type junctions may cause features in the current-voltage characteristics that are not represented by Eqns (13) and (14) (see discussion in Section B2).

Minority diffusion coefficients are related to minority carrier mobilities via the Einstein relation

$$D_{n,p} = \mu_{n,p} kT/q \tag{23}$$

Data on minority carrier mobilities have been reviewed by Klaassen [33]. Model calculations that are fitted to these data are shown in FIGURE 7 for electrons and holes. Using Monte Carlo methods minority electron mobilities in strained SiGe layers have been calculated by Kay and Tang [33a].

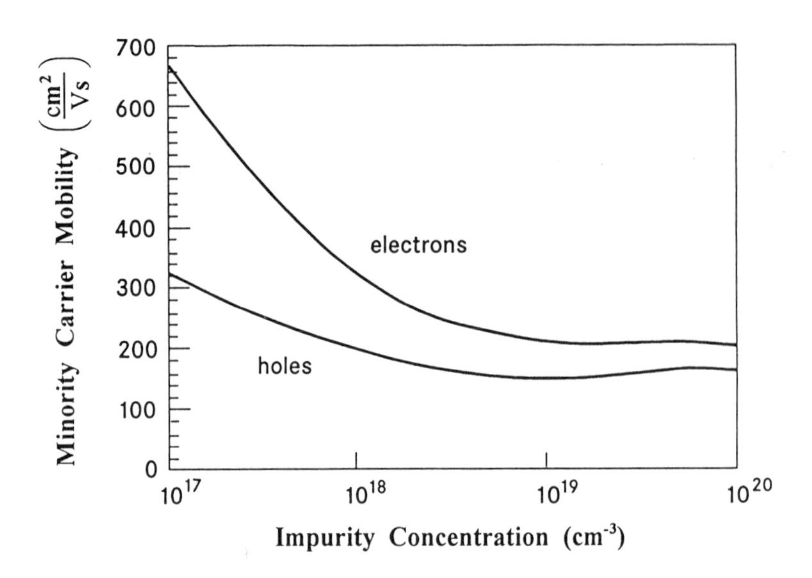

FIGURE 7 Minority carrier mobilities in Si as a function of impurity concentration [33].

Extensive studies have been made on the product of the concentrations of electrons and holes $np = n_i^2$ in (unstrained) Si ([33] and references therein) but only minor results on n_i^2 have been reported so far for strained Si or $Si_{1-x}Ge_x$ layers.

For lightly doped (unstrained) Si the np product follows from

$$n_i^2 = N_c N_v \exp(-E_{g,o}/kT) \tag{24}$$

with

$$N_c = 2(2\pi kTm_{ds,n}h^{-2})^{3/2} \tag{25}$$

and

$$N_v = 2(2\pi kTm_{ds,p}h^{-2})^{3/2} \tag{26}$$

being the effective densities of states of the conduction band and the valence band, respectively.

For heavily doped layers the np product deviates significantly from that given by Eqn (24). Firstly, Boltzmann statistics used in Eqn (24) have to be replaced by Fermi-Dirac statistics. Secondly, as the bandgap E_g shrinks by virtue of the heavy doping, bandgap narrowing has to be considered:

$$E_g = E_{g,o} - \Delta E_g \tag{27}$$

Both aspects can be condensed into a single parameter, the so-called apparent bandgap narrowing ΔE_g^a that is related to ΔE_g via [34]

$$\Delta E_g^a = \Delta E_g + \Delta E^{FD} \tag{28}$$

with

$$\Delta E^{FD} = kT \ln[\exp(-E_F/kT)F_{\frac{1}{2}}(E_F/kT)] \tag{29}$$

Thus, for heavily doped layers, the np product can be written as

$$n_i^2 = N_c N_v \exp(-E_{g,o}/kT) \exp(\Delta E_g^a/kT) \tag{30}$$

Following Klaassen et al [35] no difference exists in apparent bandgap narrowing of n- and p-type Si. FIGURE 8 shows a fit to experimental results for (unstrained) silicon [35].

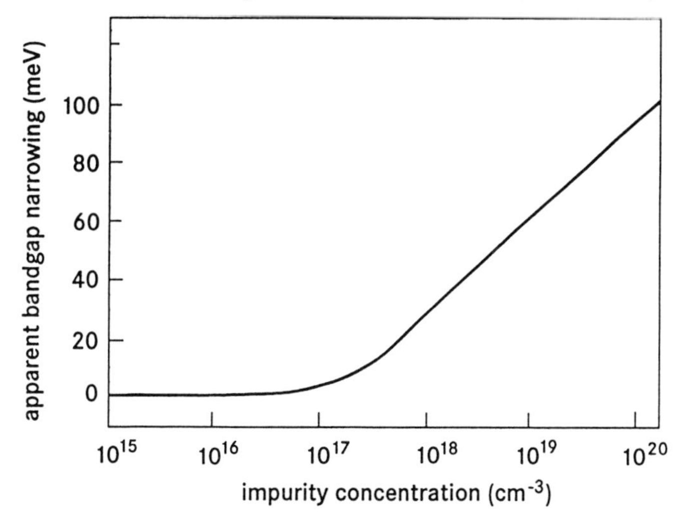

FIGURE 8 Apparent bandgap narrowing in Si (p- and n-type) as a function of impurity concentration [35].

The np product in $Si_{1-x}Ge_x$ layers deviates from that in Si

(a) by the smaller gap $E_{g,o}$ (see Chapter 4 in this book). For $Si_{1-x}Ge_x$ layers grown pseudomorphically on Si(100) the following approximate expression holds

$$E_{g,o} = (1.12 - 0.74x) \text{ eV} \tag{31}$$

(b) by the apparent bandgap narrowing ΔE_g^a. A simple expression for ΔE_g^a (Slotboom-Dunke-Swirhun fit) for use in computer simulation was suggested by Jain et al [34]

$$\Delta E_g^a = K_1 \{\ln(N/K_2) + [\ln^2(N/K_2) + 0.5]^{1/2}\} \tag{32}$$

where N is the dopant concentration. For p-type $Si_{1-x}Ge_x$ up to $x = 0.3$ the constants are $K_1 = 9.0$ meV and $K_2 = 10^{17}$ cm^{-3}. These parameters are valid up to $N = 10^{20}$ cm^{-3}.

(c) by the product of the effective densities of states $N_c N_v$. In strained $Si_{1-x}Ge_x$ layers the effective density of states, both in the conduction and in the valence band, gets smaller due to strain-induced split-off (2-4 split-off in the conduction band, lh-hh split-off in the valence band). FIGURE 9 shows the ratio of the products $N_c N_v$ in $Si_{1-x}Ge_x$ (strained) and Si (unstrained) as a function of x [36]. Thermal occupation of higher energy split-off bands decreases with increasing x and, hence, the effective density of states decreases (see also Eqn (10)). Density of states effective hole masses have been calculated by Manku and Nathan [32]. The results show a significant decrease compared to pure Si. Pruijmboom et al [37] found experimental evidence of an N_cN_v ratio of 0.4 in good agreement with the result of Prinz et al [36].

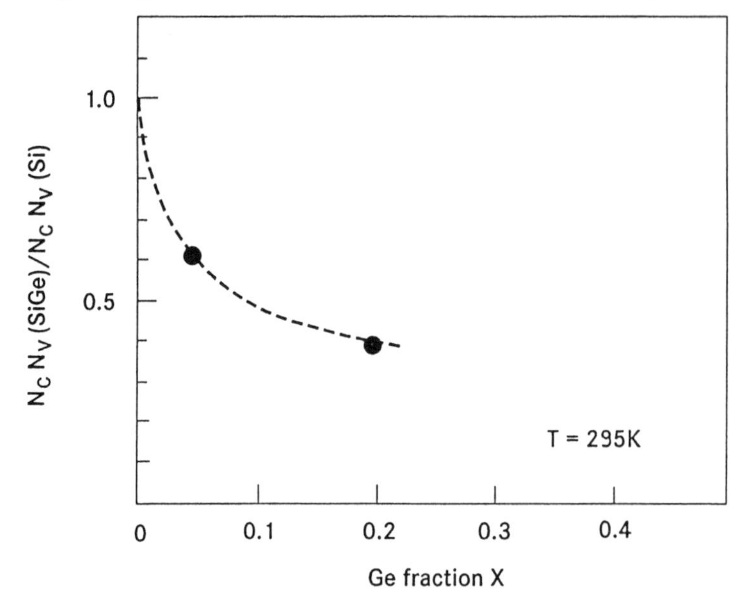

FIGURE 9 Ratio of the N_cN_v product in $Si_{1-x}Ge_x$ (strained) and Si (unstrained) vs. Ge fraction x following Prinz et al [36].

D CONCLUSION

A key feature of the $Si/Si_{1-x}Ge_x$ heterojunction system is the existence of lattice mismatch between Si and $Si_{1-x}Ge_x$ alloy layers. Depending on the Ge content x the mismatch is up to 4.2% (x = 1). Thus, in pseudomorphic (i.e. misfit dislocation free) systems - subject to the kind of substrate used - either the Si or the $Si_{1-x}Ge_x$ layer is under biaxial strain. The strain distribution decisively influences band structure and heterojunction properties and, in this way, besides other factors, injection currents.

Models used to describe injection currents in $Si/Si_{1-x}Ge_x$ heterojunctions are discussed. In the case of isotype junctions (junctions of the n-n or p-p type) thermionic emission theory is preferably used. Conversely, for anisotype (n-p and p-n type) junctions the drift-diffusion approach is more common. This is because in the latter case the total barrier potential is generally smoother as it is governed by the electrostatic potential from dopant atoms rather than by the abrupt heterostructure potential. Relevant material properties in both cases are reviewed.

REFERENCES

[1] Strain relaxed $Si/Si_{1-x}Ge_x$ heterojunctions (Si and $Si_{1-x}Ge_x$ unstrained) contain necessarily a high density of misfit dislocations at the interface. Such defects (a) induce additional barriers by Fermi level pinning at midgap states and (b) induce generation/recombination currents. Heterojunctions containing defects are not considered here.

[2] H.C. Liu, D. Landher, M. Buchanan, D.C. Houghton [*Appl. Phys. Lett. (USA)* vol.52 (1988) p.1809-11]; K.L. Wang, R.P.G. Karunasiri, J. Park, S.S. Rhee, C.H. Chen [*Superlattices Microstruct. (UK)* vol.5 (1989) p.201-6]

[3] G. Schubert, G. Abstreiter, E. Gornik, F. Schäffler, J.F. Luy [*Phys. Rev. B (USA)* vol.43 (1991) p.2280-4]

[4] H.C. Liu, J. Li, M. Buchanan, J.-M. Baribeau, J.G. Simmons [*Electron. Lett. (UK)* vol.29 (1993) p.407-9]

[5] S.S. Rhee, J.S. Park, R.P.G. Karunasiri, Q. Ye, K.L. Wang [*Appl. Phys. Lett. (USA)* vol.53 (1988) p.204-6]

[6] K. Ismail, B.S. Meyerson, P.J. Wang [*Appl. Phys. Lett. (USA)* vol.59 (1991) p.973-5]

[7] S.M. Sze [*Physics of Semiconductor Devices* (Wiley, New York, 1981)]

[8] This expression presumes $(E_z - E_F) >> kT$.

[9] U.V. Bhapkar, R.J. Mattauch [*IEEE Trans. Electron Devices (USA)* vol.40 (1993) p.1038-46]

[10] S. Collins, D. Lowe, J.R. Barker [*J. Appl. Phys. (USA)* vol.63 (1988) p.142-9]

[11] Ch. Zeller, G. Abstreiter [*Z. Phys. B, Condens. Matter (Germany)* vol.64 (1986) p.137-43]

[12] A.M. Cruz Serra, H. Abreu Santos [*J. Appl. Phys. (USA)* vol.70 (1991) p.2734-8]

[13] W. Pötz [*J. Appl. Phys. (USA)* vol.66 (1989) p.2458-66]

[14] T. Fiig, A.P. Jauho [*Appl. Phys. Lett. (USA)* vol.59 (1991) p.2245-7]

[15] B. Pejcinovic, L.E. Kay, T.-W. Tang, D.H. Navon [*IEEE Trans. Electron Devices (USA)* vol.36 (1989) p.2129-36]

[16] It should be mentioned that this approach requires a coherent injection process such as e.g. resonant tunnelling. Processes such as sequential tunnelling (S. Luryi [*Appl. Phys. Lett. (USA)* vol.47 (1985) p.490-2]) are not reproduced.

[17] A.G. Milnes, D.L. Feucht [*Heterojunctions and Metal-Semiconductor Junctions* (Academic Press, New York, 1972) ch.4]

[18] J. Smoliner, W. Demmerle, G. Berthold, E. Gornik, G. Weimann, W. Schlapp [*Phys. Rev. Lett. (USA)* vol.63 (1989) p.2116-9]

[19] E.E. Mendez, J. Nocera, W.I. Wang [*Phys. Rev. B (USA)* vol.45 (1992) p.3910-13]; H. Jorke [*J. Appl. Phys. (USA)* vol.72 (1992) p.3215-7]

[20] W. Shockley [US Patent No.2,569,347 (Sept. 25, 1951)]

[21] H. Krömer [*Proc. IREE Aust. (Australia)* vol.45/2 (1957) p.1535-7]

[22] The validity of the Shockley-type Eqns (16) and (17) is subject to some restrictions such as (a) low injection and (b) no generation/recombination in the depletion layer.

[23] H. Krömer [*Solid-State Electron. (UK)* vol.28 (1985) p.1101-3]

[24] The diffusion-equation description of transport in the base breaks down if the base length W is comparable to the minority carrier mean free path. An analytical treatment of this problem using the Boltzmann transport equation was given by Grinberg and Luryi (A.L. Grinberg, S. Luryi [*Solid-State Electron. (UK)* vol.35 (1992) p.1299-309]).

[24a] J.W. Slotboom, G. Streutker, A. Pruijmboom, D.J. Gravesteijn [*IEEE Electron Device Lett. (USA)* vol.12 (1991) p.486-8]

[25] R.L. Anderson [*Solid-State Electron. (UK)* vol.5 (1962) p.341]

[26] Ref [17], ch.2

[27] D. Fink, R. Braunstein [*Phys. Status Solidi B (Germany)* vol.73 (1976) p.361]

[28] Landolt-Börnstein [*Numerical Data and Functional Relationships in Science and Technology*, New Series, Group III (Springer-Verlag, Berlin, 1982) vol.17a p.43]

[29] W.H. Kleiner, L.M. Roth [*Phys. Rev. Lett. (USA)* vol.2 (1959) p.334-7]

[30] A more rigorous approach to determining hole injection currents may proceed from the expression

$$dJ_i = qg_i \frac{1}{\hbar} \frac{\partial E}{\partial k_z} \frac{1}{4\pi^3} dF_i dk_z f(E) T_i \qquad i = lh, hh$$

for the differential injection current taking the specific dispersion relations given by Eqns (21) and (22) (H. Jorke [*Solid-State Electron. (UK)* vol.36 (1993) p.975-9]).
[31] R. Wessel, M. Altarelli [*Phys. Rev. B (USA)* vol.40 (1989) p.12457-62]
[32] T. Manku, A. Nathan [*J. Appl. Phys. (USA)* vol.69 (1991) p.8414-6]
[33] D.B.M. Klaassen [*Solid-State Electron. (UK)* vol.35 (1992) p.953-9]
[33a] L.E. Kay, T.-W. Tang [*J. Appl. Phys. (USA)* vol.70 (1991) p.1483-8]
[34] S.C. Jain, D.J. Roulston [*Solid-State Electron. (UK)* vol.34 (1991) p.453-65]
[35] D.B.M. Klaassen, J.W. Slotboom, H.C. DeGraaff [*Solid-State Electron. (UK)* vol.35 (1992) p.125-9]
[36] E.J. Prinz, P.M. Garone, P.V. Schwartz, X. Xiao, J.C. Sturm [*Int. Electron Devices Meet. Tech. Dig. (USA)* (1989) p.639-42]
[37] A. Pruijmboom, J.W. Slotboom, D.J. Gravensteijn [*IEEE Electron Device Lett. (USA)* vol.12 (1991) p.357-9]
[38] R. People [*IEEE J. Quantum Electron. (USA)* vol.22 (1986) p.1696-710]
[39] K. Nauka, T.I. Kamins, J.E. Turner, C.A. King, J.L. Hoyt, J.F. Gibbons [*Appl. Phys. Lett. (USA)* vol.60 (1992) p.195-7]

5.3 Magnetotransport in SiGe/Si structures

G. Stöger and G. Bauer

January 1994

A INTRODUCTION

From magnetotransport measurements one obtains information on properties like carrier concentration, effective masses, g-factor and scattering times. In this Datareview we report on Shubnikov de Haas and quantum Hall effect for two dimensional electron and hole gases, on fractional quantum Hall effect observed in n-type heterojunctions, on quantum corrections to the magnetoresistance due to weak localisation and on resonant tunnelling of holes in $Si/Si_{1-x}Ge_x$ structures.

B SHUBNIKOV DE HAAS EFFECT AND QUANTUM HALL EFFECT

The oscillatory magnetoresistance yields information on the Fermi energy and thus on carrier concentration and on g-factor, but from its damping with magnetic field it also yields information on Landau-level broadening and therefore on the role of scattering mechanisms. The effective mass is obtained from the temperature dependence of the Shubnikov de Haas (SdH) oscillations. From the oscillation period of the SdH effect and the filling factor derived from the plateaux observed in the quantum Hall effect (QHE) the degeneracy of conduction or valence band can be determined. Measurements of the SdH effect in $Si/Si_{1-x}Ge_x$ structures have been performed for n- and p-type heterojunctions and also for multi-quantum wells [1].

B1 Two Dimensional Electron Gas

The damping of the SdH oscillations with magnetic field is determined by the single particle scattering time, which is correlated with the total scattering rate. For the momentum relaxation (transport) time on the other hand, which is calculated from the mobility, the scattering events are weighted by the scattering angle [2]. Thus the ratio of these two scattering times is closely related to the scattering mechanism. For n-type MBE samples grown on graded $Si_{1-x}Ge_x$ buffers with spacer layers in the range of 100 Å to 200 Å and channel widths of about 100 Å, mobilities as high as 180000 cm^2/V s [3] and 175000 cm^2/V s [2] were reported. The ratio of the total scattering rate to the large angle scattering rate was found to be in the range from 4 to 14 [3] and 11 to 26 [2], respectively. This ratio being much larger than unity clearly indicates that in high quality $Si/Si_{1-x}Ge_x$ heterostructures the mobility is limited by remote impurity scattering as predicted by theoretical calculations [4].

The sixfold degenerate valleys of the bulk Si conduction band are split due to the biaxial tensile strain in the Si channel into twofold and fourfold degenerate valleys with the twofold degenerate ones having the lower energy. The degeneracy factor can be determined experimentally from the period of the SdH oscillations and the filling factor ν obtained from the QHE. The filling factor is calculated from the value of the Hall resistance ρ_{xy} in the plateaux by

$$\rho_{xy} = \frac{h}{e^2 \nu}$$

As an example, the magnetoresistance ρ_{xx} and Hall resistance ρ_{xy} for a $Si/Si_{1-x}Ge_x$ heterostructure with a carrier concentration of 0.9 x 10^{12} cm^{-2} and a mobility of 77300 cm^2/V s are shown in FIGURE 1 [2]. For magnetic fields B < 3 T, the magnetoresistance oscillations correspond to filling factors which are multiples of 4 (since neither the twofold spin splitting nor the twofold valley splitting is resolved). For increasing magnetic field the spin degeneracy and finally the valley degeneracy is lifted [2,3,5]. As displayed in FIGURE 1, the clearly resolved dip in ρ_{xx} between the minima corresponding to filling factors of $\nu = 8$ and $\nu = 12$ shows the effect of spin splitting. The appearance of the dip in the magnetoresistance for the odd filling factor $\nu = 5$ indicates the lifting of the twofold valley degeneracy. Well defined quantum Hall plateaux are resolved for filling factors as large as $\nu = 20$, plateaux corresponding to $\nu = 10$ and 6 indicate the lifted spin degeneracy, and the plateau corresponding to $\nu = 5$ indicates the lifted valley degeneracy.

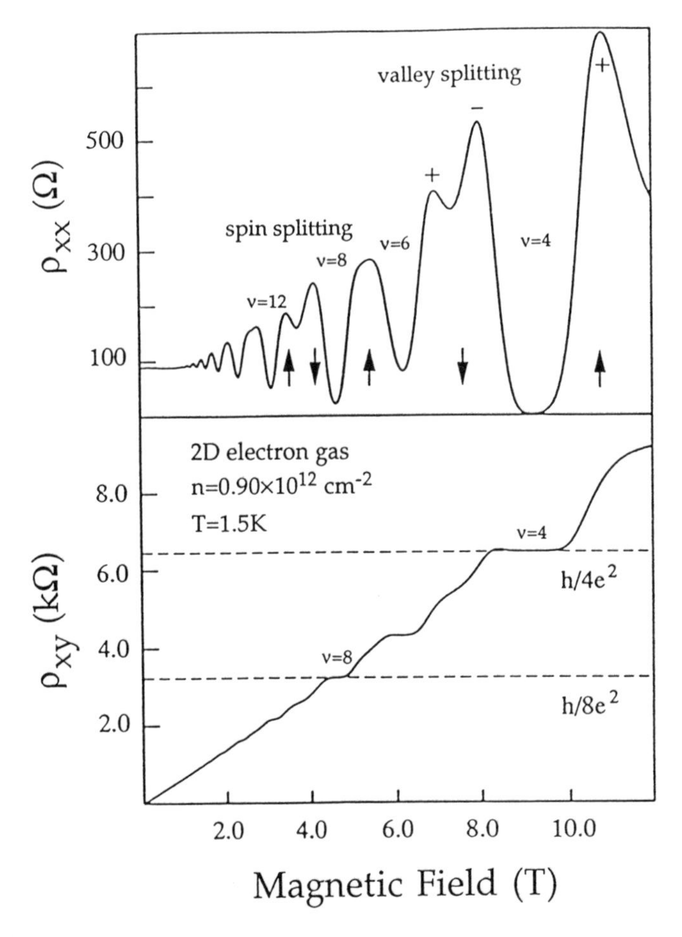

FIGURE 1 SdH oscillations and quantum Hall effect for a 2DEG at T = 1.5 K [2]. Spin split states (↑/↓) of the third Landau (between $\nu = 12$ and $\nu = 8$) and valley splitting (+/-) for $\nu \leq 5$ are clearly resolved. Hall plateaux are indicated for filling factors $\nu = 4$ and 8.

From the damping of the SdH oscillation amplitude with increasing lattice temperature an effective mass of $m^* \sim 0.17 \pm 0.02\ m_0$ [3] is estimated. Within the experimental uncertainty this value is in excellent agreement with the transverse effective mass in bulk Si ($m^* = 0.19$).

B2 Two Dimensional Hole Gas

Two dimensional hole gases (2DHG) can be fabricated by growing a $Si_{1-x}Ge_x$ layer on pure Si in a pseudomorphically strained fashion. In this situation nearly the entire band offset between Si and the alloy occurs in the valence band. Since the bandgap of the alloy is smaller than the bandgap of Si, the holes are confined to the $Si/Si_{1-x}Ge_x$ layer. In addition, the strain lifts the degeneracy of the light- and the heavy-hole valence band; for the compressively strained $Si_{1-x}Ge_x$ layer on Si the heavy-hole band is lowest in energy [6,7]. A novel method for the fabrication of high mobility 2DHG structures is the growth of a Ge channel on a relaxed graded $Si_{1-x}Ge_x$ ($x \sim 0.6$ to 0.7) buffer [3,8], the Ge channel being under an in-plane biaxial compressive strain. For this layer design hole mobilities at 4.2 K as high as 55000 cm^2/V s were observed.

An evaluation of the periods of the SdH oscillations yields a valley degeneracy factor of one thus indicating that the separation of the heavy-hole and light-hole states is larger than the Fermi energy [9]. Indeed, for a $Si/Si_{0.8}Ge_{0.2}$ heterostructure an energy difference between the heavy-hole and light-hole states of about 23 meV is expected [10].

The shape of the SdH oscillations for p-type heterostructures depends somehow on specific sample parameters. For a $Si/Si_{0.88}Ge_{0.12}$ heterostructure the spin splitting was interpreted to be larger than half of the separation of the Landau levels [9] since (with the exception of the even filling factor $\nu = 2$ at high magnetic field) only plateaux corresponding to odd filling factors ($\nu = 1, 3, 5, 7$) were observed in the QHE. A similar behaviour was observed in a p-type $Si/Si_{0.72}Ge_{0.28}$ heterostructure [11] with a hole concentration and a mobility comparable to the samples of [9]. On the other hand for a $Si/Si_{0.8}Ge_{0.2}$ double modulation-doped heterostructure grown by rapid thermal chemical vapour deposition only filling factors which are multiples of 4 were observed in the QHE [12]. The additional factor of 2 which enters the filling factor is due to the two parallel 2D hole gases on either side of the quantum well but the result clearly shows that spin splitting is not resolved in these samples. High quality $Si/Si_{1-x}Ge_x/Ge$ heterostructures [3] show minima in the SdH oscillations as well as plateaux in the QHE regime corresponding both to even and odd filling factors. FIGURE 2 shows the magnetoresistance and Hall resistance for a $Si/Si_{1-x}Ge_x/Ge$ heterostructure with a hole concentration of about 5×10^{11} cm^{-2} and a mobility of 18000 cm^2/V s [3]. For filling factors between $\nu = 4$ and $\nu = 6$ the structure in the magnetoresistance is attributed to the spin splitting by the present authors. For magnetic fields $B < 3.5$ T, the magnetoresistance oscillations correspond to filling factors which are multiples of 2 indicating only spin degeneracy. Quantum Hall plateaux are observed for filling factors up to $\nu = 12$. In contrast to QHE measurements performed on p-type $Si/Si_{1-x}Ge_x$ heterostructures [9,11,12], for this particular $Si/Si_{1-x}Ge_x/Ge$ heterostructure plateaux corresponding to even and odd filling factors are clearly resolved. For the filling factors displayed in FIGURE 2, the present authors have corrected the value $\nu = 7$ given in [3] to $\nu = 8$.

From the temperature dependent SdH oscillation amplitude data for the $Si/Si_{1-x}Ge_x$ heterostructures, effective hole masses between $m^* \sim 0.23 \pm 0.02\ m_0$ ($x = 0.13$) [13], $m^* \sim 0.30 \pm 0.02\ m_0$ ($x = 0.2$) [14] and $m^* \sim 0.44 \pm 0.03\ m_0$ ($x = 0.15$) [15] were evaluated. For the $Si/Si_{1-x}Ge_x/Ge$ heterostructures [8] the effective hole mass in the Ge channel was found to be $\leq 0.1\ m_0$.

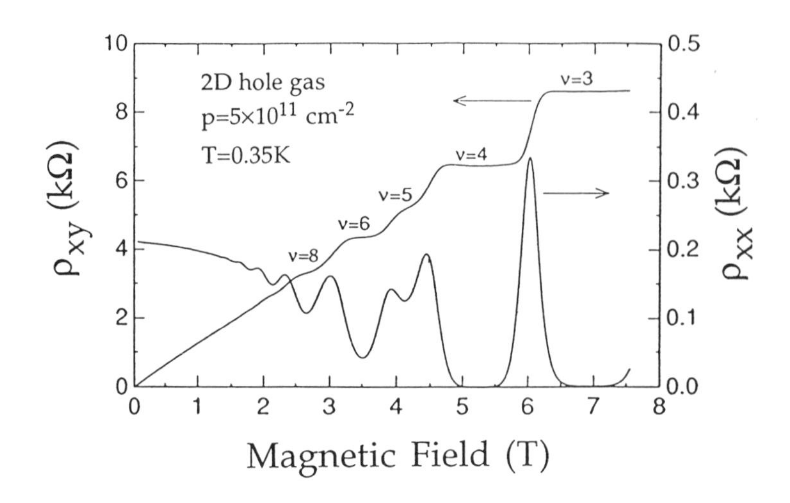

FIGURE 2 SdH oscillations and quantum Hall effect for a 2DHG at T = 0.35 K [3]. Hall plateaux are indicated for filling factors ν = 3, 4, 5, 6 and 8.

The anisotropy of the valence band g-factor can be obtained from measurements of the SdH oscillations in tilted magnetic fields [9]. For a strained $Si_{0.8}Ge_{0.2}$ layer values of $g\perp = 4.9 \pm 0.1$ (perpendicular to the hole gas) and for parallel fields values of $g\| = 1.8 \pm 0.1$ were determined. For an unstrained $Si_{0.8}Ge_{0.2}$ layer grown on a biaxially tensilely strained Si layer values of $g\perp = 4.8 \pm 0.1$ and $g\| = 4.9 \pm 0.1$ were obtained, which is isotropic within experimental accuracy.

C FRACTIONAL QUANTUM HALL EFFECT

Non-integer filling factors have been observed in various n-type samples grown by MBE [16] and ultra high vacuum chemical vapour deposition (UHV-CVD) [17]. Hints of the appearance of a plateau in ρ_{xy} related to a non-integer filling factor were also reported in [5].

In [17] the temperature dependences of the dips in ρ_{xx} corresponding to fractional filling factors of $\nu = 4/3$ and $\nu = 2/3$ were studied. An activation analysis for $\nu = 4/3$ using $\rho_{xx} = \rho_0 \exp(-E_g/2k_BT)$ in the temperature range from 0.5 to 1.4 K yields an energy gap of 0.85 ± 0.17 K ($\sim 0.073 \pm 0.015$ meV) and for $\nu = 2/3$ an energy gap of 1.16 ± 0.45 K ($\sim 0.10 \pm 0.040$ meV). FIGURE 3 shows the magnetoresistance and Hall resistance for a 2DEG with a carrier concentration of 4×10^{11} cm^{-2} and a mobility of 85000 cm^2/V s [17]. The fractional filling factor $\nu = 4/3$ is clearly visible as a plateau in the Hall resistance and a minimum in the magnetoresistance.

In [16] a minimum in ρ_{xx} with a fractional filling factor of $\nu = 2/3$ was observed in the temperature range from 0.3 K to 1.2 K. The activation energy was determined to be of the order of 0.02 meV (~ 0.23 K).

Neither the number of papers on fractional quantum Hall effect (FQHE) in $Si/Si_{1-x}Ge_x$ nor the number of fractions so far observed is comparable to the numbers found for $GaAs/Ga_{1-x}Al_xAs$.

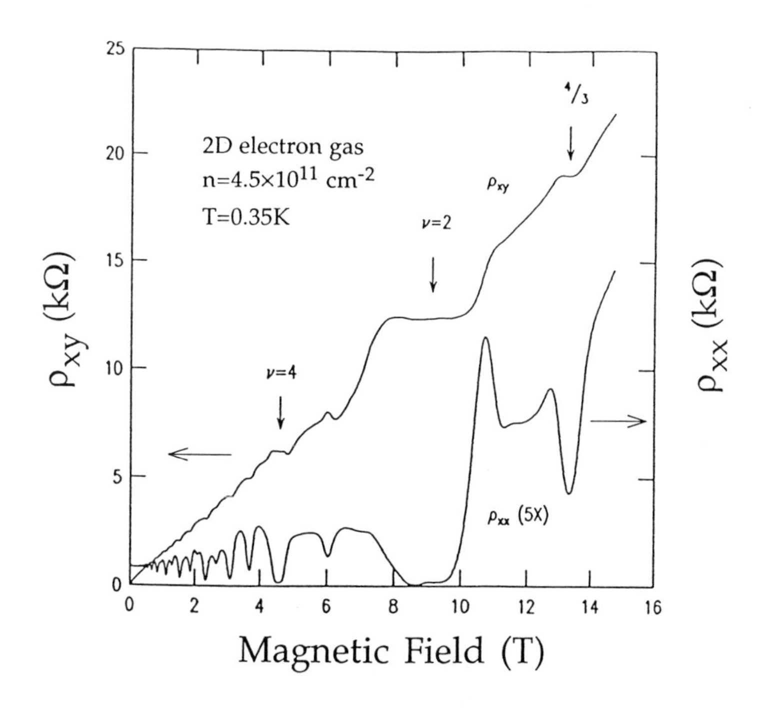

FIGURE 3 SdH oscillations and quantum Hall effect for a 2DEG at T = 0.35 K [17]. The fractional filling factor $\nu = 4/3$ is clearly resolved as a dip in ρ_{xx} and a plateau in ρ_{xy}.

D WEAK LOCALISATION

From the temperature dependences of the resistivity and the Hall coefficient for a 2DHG in a $Si_{0.8}Ge_{0.2}$ quantum well the presence of weak localisation and hole-hole interaction effects was deduced [7,18]. The conductivity at zero magnetic field exhibits a logarithmic increase with increasing temperature which is an indication for 2D weak localisation. Fitting the magnetoresistance, which shows a negative slope for $B < 0.3$ T, yields a phase coherence length of $l_\phi \sim 180$ nm.

In n-type heterostructures the suppression of the weak localisation with increasing electric field and lattice temperature has been investigated [19]. The phase coherence time was deduced to be in the order of 1 to 2 ps, which corresponds to a value of l_ϕ of about 90 to 120 nm.

E RESONANT TUNNELLING

To our knowledge, so far only the resonant tunnelling of holes has been studied in $Si/Si_{1-x}Ge_x$ structures.

For $Si/Si_{1-x}Ge_x$ ($x \sim 0.2$) double barrier resonant tunnelling structures the observed resonances can be explained by the tunnelling of the incoming heavy holes through both light-hole and heavy-hole states in the well [20]. In addition, the resonances were studied with magnetic field applied parallel to the growth direction, i.e. parallel to the current direction. The derivatives of the tunnelling characteristics reflect the density of states associated with Landau levels. From the splitting of the Landau levels with magnetic field an in-plane effective mass of $m^* \sim 0.20m_0$ in a $Si_{0.77}Ge_{0.23}$ well is obtained.

Tunnelling through double barrier heterostructures with applied magnetic field perpendicular to the growth direction and thus parallel to the interfaces was studied in [21]. By rotating the magnetic field in the plane of the interfaces it is possible to map out the energy surface of the quantum well levels, since this angle-resolved magnetotunnelling spectroscopy does not average over all momentum (k) space but directly probes states with specific k values. It was found that the warping of the heavy-hole band and the light-hole band are 45° out of phase with respect to each other as in the case of bulk Si. This band warping was observed for samples containing a strained $Si_{0.67}Ge_{0.33}$ well and an unstrained Si barrier grown on a strained graded $Si_{1-x}Ge_x$ buffer as well as for tunnelling structures with a relaxed $Si_{0.5}Ge_{0.5}$ well and strained Si barriers. Due to the structural quality, the anisotropy is more pronounced in the case of the strained quantum well.

F CONCLUSION

The structural quality and thus consequently also the electric properties of $Si/Si_{1-x}Ge_x$ structures have been improved substantially in recent years. Thus investigations of magnetotransport phenomena of low-dimensional Si-based structures have become possible which were previously restricted to $GaAs/Ga_{1-x}Al_xAs$.

Low and high field magnetotransport phenomena yield relevant information on various scattering times, effective masses and their nonparabolicity, and on g-factors. QHE, FQHE, weak localisation and resonant tunnelling have been observed.

REFERENCES

[1] H.-J. Herzog, H. Jorke, F. Schäffler [*Thin Solid Films (Switzerland)* vol.184 (1990) p.237]

[2] D. Többen, F. Schäffler, A. Zrenner, G. Abstreiter [*Phys. Rev. B (USA)* vol.46 (1992) p.4344]

[3] Y.H. Xie, E.A. Fitzgerald, D. Monroe, P.J. Silverman, G.P. Watson [*J. Appl. Phys. (USA)* vol.73 (1993) p.8364]

[4] F. Stern, S.E. Laux [*Appl. Phys. Lett. (USA)* vol.61 (1992) p.1110]

[5] D. Többen, F. Schäffler, A. Zrenner, G. Abstreiter [*Thin Solid Films (Switzerland)* vol.222 (1992) p.15]

[6] C.J. Emeleus, T.E. Whall, D.W. Smith, R.A. Kubiak, E.H.C. Parker, M.H. Kearney [*J. Appl. Phys. (USA)* vol.73 (1993) p.3852]

[7] C.J. Emeleus et al [*Phys. Rev. B (USA)* vol.47 (1993) p.10016]

[8] Y.H. Xie, D. Monroe, E.A. Fitzgerald, P.J. Silverman, F.A. Thiel, G.P. Watson [*Appl. Phys. Lett. (USA)* vol.63 (1993) p.2263]

[9] F.F. Fang, P.J. Wang, B.S. Meyerson, J.J. Nocera, K. Ismail [*Surf. Sci. (Netherlands)* vol.263 (1992) p.175]

[10] R. People [*IEEE J. Quantum Electron. (USA)* vol.22 (1986) p.1696]

[11] J.F. Nützel, F. Meier, E. Friess, G. Abstreiter [*Thin Solid Films (Switzerland)* vol.222 (1992) p.150]

[12] V. Venkataraman, P.V. Schwartz, J.C. Sturm [*Appl. Phys. Lett. (USA)* vol.59 (1991) p.2871]

[13] N.L. Mattey et al [*Jpn. J. Appl. Phys. (Japan)* vol.33 (1994) p.2348]

[14] R. People et al [*Appl. Phys. Lett. (USA)* vol.45 (1984) p.1231]

[15] P.J. Wang, F.F. Fang, B.S. Meyerson, J. Nocera, B. Parker [*Appl. Phys. Lett. (USA)* vol.54 (1989) p.2701]

[16] D. Monroe, Y.H. Xie, E.A. Fitzgerald, P.J. Silverman [*Phys. Rev. B (USA)* vol.46 (1992) p.7935]

[17] S.F. Nelson et al [*Appl. Phys. Lett. (USA)* vol.61 (1992) p.64]

[18] C.J. Emeleus, T.E. Whall, D.W. Smith, R.A. Kubiak, E.H.C. Parker, M.J. Kearney [*Thin Solid Films (Switzerland)* vol.222 (1992) p.24]

[19] G. Stöger et al [*Semicond. Sci. Technol. (UK)* vol.9 (1994) p.765]; G. Stöger et al [*Phys. Rev. B (USA)* vol.49 (1994) p.10417]

[20] G. Schuberth, G. Abstreiter, E. Gornik, F. Schäffler, J.F. Luy [*Phys. Rev. B (USA)* vol.43 (1991) p.2280]

[21] U. Gennser, V.P. Kesan, D.A. Syphers, T.P. Smith III, S.S. Iyer, E.S. Yang [*Phys. Rev. Lett. (USA)* vol.67 (1991) p.3828]

CHAPTER 6

SURFACE PROPERTIES

6.1 Reconstruction and bonding configurations of the SiGe(100) surface

P.C. Kelires and G. Theodorou

January 1994

A INTRODUCTION

Reconstructed semiconductor surfaces have attracted considerable attention over the years, primarily because of their crucial role in epitaxial growth techniques. The surface properties of elemental systems are rather well understood. However, alloy surfaces are more complex and inhomogeneous, not because of the existence of new types of reconstructions, but because the composition profile of the surface differs from that of the bulk system, leading to a variety of bonding configurations. A prototypical example is offered by the (100) surface of the Si-Ge alloy. Here, we discuss certain issues related to this surface, namely the primary reconstruction and its energetics (Section B), surface and sub-surface stresses (Section C), and the composition profile (Section D).

B RECONSTRUCTION

It is by now well established that the primary reconstruction of the Si(100) surface is the 2x1 dimerisation geometry [1]. Accurate total-energy calculations [2,3] give the reconstruction energy as ~1.0 eV per surface atom and the dimer bond length as 2.23 Å (evidence for n-bonding character). Dimerisation occurs because it reduces the number of dangling (unsaturated) bonds in the surface atoms from two to one, leading to bonding (σ) and antibonding (σ^*) states, while the remaining dangling bonds form π and π^* surface states in the gap. Extended defects (missing dimer rows) enhance the stability of the dimer structure [4]. A similar (2x1) reconstruction, along with local c(2x2) and p(4x2) symmetry, has been observed on the Ge(100) surfaces [5]. Thus, dimerisation is also the essential feature of this surface. Due to the close similarity of Si and Ge, one expects the same feature to hold for the $Si_{1-x}Ge_x$ alloy as well. Indeed, a low-energy electron-diffraction study [6] of the (100) surface shows a 2x1 unit cell, suggesting that the surface atoms dimerise.

It is important to examine the composition of the surface layer. Kelires and Tersoff [7] found, from Monte Carlo (MC) simulations, that there is strong segregation of Ge atoms to the top surface layer. The Ge surface energy for a (100) 2x1 reconstruction is about 0.07 eV/atom lower than for Si, so segregation of a layer of Ge to the surface reduces the enthalpy. An almost complete coverage with Ge of the surface of stoichiometric (50-50%) alloys is predicted for temperatures up to ~600 K. There has been experimental evidence for such behaviour. A high-resolution electron-energy-loss spectroscopy study [6] found that annealing a $Si_{0.60}Ge_{0.40}$ alloy at 700 K produced an almost pure Ge surface. For a composition $Si_{0.80}Ge_{0.20}$, the surface was found to be ~75% Ge. The lower surface energy of Ge can be easily understood if we consider that it has weaker bonds (~-1.92 eV) than Si (-2.31 eV), and that it is energetically favourable to have Ge rather than Si unsaturated bonds at the surface.

Theoretical calculations of Ciraci and Batra [8] found that the total energy of the reconstructed Si(100) surface covered with Ge is ~0.5 eV per surface Ge atom lower than that if the Ge atoms continue the ideal 1x1 structure. The difference in energy between Si-Si and Ge-Ge dimer bonds is found to be ~0.5 eV. Taking into account the lower surface energy (~0.07 eV/atom) of Ge discussed above, one can estimate the energy difference between unsaturated (dangling) Si and Ge bonds to be (2 x 0.07 + 0.5)/2 = 0.32 eV. These values refer to alloys grown epitaxially on Si substrates ($a_{\|}$ = 5.43 Å), but similar values and trends are expected when the alloys take their natural lattice constant. Note that the relative fractions of Si-Si, Ge-Ge and Si-Ge dimer bonds depend on the Ge concentration x of the alloy. For example, at low Ge content it is favourable to replace a Ge dimer bond with a Si-Ge bond. Recalling that the excess bond energy $\Delta V_{Si\text{-}Ge}$,

$$\Delta V_{Si\text{-}Ge} = V_{Si\text{-}Ge} - [V_{Si\text{-}Si} + V_{Ge\text{-}Ge}]/2,$$

equals ~4.5 meV (as found by pseudopotential calculations [9]), this replacement saves ~0.18 eV [9].

C SURFACE AND SUB-SURFACE STRESSES

The knowledge of values and signs of stresses which are present at the surface and sub-surface layers is important in determining the stability of the reconstructed surface, as well as the composition profile in the surface region. The surface stress tensor is defined as

$$\sigma_{ij}^{surf} = \frac{1}{A}\frac{dE^{surf}}{d\varepsilon_{ij}} \tag{1}$$

where E^{surf} is the surface energy, A the surface area, and $\{\varepsilon_{ij}\}$ the two-dimensional strain tensor; a positive value corresponds to tensile stress. Theoretical estimates of stresses for the ideal Si(100) 2x1 surface have appeared in the literature. TABLE 1 gives results for $\sigma_{\|}$ and $\sigma_{\perp}$, the stress components parallel and perpendicular to the surface dimers, from quantum-mechanical calculations of Payne et al [3] and of Meade and Vanderbilt [10]. Values, and especially signs, should not be altered significantly for the alloyed SiGe(100) 2x1 surface. While the two calculations differ quantitatively (values and net tension), they show the same trend. The surface is under a tensile stress parallel to the dimer bond, and a compressive stress perpendicular to the dimer bond. They also predict a stress anisotropy $\sigma_{\|} - \sigma_{\perp}$, which is crucial in determining the behaviour of steps on Si(100) [11], of about 2.5 eV/a^2, while experimental measurements [12] give 1.0 eV/a^2. Tersoff proposed [13] that the discrepancy between theory and experiment can be accounted for, if a density of ~4% missing dimers is considered.

TABLE 1 Surface stress for the Si(100) 2x1 surface, in eV/(1x1) cell, parallel and perpendicular to the dimers ($\sigma_{\|}$ and $\sigma_{\perp}$), and their sum and difference in units of eV/a^2.

Ref	$\sigma_{\|}$	$\sigma_{\perp}$	$\sigma_{\|} + \sigma_{\perp}$	$\sigma_{\|} - \sigma_{\perp}$
[3]	0.7	-2.0	-1.3	2.7
[10]	1.6	-0.9	0.7	2.5

To estimate sub-surface stresses induced by the 2x1 reconstruction, the concept of an atomic stress tensor, or its trace, which defines a local compression, was used [7]. This is made possible by partitioning the total energy of the system into atomic contributions E_i. Then, considering a uniform expansion (contraction) of the system, and by analogy with the macroscopic pressure, one can define an atomic compression

$$C_i = -V\, dE_i/dV \sim p\Omega \qquad (2)$$

where E_i is the energy of atom i and V is the volume. This compression can be converted into units of pressure by dividing by an appropriate atomic volume Ω. The results of such an analysis of the (100) 2x1 surface are as follows:

(a) For the surface dimer layer, the compression is very weak, consistent with the relatively unconstrained geometry.

(b) The second layer is under a large compression, about 0.4 eV/atom, corresponding to a pressure of ~30 kbar.

(c) In the third and fourth layer, the two atoms per cell are not equivalent. One atom is directly below the dimer, and is under a compression of about 0.3 eV. The other is between dimers and is under a tension of similar magnitude. We discuss the consequences of these observations in Section D.

D COMPOSITION PROFILE

We showed in Section B that under equilibrium conditions the SiGe(100) 2x1 surface is Ge-rich. It is evident that the surface composition is driven by the reduction of surface energy associated with the dangling bond, and not by atomic compression (which is very weak). To the contrary, the composition of layers just below the surface is strongly driven by the sub-surface atomic stresses discussed above. MC calculations [7] led to the following conclusions (the equilibrium structure of the top four surface layers at 300 K is shown in FIGURE 1):

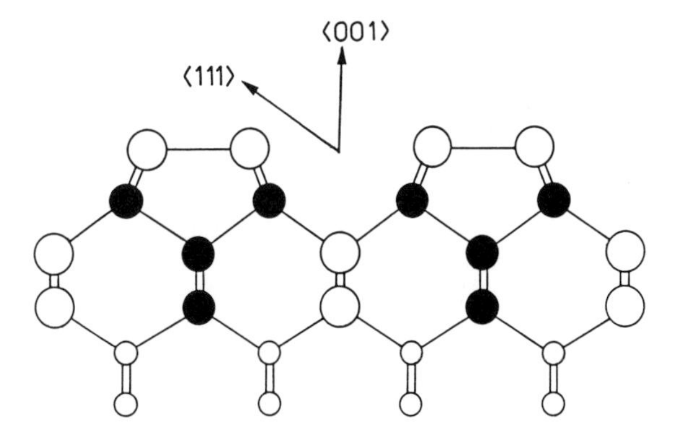

FIGURE 1 Arrangement of atoms in surface layers. At the top are surface dimers (Ge). Large filled circles denote sites under compression (Si). Large open circles denote sites under tension (Ge). Small circles denote sites with bulk composition.

(a) Substituting a Ge atom, which is larger than Si, would raise the energy of a site that is under compression, but would lower the energy of a site under tension.

(b) The second layer is under compression. The energy loss by substituting Ge for Si equals $\Delta E = C_i \Delta \ln V = 0.4 \times 0.12 \sim 50$ meV/atom (see Eqn (2)). Thus the Ge concentration is strongly reduced in the second layer. Khor and Das Sarma [14] also arrived at similar estimates for this energy cost.

(c) In the third and fourth layer, the site which is under compression is driven towards being occupied by Si (there is an energy cost of ~30 meV/atom for substituting Ge in this site), while the site under tension is driven towards being occupied by Ge.

(d) The oscillatory behaviour of the concentration profile at the surface persists up to high temperatures (~1200 K), where equilibration of the surface is experimentally more easily achieved.

The equilibrium structure of the SiGe(100) 2x1 surface (shown in FIGURE 1) has been the basis for a growth mechanism that could explain the observation of long-range ordering in bulk $Si_{1-x}Ge_x$ alloys. The interested reader should refer to the work of LeGoues et al [15].

E CONCLUSION

We discussed in this Datareview the 2x1 reconstruction of the SiGe(100) surface, its energetics, the stresses associated with the reconstruction, and the equilibrium composition profile. It is seen that two factors play a crucial role in determining the distribution of species and the bonding configurations: (a) the lower surface energy of Ge, which drives the surface towards being Ge-rich, and (b) the sub-surface atomic stresses which give rise to a non-monotonic concentration profile. There is experimental evidence for the enrichment of the top surface layer with Ge. However, more experimental work is needed to study the equilibrium concentration profile of the reconstructed surface.

ACKNOWLEDGEMENT

This work has been supported in part by the ESPRIT Basic Research Action No 7128.

REFERENCES

[1] R.E. Schlier, H.E. Farnsworth [*J. Chem. Phys. (USA)* vol.30 (1959) p.917]
[2] M.T. Yin, M.L. Cohen [*Phys. Rev. B (USA)* vol.24 (1981) p.2303]
[3] M.C. Payne, N. Roberts, R.J. Needs, M. Needels, J.D. Joannopoulos [*Surf. Sci. (Netherlands)* vol.221 (1989) p.1]
[4] R.M. Tromp, R.J. Hamers, J.E. Demuth [*Phys. Rev. Lett. (USA)* vol.55 (1985) p.1303]
[5] J.A. Kubby, J.E. Griffith, R.S. Becker, J.S. Vickers [*Phys. Rev. B (USA)* vol.36 (1987) p.6079]
[6] J.A. Schaefer, J.Q. Broughton, J.C. Bean, H.H. Farrel [*Phys. Rev. B (USA)* vol.33 (1986) p.2999]
[7] P.C. Kelires, J. Tersoff [*Phys. Rev. Lett. (USA)* vol.63 (1989) p.1164]
[8] S. Ciraci, I.P. Batra [*Phys. Rev. B (USA)* vol.38 (1988) p.1835]
[9] J.L. Martins, A. Zunger [*Phys. Rev. Lett. (USA)* vol.56 (1986) p.1400]
[10] R.D. Meade, D. Vanderbilt [in *Proc. 20th Int. Conf. on the Physics of Semiconductors*, Eds E.M. Anastassakis, J.D. Joannopoulos (World Scientific, Singapore, 1990) p.123]
[11] E. Pehlke, J. Tersoff [*Phys. Rev. Lett. (USA)* vol.67 (1991) p.465 and p.1290]

[12] M.B. Webb, F.K. Men, B.S. Swartzentraber, R. Kariotis, M.G. Lagally [*Surf. Sci. (Netherlands)* vol.242 (1991) p.23]
[13] J. Tersoff [*Phys. Rev. B (USA)* vol.45 (1992) p.8833]
[14] K.E. Khor, S. Das Sarma [*Phys. Rev. B (USA)* vol.43 (1991) p.9992]
[15] F.K. LeGoues, V.P. Kesan, S.S. Iyer, J. Tersoff, R. Tromp [*Phys. Rev. Lett. (USA)* vol.64 (1990) p.2038]

6.2 Atomic steps on growing and annealed (001) surfaces of Si and Ge

G. Theodorou and E. Karra

September 1993

A INTRODUCTION

Perfectly flat surfaces are almost impossible to produce, and the formation of steps cannot be avoided. The behaviour of steps on semiconductor surfaces is an issue of fundamental importance. From an applied viewpoint, the control of step density and meander is of great practical importance for growth of compound semiconductor overlayers and reduced-dimensional devices. We discuss here the types of step on (001) surfaces of Si and Ge (Section B) and their energies (Section C), and also present some specific cases (Section D).

B TYPES OF STEP

The Si(001) and Ge(001) surfaces reconstruct to form rows of dimerised atoms. Because of the symmetry of the diamond lattice, dimer rows are perpendicular to each other on terraces that are separated by an odd number of monatomic steps. Depending on the growth conditions and the treatment of the crystal [1], two kinds of step appear on the (001) surfaces of Si and Ge: the single-layer (S) and the double-layer (D) steps. Single-layer steps are more commonly seen on 'normal' (non-tilted) (001) surfaces, while double-layer steps are more likely to occur on tilted surfaces. Each of the above-mentioned types of step can be divided into two others, labelled by S_A, S_B and D_A, D_B respectively. Following the literature [2] the subscripts A and B denote the dimerisation direction of the upper terrace near the step, normal (A) or parallel (B) to the step edge. Terraces are also characterised as type A or B. In type A the dimerisation direction is along the step edge and in type B normal to it. FIGURE 1 shows the structure of terraces of type A and B and the steps S_A, S_B and D_B. In the formation of a single step, both types of terrace are present and the rows on alternate terraces run at right

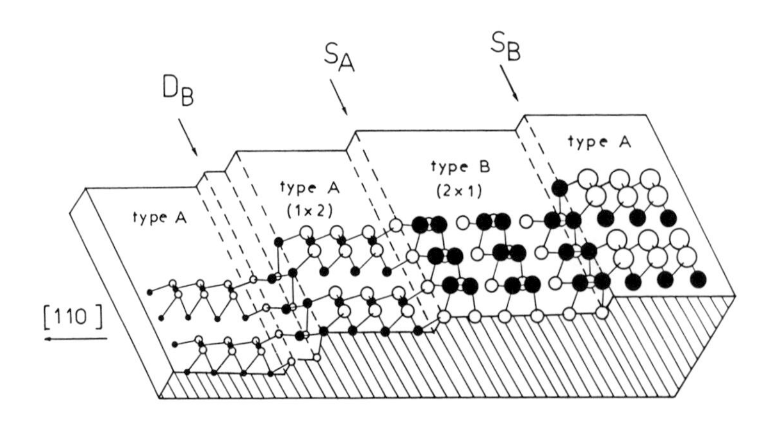

FIGURE 1 Structure of terraces of type A and B and steps of type S_A, S_B and D_B.

angles to each other (double domain surface). Steps on 'flat' surfaces as well as on surfaces tilted towards the [110] axis have been observed.

C ENERGIES

The formation energies per unit length relative to that of a fully relaxed Si(001) surface for the four types of step have been estimated to be approximately equal to [2]:

$$\lambda(S_A) \sim (0.01 \pm 0.01 \text{ eV})/a$$

$$\lambda(S_B) \sim (0.15 \pm 0.03 \text{ eV})/a$$

$$\lambda(D_A) \sim (0.54 \pm 0.10 \text{ eV})/a$$

$$\lambda(D_B) \sim (0.05 \pm 0.02 \text{ eV})/a$$

The single layer step S_A has the lowest formation energy, because it is the only step that does not lead to large strains or to extra dangling bonds, making it likely to occur on well aligned (001) surfaces. In this case each step alternately raises or lowers the terrace height by one interplanar lattice spacing, so on the average the surface remains untilted. The preference for S_A steps leads to smaller mean terrace widths along the dimerisation direction than normal to it. For tilted surfaces it is not possible to have only S_A steps. The occurrence of an S_A step makes the existence of an S_B step across some boundary between terraces unavoidable. Even though the above-mentioned estimates for the step-energies imply $\lambda(S_A) + \lambda(S_B) - \lambda(D_B) \sim 0.1$ eV/a, making the D_B steps favourable over the pair of single steps $(S_A + S_B)$, Scanning-Tunnelling-Microscopy (STM) studies of Si(001) and Ge(001) surfaces have revealed the presence of both S_A and S_B monatomic steps. More recent calculations indicate that the elastic interaction between steps makes the single steps $(S_A + S_B)$ lower in energy than the double step D_B for a tilted surface with a miscut angle, θ, smaller than a critical angle θ_c, and higher for $\theta > \theta_c$ [3-5]. Experimental results [6,7] indicate a value for the critical angle θ_c of approximately 2°. The high energy of formation of a D_A step relative to a D_B has as a result that the double steps that appear on an annealed surface are always of the D_B-type, with the surface dimerisation axis parallel to the step edge.

D SPECIFIC CASES

D1 Steps on a Si(001) Well-Oriented Surface

It has been observed that alternating surface reconstruction between 2x1 and 1x2 appears during growth on a single-domain surface [1,8-10]. In this well oriented case, both of the surface reconstructions can stably exist at the lower growth temperature range (400°C to 500°C), since the influence of the steps on the surface reconstruction in the large size terraces is considered to be weak. High temperature annealing, following a buffer layer growth, is found to be very effective in obtaining an atomically smooth surface [1,8-11]. During this annealing process a change of the surface reconstruction takes place; the originally double domain structure (2x1 + 1x2) changes to a (2x1) single-domain structure [12-14]. The annealing surface consists of large terraces and surrounding stepbands which contain an even number of monatomic layer steps [15].

D2 Steps on a Si(001) Surface Tilted by a Large Angle (4°)

It has been discussed in Section C that for vicinal surfaces with a tilt larger than a critical angle θ_c (~2°), the most stable steps are the double steps D_B. In a cut with a tilt of a few degrees towards the [110] direction, Si(001) and Ge(001) surfaces can lower their energy by forming double steps. As a result, a widely employed strategy for eliminating the single steps is the use of vicinal surfaces, with a tilt toward the (110) plane. RHEED patterns show [1] that the vicinal Si(001) surface tilted by 4° toward the [110] azimuth has the (1x2) single domain structure with ordered double-layer D_B steps, which separate the type A terraces. At a substrate temperature above 580°C, RHEED pattern [1] and STM images [16] show the occurrence of double-layer steps from the initial stages of the growth, while results for temperatures below 450°C show intensity oscillations at the initial stage that die out rapidly and the pattern changes to a streaky one, characteristic of double steps. These results indicate that for a substrate temperature below 450°C the growth at the initial stages occurs in a monolayer-by-monolayer fashion. The steps on a surface tilted by a large angle (3.5° - 7°) become sharper after a thermal treatment, consisting of flashes of 1000 - 1200°C, followed by fifteen minute anneals at 950°C. A similar thermal treatment of a 2.5°-off surface leads to a mostly double step structure [17].

D3 Steps on a Si(001) Surface Tilted by a Small Angle (0.3° - 1°)

The surface step structure of samples tilted by a small angle (0.3° - 1°) is quite different from that of the well oriented substrate, or 4°-off substrate [1,10]. In this range STM images show the existence of straight as well as kinked step edges. The straight edges correspond to S_A steps, while the kinked edges correspond to S_B steps. The S_A type steps, because of their small energy, are kinked and very active in thermal equilibrium conditions [18-22]. It is found that annealing at about 1200°C is necessary to obtain a clean surface with a regular step distribution [23,24].

E CONCLUSION

In this Datareview we have discussed the atomic steps on growing and annealed (001) surfaces of Si and Ge. The general conclusions are that the steps with the lowest formation energy are the S_A, followed by D_B and S_B, while the D_A steps practically never appear. Also the single steps can be eliminated by growing on a vicinal surface with a tilt of ~4° toward the (110) plane.

ACKNOWLEDGEMENT

We wish to acknowledge valuable discussions with Prof. J. Stoemenos. This work has been supported in part by the ESPRIT Basic Research Action No 7128.

REFERENCES

[1] T. Sakamoto, K. Sakamoto, K. Miki, H. Okumura, S. Yoshida, H. Tokumoto [in *Kinetics of Ordering and Growth at Surfaces*, Ed. M.G. Lagally (Plenum Press, New York, 1990)]

[2] D.J. Chadi [*Phys. Rev. Lett. (USA)* vol.59 (1987) p.1691]

[3] T.W. Poon, S. Yip, P.S. Ho, F.F. Abraham [*Phys. Rev. Lett. (USA)* vol.65 (1990) p.2161]

[4] E. Pehlke, J. Tersoff [*Phys. Rev. Lett. (USA)* vol.67 (1991) p.1290]

[5] O.L. Alerhand, A.N. Berker, J.D. Joannopoulos, D. Vanderbilt [*Phys. Rev. Lett. (USA)* vol.64 (1990) p.2406]

[6] J.E. Griffith, J.A. Kubby, P.E. Wierenga, G.P. Kochanski [*Mater. Res. Soc. Symp. Proc. (USA)* vol.116 (1988) p.27]

[7] X. Tong, P.A. Bennett [*Phys. Rev. Lett. (USA)* vol.67 (1991) p.101]

[8] T. Sakamoto, G. Hashiguchi [*Jpn. J. Appl. Phys. (Japan)* vol.25 (1986) p.178]

[9] T. Sakamoto, T. Kawamura, G. Hashiguchi [*Appl. Phys. Lett. (USA)* vol.48 (1986) p.1612]

[10] N. Aizaki, T. Tatsumi [*Surf. Sci. (Netherlands)* vol.174 (1986) p.658]

[11] T. Sakamoto, T. Kawamura, S. Nago, G. Hashiguchi, K. Sakamoto, K. Kuniyoshi [*J. Cryst. Growth (Netherlands)* vol.81 (1987) p.59]

[12] N. Inoue, Y. Tanishiro, K. Yagi [*Jpn. J. Appl. Phys. (Japan)* vol.26 (1987) p.L298]

[13] T. Doi, M. Ichikawa [*J. Cryst. Growth (Netherlands)* vol.95 (1989) p.468]

[14] Y. Enta, S. Suzuki, S. Kono, T. Sakamoto [*Phys. Rev. B (USA)* vol.39 (1989) p.5524]

[15] R.M. Feenstra, J.A. Stroscio [*Phys. Rev. Lett. (USA)* vol.59 (1987) p.2173]

[16] P.E. Wierenga, J.A. Kubby, J.E. Griffith [*Phys. Rev. Lett. (USA)* vol.59 (1987) p.2169]

[17] C.E. Aumann, D.E. Savage, R. Kariotis, M.G. Lagally [*J. Vac. Sci. Technol. A (USA)* vol.6 (1988) p.1963]

[18] K. Miki, H. Tokumoto, T. Sacamoto, K. Kajimura [*Jpn. J. Appl. Phys. (Japan)* vol.28 (1989) p.L1483]

[19] A.J. Hoeven, E.J. van Loenen, D. Dijkkamp, J.M. Lenssinck, J. Dieleman [*Thin Solid Films (Switzerland)* vol.183 (1989) p.263]

[20] A.J. Hoeven, J.M. Lenssink, D. Dijkkamp, E.J. van Loenen, J. Dieleman [*Phys. Rev. Lett. (USA)* vol.63 (1989) p.1830]

[21] R.J. Hamers, U.K. Kohler, J.E. Demuth [*J. Vac. Sci. Technol. A (USA)* vol.8 (1990) p.195]

[22] A.J. Hoeven, D. Dijkkamp, J.M. Lenssink, E.J. van Loenen [*J. Vac. Sci. Technol. A (USA)* vol.8 (1990) p.3657]

[23] D. Dijkkamp, A.J. Hoenen, E.J. van Loenen, J.M. Lenssinck, J. Dieleman [*Appl. Phys. Lett. (USA)* vol.56 (1990) p.39]

[24] D. Dijkkamp, E.J. van Loenen, A.J. Hoenen, J. Dieleman [*J. Vac. Sci. Technol. A (USA)* vol.8 (1990) p.218]

6.3 Segregation of Ge and dopant atoms during growth of SiGe layers

H. Jorke

January 1994

A INTRODUCTION

Surface segregation during MBE growth, which is well known from studies on dopant incorporation in Si MBE growth ([1] and references therein), has attracted attention for Ge on Si only in the last few years. The first evidence of surface segregation was found in an asymmetric intermixing phenomenon that disturbs interfacial sharpness in Si/Ge heterostructures [2,3] and in $Si/Si_{1-x}Ge_x$ heterostructures [4].

The sandwich layer setup commonly used to study surface segregation in epitaxial growth is sketched in FIGURE 1. After growth of a Si buffer layer (to generate a well-defined surface free of preparation induced defects) impurity atoms (Ge or dopant atoms) are deposited (typically in the order of one monolayer). Then, to form a Si cap layer, Si growth is continued. Surface segregation becomes manifest in a redistribution of impurity atoms into the Si cap layer (in contrast, solid-state diffusion would cause redistribution into cap and buffer layer).

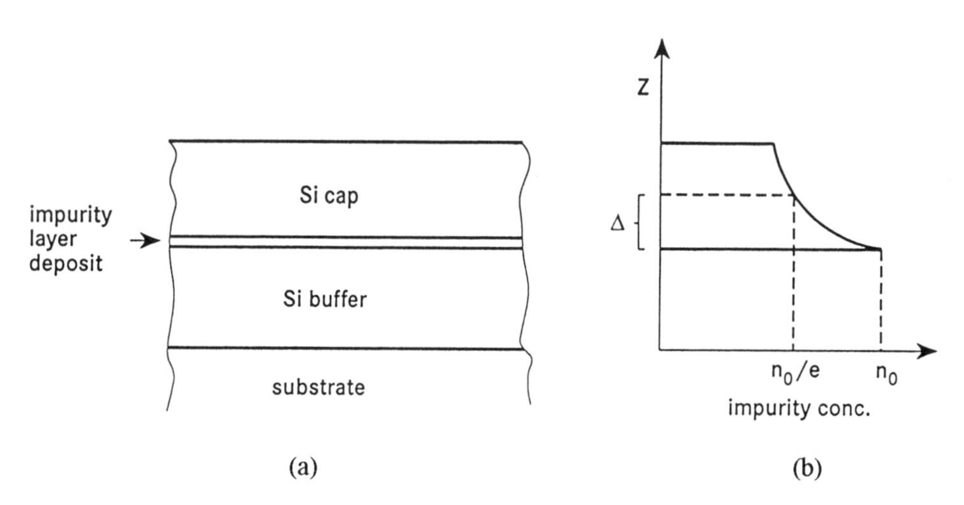

FIGURE 1 Sandwich structure commonly used to investigate impurity redistribution caused by surface segregation (a). Redistributions are quantified by the 1/e decay length Δ (b).

Quantitatively, surface segregation is described by the segregation coefficient r_s [5] that expresses the ratio of the impurity surface concentration n_s to the bulk concentration n

$$r_s = \frac{n_s}{n} \tag{1}$$

With experiments where (a) solid-state diffusion is quenched and (b) desorption of impurity atoms from the growing surface is negligible there is a second relationship existing between n and n_s

$$n_s = n_{s,o} - \int_0^z n(z')dz' \tag{2}$$

where $n_{s,o}$ is the initial impurity coverage (on the buffer layer). From Eqns (1) and (2) the integral equation

$$r_s n(z) = n_{s,o} - \int_0^z n(z')dz' \tag{3}$$

follows.

In general, r_s may not solely depend on growth parameters (temperature and cap layer growth rate) but also on the surface coverage n_s itself (self-limitation in surface segregation [6]) and, occasionally, on time [7]

$$r_s = f(T,R,n_s,t)$$

In this most universal case Eqn (3) is a complex integral equation of n(z). If (as presumably in the case of sufficiently low impurity coverages) the dependences on n_s and t can be neglected, i.e.

$$r_s = f(T,R)$$

then Eqn (3) can be easily integrated

$$n(z) = \frac{n_{s,o}}{r_s} \exp\left(-\frac{z}{r_s}\right) \tag{4}$$

Thus, in the low coverage limit (no interaction between individual impurity atoms) the impurity density decay is exponential and the segregation coefficient r_s is identical to the 1/e decay length Δ defined by (FIGURE 1(b))

$$\frac{n(\Delta)}{n(0)} = 0.368 \tag{5}$$

Conversely, deviations of the impurity profile from a simple exponential decay are indicative of impurity-impurity interactions [6,8].

Questions related to interactions between impurities of different nature are studied in the field of surfactant assisted growth [4,9-14]. In these studies Ge is deposited together with a (strongly segregating) impurity species. Results show that Ge segregation is quenched by having such a co-deposited impurity layer on the surface of the growing structure.

B Ge SURFACE SEGREGATION ON Si

Quantitative studies on Ge surface segregation on Si have used either secondary ion mass spectrometry (SIMS) to profile Ge redistributed into the Si cap layer [4,6,8,10,11,13] or X-ray photoemission spectroscopy (XPS) to determine the evolution of the Ge surface concentration during Si overlayer growth [10,15]. At a typical initial Ge coverage of a few monolayers the Ge signal (SIMS and XPS) is found to decay non-exponentially in the growth direction (FIGURE 2). Following an initially steep decay (Δ_1) the decay weakens, approaching an exponential decay with (lower) slope Δ_2. Fukatsu et al have explained this change in slope by a self-limitation mechanism that becomes operative at high Ge coverage [6]. Actually, the steep slope part disappears at Ge coverages below 0.01 monolayers [8]. TABLE 1 summarises values of the 1/e decay length for Ge on Si(100) and Si(111) as found from concentration vs. thickness plots given in [4,6,8,10,11,13] and [11,13], respectively.

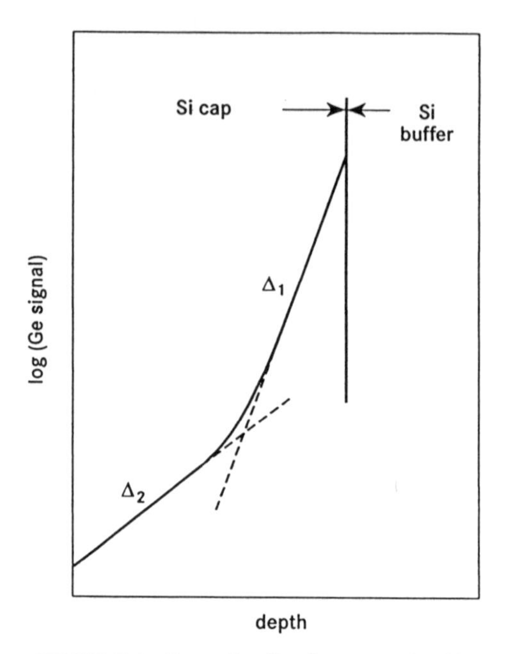

FIGURE 2 Ge redistribution onto the Si cap layer due to surface segregation. Following an initially steep decay (Δ_1) the decay flattens, approaching an exponential decay with (lower) slope Δ_2.

Regarding the slope in the heel (Δ_2) there is a reasonable conformity between various SIMS data [4,6,8] and between SIMS [4,6,9] and XPS data [15].

With regard to the slope of the initial decay, however, SIMS data show some scattering. This is because the value of Δ_1 is more sensitive than that of Δ_2 to the specific choice of SIMS process parameters (depending on the primary ion energy the original distribution is smeared by knock-on processes [10]). Therefore, a precise comparison of Δ_1 values would require use of identical SIMS process parameters.

Values of the decay length in the low-coverage limit (Δ_2) have also been calculated [6] using a two-state exchange scheme [1,16]. FIGURE 3 shows a comparative plot of modelling results

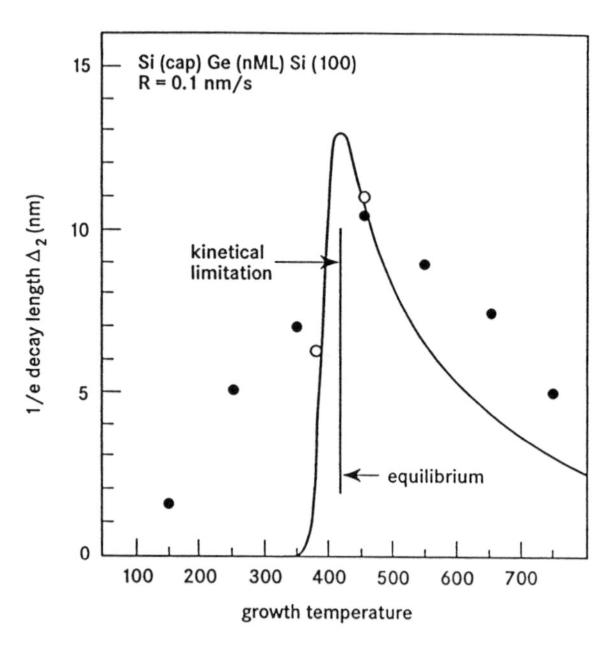

FIGURE 3 1/e decay length Δ_2 (see FIGURE 2) of Ge on Si(100). Closed symbols are from XPS measurements [15], open symbols from SIMS data [6,8].

(solid line) and experimental data (TABLE 1) for Ge on Si(100). Whereas in the high temperature (equilibrium) range data do not strongly differ, modelling grossly underestimates surface segregation in the low temperature (kinetically-limited) regime.

TABLE 1 1/e decay length Δ_1, Δ_2 (for definition see FIGURE 2) for Ge on Si with different growth parameters (T_g = growth temperature, R = cap layer growth rate) and surface orientations.

T_g (°C)	R (nm/s)	Orientation	Δ_1 (nm)	Δ_2 (nm)	Ref	Method
560	0.1	(100)	-	7.4	[4]	SIMS
470	0.1	(100)	-	7.4	[4]	SIMS
380	0.1	(100)	1.1	6.1	[6]	SIMS
450	0.1	(100)	5.1	11	[8]	SIMS
500	0.055 - 0.2	(100)	1.9	-	[10]	SIMS
500	0.055 - 0.2	-	-	4.4	[10]	XPS
300	-	(100)	0.9	-	[11]	SIMS
350	-	(100)	1.5	-	[11]	SIMS
400	-	(100)	1.9	-	[11]	SIMS
450	-	(100)	3.3	-	[11]	SIMS
500	-	(100)	2.6	-	[11]	SIMS
550	-	(100)	2.8	-	[11]	SIMS
600	-	(100)	2.3	-	[11]	SIMS
400	-	(111)	1.4	-	[11]	SIMS
500	-	(111)	2.1	-	[11]	SIMS
600	-	(111)	1.8	-	[11]	SIMS
150	0.1	(100)	-	1.6	[15]	XPS
250	0.1	(100)	-	5.0	[15]	XPS
350	0.1	(100)	-	7.0	[15]	XPS
450	0.1	(100)	-	10.5	[15]	XPS
550	0.1	(100)	-	9.0	[15]	XPS
650	0.1	(100)	-	7.5	[15]	XPS
750	0.1	(100)	-	5.0	[15]	XPS
350	0.1	(111)	-	3.0	[15]	XPS
450	0.1	(111)	-	4.0	[15]	XPS
550	0.1	(111)	-	4.5	[15]	XPS
650	0.1	(111)	-	4.8	[15]	XPS
750	0.1	(111)	-	4.3	[15]	XPS

C DOPANT SURFACE SEGREGATION ON $Si_{1-x}Ge_x$

In comparison with Ge on Si some dopants (e.g. Sb and Ga) exhibit a much stronger surface segregation tendency (under particular growth conditions) [1,17,18,22]. Use of the sandwich layer setup (FIGURE 1) allows for a reliable determination of the 1/e decay length Δ only if

$$\Delta \leq \text{cap layer thickness}$$

Dopant redistributions that fulfil this condition have been probed by (bulk sensitive) SIMS analysis [19-21], and by (surface sensitive) XPS [22] and Auger electron spectroscopy (AES) measurements [23].

In the case of decay lengths in excess of the cap layer thickness an accurate determination of the decay length is not feasible. Under this condition the dopant concentration within the cap is essentially homogeneous and, instead, surface segregation can be characterised by the segregation coefficient

$$r_s = \frac{n}{n_{s,o}}$$

with $n_{s,o}$ being the pre-deposited dopant surface concentration. n values were determined by SIMS [20] and Hall measurements [1]. In the medium range where

$$\Delta \sim \text{cap layer thickness}$$

both techniques are applicable and showed almost identical results in the case of Sb and B on Si(100) [20,7]. As mentioned before, this finding is indicative of the absence of dopant desorption and solid-state diffusion during cap layer growth.

C1 Sb Segregation on $Si_{1-x}Ge_x$

Extensive experimental studies have been made of Sb on Si(100) [1,19,21,22,24-28] and Si(111) [22,29] but little work has been done on Sb on $Si_{1-x}Ge_x$ [20] and Ge [23]. FIGURE 4 shows a plot of the decay length (segregation coefficient) vs. temperature of Sb on Si, $Si_{0.6}Ge_{0.4}$ and Ge for the (100) orientation. The 1/e decay lengths for Sb on Si in the low temperature range (T < 550°C) determined by SIMS [19] and XPS [22] do reasonably agree.

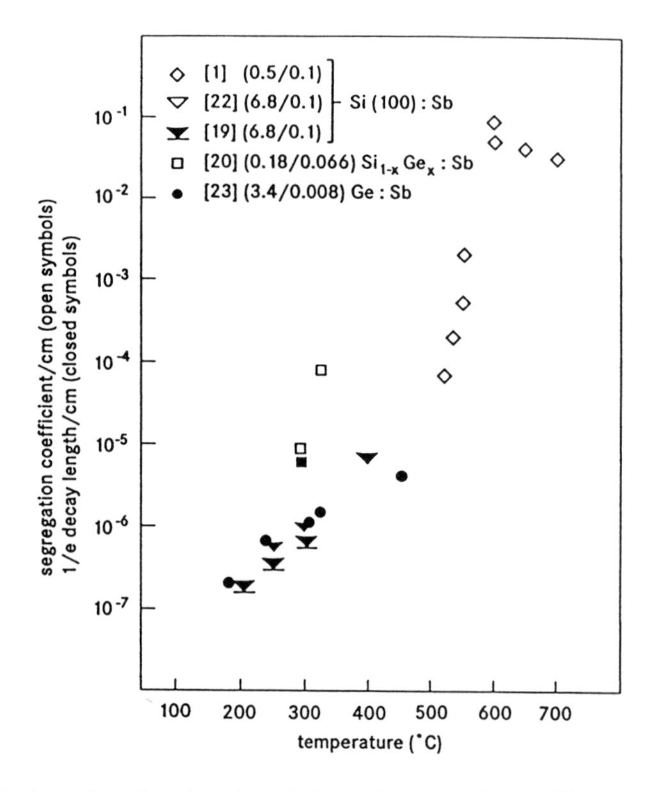

FIGURE 4 1/e decay lengths (closed symbols) and segregation coefficients (open symbols) of Sb on Si, $Si_{1-x}Ge_x$ and pure Ge vs. growth temperature in cap/Sb/buffer structures (FIGURE 1). The first number in parentheses denotes the initial Sb surface coverage (in units of 10^{14} atoms/cm^2), the second the cap layer growth rate (in units of nm/s).

Surprisingly, data for Sb on Ge(100) [23] also do not differ strongly in that temperature range. In contrast, much stronger segregation is observed for Sb on $Si_{0.6}Ge_{0.4}$. This may be partially attributed to the lower Sb coverage used. Similarly, use of a lower Sb coverage on Si (0.1 monolayer instead of 1 monolayer) also results in an increase of Δ of about one order of magnitude. This behaviour resembles that of Ge on Si(100) (see Section B). In may, thus, also be caused by a self-limitation mechanism [21]. This explanation is supported by the appearance of a two-slope decay in Sb SIMS profiles where the initial steep decay is the less pronounced the lower the initial Sb coverage [19,20]. Nevertheless, if the phenomenon of a coverage dependent segregation coefficient is taken into account, the surface segregation tendency of Sb on $Si_{0.6}Ge_{0.4}$ appears to be more distinct than that of Sb on Si and Ge.

C2 B Segregation on $Si_{1-x}Ge_x$

Despite the difficulties in evaporating elemental boron [30] it is the preferred p-type dopant in Si MBE growth. This is due to the favourable incorporation behaviour compared to that of other p-type dopants. However, some surface segregation effects during growth also exist with boron. TABLE 2 lists recent publications that address questions of boron segregation in Si and $Si_{1-x}Ge_x$ MBE growth.

TABLE 2 Publications that deal with surface segregation of boron on Si(100), Si(111) and $Si_{1-x}Ge_x$ layers.

System	Authors	Ref	Method
Si(100):B	Tatsumi	[31]	AES
Si(100):B	Mattey et al	[32]	SIMS
Si(100):B	Jorke, Kibbel	[7]	SIMS
Si(100):B	Weir et al	[33]	X-ray diffraction
Si(100):B	Parry et al	[34]	SIMS
Si(100):B	Parry et al	[35]	SIMS
Si(100):B	Sardela et al	[36]	AES
Si(100):B	Sardela et al	[37]	AES
Si(100):B	Sardela et al	[38]	AES
Si(100):B	Murakami et al	[39]	AES
Si(111):B	Fresart et al	[40]	AES
Si(111):B	Sardela et al	[36]	AES
$Si_{1-x}Ge_x$(100):B	Jorke, Kibbel	[7]	SIMS
$Si_{1-x}Ge_x$(100):B	Parry et al	[34]	SIMS
$Si_{1-x}Ge_x$(100):B	Sardela et al	[37]	AES

Results on studies of B segregation on (100) oriented surfaces that have used a sandwich structure (FIGURE 1) are shown in FIGURE 5. Scatter in data is largely due to differing process parameters (cap layer growth rate, B coverage) used in different studies. Obviously - as with Ge and Sb on Si (see Section B) - a transition exists around 600°C that separates a high temperature regime, with decay lengths of a few tenths of nm, from a low temperature range with significantly reduced decay lengths.

Regarding incorporation of B into $Si_{1-x}Ge_x$ layers, experiments with $x = 0.2$ showed consistently lower Δ values compared to that of pure Si [7,34]. With $x = 0.1$, however, data are conflicting. Whereas in the higher temperature regime ($T > 600$°C) decay lengths are found to be reduced also in $Si_{0.9}Ge_{0.1}$ layers [34] the situation at $T < 600$°C is not clear: Parry

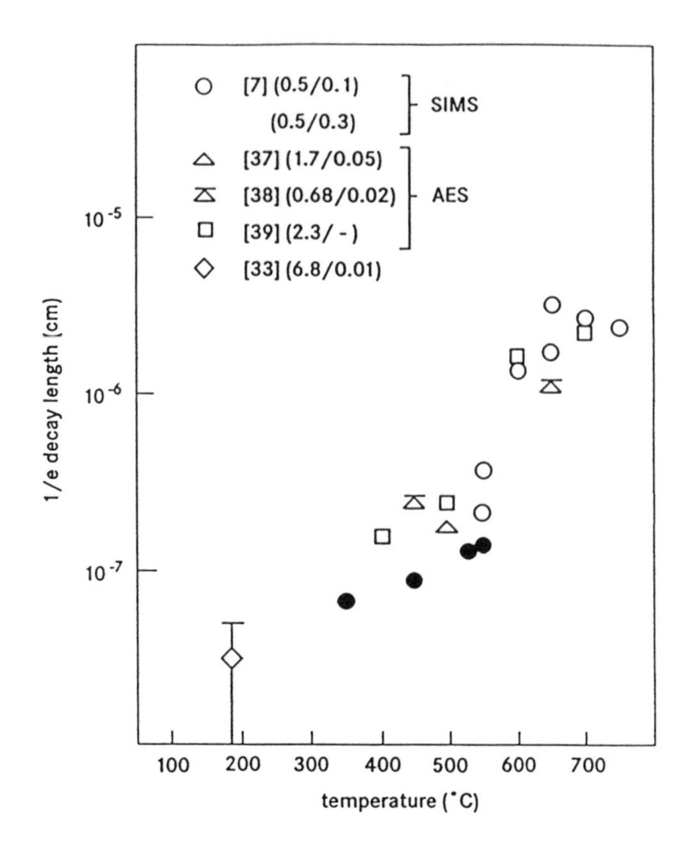

FIGURE 5 1/e decay length of B on Si(100) (open symbols) and $Si_{0.8}Ge_{0.2}(100)$ (closed symbols). The first number in parentheses denotes the initial B surface coverage (in units of 10^{14} atoms/cm^2), the second the cap layer growth rate (in units of nm/s).

et al [34] reported an increase of Δ compared to Δ of Si. In contrast, Sardela et al [37] observed a decrease. The latter group also investigated segregation of Ge during $Si_{1-x}Ge_x$ layer growth. Their AES data for B and Ge indicate that both segregation effects are related, and the Ge segregation reduces B segregation in comparison with the case of doping in pure silicon.

D SURFACTANT-ASSISTED GROWTH

The presence of a strongly segregating species at the surface of growing Si/Ge heterostructures influences the growth process in two ways. First, by virtue of a surfactant (surface-active species) island formation is suppressed (without a surfactant Ge grows on Si in the Stransky-Krastanov mode: a few monolayers grow layer-by-layer followed by islanding growth). This phenomenon was first described by Copel et al [9] using As as a surfactant. Secondly, intermixing phenomena due to surface segregation are also effectively suppressed with surfactant-assisted growth. The first evidence of the latter effect was found by Zalm et al [4] using Ga as a surfactant.

Subsequent publications have dealt also with other surfactants such as Sb [8,10,41-44], Bi [13,45], Sn [12], Te [46] and H [14]. In the following, results on Ge surface segregation in surfactant-assisted Si/Ge heterostructure growth are briefly summarised for some species.

D1 Antimony

Fujita et al [10] demonstrated that Ge surface segregation is suppressed by depositing Sb atoms on Ge layers prior to Si overgrowth and that Ge layers are confined to within 0.8 nm. The lower limit of the amount of Sb needed to realise sharp Si/Ge interfaces was found to be 0.75 monolayers [8]. In a comparative study at $T_g = 400°C$ the 1/e decay length of Ge was 5.6 monolayers (= 0.76 nm) without Sb and 3.8 monolayers (= 0.52 nm) with Sb [43]. In a microscopic study using high resolution photoemission [44] Sb on Si(100) and Ge(100) was found to fully saturate surface dangling bonds. During growth Sb atoms form a new ordered layer while leaving the uniform epitaxial Si/Ge layer behind [44].

D2 Bismuth

In a comparative investigation using Sb and Bi, Sakamoto et al [13] showed that both species produce abrupt Si/Ge/Si(100) interfaces. They found that at a growth temperature of $T_g = 400°C$ a significant amount of Sb is incorporated. On the other hand, the amount of Bi incorporated was found to be smaller than the detection limit of SIMS ($< 5 \times 10^{16}$ cm^{-3}).

D3 Gallium

The presence of about one monolayer of Ga on the surface at $T_g = 560°C$ (that yields a doping level of around 10^{18} cm^{-3}) was found to suppress Ge segregation in a $Si/Si_{1-x}Ge_x/Si$ heterostructure [4]. In a subsequent study using Sb and Ga Fukatsu et al [11] showed that surfactant-assisted growth improves the Si/Ge interface decay length down to 0.6 nm. Results showed also that surfactant-assisted growth using Sb and Ga is most effective at 500°C and 400°C growth temperature, respectively [11].

D4 Tin

In a comparative investigation using Sb and Sn Dondl et al [12] showed that Sn is more strongly segregating than Sb. This feature would favour tin as it is incorporated to a lesser extent during growth than antimony. However, due to an unwanted islanding tendency tin is not usable for surfactant-assisted growth [12].

D5 Hydrogen

The presence of surface hydrogen is an inherent feature of gas source MBE and related methods. Ohtani et al [14] have investigated Ge surface segregation during gas source MBE growth and have found that Ge surface segregation can be minimised below 600°C. Comparison with solid source growth results suggests that hydrogen significantly suppresses the Ge segregation.

E CONCLUSION

Surface segregation during Si MBE growth, known from studies on dopant incorporation, is observed also for Ge on Si. The phenomenon degrades the interfacial sharpness in Si/Ge and $Si/Si_{1-x}Ge_x$ heterostructures and superlattices. At high Ge coverages surface segregation becomes self-limited as surface Ge prevents sub-surface Ge from being segregated to the surface.

Interactions between impurities of different nature are studied in the field of surfactant-assisted growth where Ge is deposited together with a (strongly segregating) impurity species (Sb, Ga, Bi, Sn, H). Quenching of Ge segregation is observed with Sb and Ga. Unfortunately, a significant amount of these surfactants is incorporated (at least in the low temperature growth regime). Low surfactant incorporation but still strong quenching is observed with Bi. Experiments with hydrogen also show promising results. Sn was found to be unusable for surfactant-assisted growth.

Doping of $Si_{1-x}Ge_x$ with Sb (n-type) and B (p-type) shows that (i) the surface segregation tendency of Sb on $Si_{1-x}Ge_x$ ($x = 0.4$) appears to be more distinct than that of Sb on Si and Ge and (ii) Ge segregation during growth of $Si_{1-x}Ge_x$ layers reduces B segregation as compared to the case of B doping in pure silicon.

REFERENCES

[1] H. Jorke [*Surf. Sci. (Netherlands)* vol.193 (1988) p.569-78]

[2] K. Eberl, G. Krötz, R. Zachai, G. Abstreiter [*J. Phys. Colloq. (France)* vol.48 (1987) p.329]

[3] S.S. Iyer, J.C. Tsang, M.W. Copel, P.R. Pukite, R.M. Tromp [*Appl. Phys. Lett. (USA)* vol.54 (1989) p.219-21]

[4] P.C. Zalm, G.F.A. van de Walle, D.J. Gravesteijn, A.A. van Gorkum [*Appl. Phys. Lett. (USA)* vol.55 (1989) p.2520-2]

[5] S.A. Barnett, J.E. Greene [*Surf. Sci. (Netherlands)* vol.151 (1985) p.67]

[6] S. Fukatsu, K. Fujita, H. Yaguchi, Y. Shiraki, R. Ito [*Appl. Phys. Lett. (USA)* vol.59 (1991) p.2103-5]

[7] H. Jorke, H. Kibbel [*Appl. Phys. Lett. (USA)* vol.57 (1990) p.1763-5]

[8] S. Fukatsu, K. Fujita, H. Yaguchi, Y. Shiraki, R. Ito [*Mater. Res. Soc. Symp. Proc. (USA)* vol.220 (1991) p.217-22]

[9] M. Copel, M.C. Reuter, E. Kaxiras, R.M. Tromp [*Phys. Rev. Lett. (USA)* vol.63 (1989) p.632-5]

[10] K. Fujita, S. Fukatsu, H. Yaguchi, Y. Shiraki, R. Ito [*Jpn. J. Appl. Phys. (Japan)* vol.29 (1990) p.L1981-3]

[11] S. Fukatsu, N. Usami, K. Fujita, H. Yaguchi, Y. Shiraki, R. Ito [*J. Cryst. Growth (Netherlands)* vol.127 (1993) p.401-5]

[12] W. Dondl, G. Lütjering, W. Wegscheider, J. Wilhelm, R. Schorer, G. Abstreiter [*J. Cryst. Growth (Netherlands)* vol.127 (1993) p.440-2]

[13] K. Sakamoto, H. Matsuhata, K. Kyoya, K. Miki, T. Sakamoto [*Int. Conf. Solid State Devices and Materials (SSDM '93)*, Chiba, Japan, 29 Aug. - 1 Sept. 1993, Extended Abstracts p.246-8]

[14] N. Ohtani, S.M. Mokler, H.H. Xie, J. Zhang, B.A. Joyce [ibid. [13] p.249-51]

[15] K. Nakagawa, M. Miyao [*J. Appl. Phys. (USA)* vol.69 (1991) p.3058-62]

[16] J.J. Harris, D.E. Ashenford, C.T. Foxon, P.J. Dobson, B.A. Joyce [*Appl. Phys. A (Germany)* vol.33 (1984) p.87-92]

[17] S.S. Iyer, R.A. Metzger, F.G. Allen [*J. Appl. Phys. (USA)* vol.52 (1981) p.5608-13]

[18] F. Schäffler, H. Jorke [*Thin Solid Films (Switzerland)* vol.184 (1990) p.75-83]

[19] H. Jorke, H. Kibbel, F. Schäffler, A. Casel, H.-J. Herzog, E. Kasper [*Appl. Phys. Lett. (USA)* vol.54 (1989) p.819-21]

[20] H. Presting, H. Kibbel, E. Kasper, H. Jorke [*J. Appl. Phys. (USA)* vol.68 (1990) p.5653-9]

[21] K.D. Hobart, D.J. Godbey, P.E. Thompson, D.S. Simons [*Appl. Phys. Lett. (USA)* vol.63 (1993) p.1381-3]

[22] K. Nakagawa, M. Miyao [*Thin Solid Films (Switzerland)* vol.183 (1989) p.315-22]

[23] J. Wilhelm, W. Wegscheider, G. Abstreiter [*Surf. Sci. (Netherlands)* vol.267 (1992) p.90-3]

[24] U. König, H. Kibbel, E. Kasper [*J. Vac. Sci. Technol. (USA)* vol.16 (1979) p.985-9]

[25] U. König, E. Kasper, H.-J. Herzog [*J. Cryst. Growth (Netherlands)* vol.52 (1981) p.151]

[26] J.C. Bean [*Appl. Phys. Lett. (USA)* vol.33 (1978) p.654-6]

[27] M. Tabe, K. Kajiyama [*Jpn. J. Appl. Phys. (Japan)* vol.22 (1983) p.423]

[28] W.-X. Ni et al [*Phys. Rev. B (USA)* vol.40 (1989) p.10449]

[29] R.A. Metzger, F.G. Allen [*J. Appl. Phys. (USA)* vol.55 (1984) p.931-40]

[30] In comparison with other p-type dopants (Al, Ga) the vapour pressure of elemental boron is rather low and, hence, requires relatively high evaporation cell temperatures. This difficulty was first overcome by Kubiak et al (R.A.A. Kubiak, W.Y. Leong, E.H.C. Parker [*Appl. Phys. Lett. (USA)* vol.44 (1984) p.878-80]).

[31] T. Tatsumi [*Thin Solid Films (Switzerland)* vol.184 (1990) p.1]

[32] N.L. Mattey et al [*Thin Solid Films (Switzerland)* vol.184 (1990) p.15]

[33] B.E. Weir et al [*Phys. Rev. B (USA)* vol.46 (1992) p.12861-4]

[34] C.P. Parry, R.A.A. Kubiak, S.M. Newstead, T.E. Whall, E.H.C. Parker [*Mater. Res. Soc. Symp. Proc. (USA)* vol.220 (1991) p.79-84]

[35] C.P. Parry, R.A.A. Kubiak, S.M. Newstead, E.H.C. Parker, T.E. Whall [*Mater. Res. Soc. Symp. Proc. (USA)* vol.220 (1991) p.103-8]

[36] M.R. Sardela Jr., W.-X. Ni, J.O. Ekberg, J.-E. Sundgren, G.V. Hansson [*Mater. Res. Soc. Symp. Proc. (USA)* vol.220 (1991) p.109-14]

[37] M.R. Sardela Jr., W.-X. Ni, H. Radpisheh, G.V. Hansson [*Thin Solid Films (Switzerland)* vol.222 (1992) p.42-5]

[38] M.R. Sardela Jr., H. Radpisheh, L. Hultman, G.V. Hansson [ibid. [13] p.222-4]

[39] E. Murakami, H. Kujirai, S. Kimura [ibid. [13] p.216-8]

[40] E. de Fresart, K.L. Wang, S.S. Rhee [*Appl. Phys. Lett. (USA)* vol.53 (1988) p.48-50]

[41] M. Copel, M.C. Reuter, M. Horn von Hoegen, R.M. Tromp [*Phys. Rev. B (USA)* vol.42 (1990) p.11682-9]

[42] H.J. Osten, J. Klatt, G. Lippert, E. Bugiel, S. Hinrich [*Appl. Phys. Lett. (USA)* vol.60 (1992) p.2522-4]

[43] K. Sakamoto et al [*Thin Solid Films (Switzerland)* vol.222 (1992) p.112-5]

[44] R. Cao, X. Yang, J. Terry, P. Pianetta [*Appl. Phys. Lett. (USA)* vol.61 (1992) p.2347-9]

[45] A. Kawano, I. Konomi, H. Azuma, T. Hioki, S. Noda [*Appl. Phys. Lett. (USA)* vol.74 (1993) p.4265-7]

[46] S. Higuchi, Y. Nakanishi [*Surf. Sci. (Netherlands)* vol.254 (1991) p.L465]

CHAPTER 7

SOME DEVICE RELATED STRUCTURES ON Si SUBSTRATES. A COLLECTION OF THE MOST IMPORTANT DATA

7.1 Si/SiGe/Si heterojunction bipolar transistors

J.C. Sturm

June 1994

A INTRODUCTION

In this Datareview, the properties of $Si/Si_{1-x}Ge_x/Si$ Heterojunction Bipolar Transistors (HBTs) grown commensurately (strained) on silicon(100) substrates will be reviewed. These devices were first reported in 1988 [1-6] and have now advanced to the point where speeds are in excess of 110 GHz [7]. Although both n-p-n and p-n-p [8,9] devices have been reported, nearly all of the technological and scientific interest is currently directed at n-p-n devices, and only n-p-n devices will be addressed here (except for the table in Section D), with a focus on room temperature performance. The Datareview will be divided into three parts: a review of the device principle, followed by DC properties and then AC properties.

B DEVICE PRINCIPLE

As in all n-p-n bipolar transistors, the output current (collector current I_C) relies on injection of electrons from the n-type emitter region into a p-type base by a forward biased base-emitter junction. In the base, the minority carrier electrons move by a combination of diffusion and drift to the collector-base junction, which is usually zero- or reverse-biased, where they are swept into the n-type collection region by the high field of the junction. The base current (I_B) has two main sources: (1) holes which are injected from the base to emitter at the forward biased base-emitter junction, and (2) recombination of some of the excess minority carrier electrons in the base with holes. In modern devices the minority carrier lifetimes are high and the base transit times are relatively short, so that in nearly all cases the first of these two sources dominates. The ratio of the collector current to base current (I_C/I_B) is known as the current gain β. $Si/Si_{1-x}Ge_x/Si$ HBTs (i.e. Si emitter and collector, and $Si_{1-x}Ge_x$ base) have two main advantages over their all-Si homojunction counterparts. First, the narrow bandgap in the base leads to a much higher collector current for the same V_{BE} (base-emitter bias) because of the lower conduction band barrier in the narrow bandgap $Si_{1-x}Ge_x$. Because of the small conduction band offset, to first order the barrier for holes being injected back into the emitter (the dominant source of I_B) is unchanged, so that the emitter efficiency and hence current gain $\beta = I_C/I_B$ is greatly enhanced. This in itself is of marginal benefit for the performance of most circuits, but one can trade this increased gain for a higher doping level (N_A) in the base. A higher base doping is desirable for circuits because it leads to a lower intrinsic base sheet resistance $R_{B,i}$. If the intrinsic base resistance is a considerable fraction of the total base resistance R_B (which also includes components due to the extrinsic region of the device, contacts, etc.), a lower $R_{B,i}$ will then lead to reduced R_BC delay times in circuits and reduced noise in microwave applications. Higher base doping will also prevent punchthrough, which limits the maximum collector voltage in some narrow-base devices, causes an increased Early voltage (output resistance), and leads to higher current-handling capability. The second advantage of $Si/Si_{1-x}Ge_x/Si$ HBTs is that the bandgap may be graded across the base (narrower at the collector) so that a 'built-in' field is provided to enhance the speed at which the minority carrier electrons cross the base. The base transit time, which is a fundamental delay inside a bipolar transistor, is therefore reduced, leading to a higher intrinsic vertical speed of electron transport across the device. Having a smaller bandgap on the collector side of the base compared to that on the emitter side also dramatically increases the Early voltage, even if the base is not heavily doped [10].

C DC PROPERTIES

For DC properties, one in principle can characterise both I_C and I_B for various structures. However, as explained above, the hole current which comprises I_B should not be affected by a $Si_{1-x}Ge_x$ base, and I_B is predominantly dependent on the emitter structure, unrelated to the base. Therefore it will not be discussed further in this Datareview. Furthermore, since I_B depends on factors other than the $Si_{1-x}Ge_x$ such as the emitter technology, parasitic B-E space-charge-region recombination, etc., the dependence of I_C on the Ge fraction x etc. will be documented in this Datareview, instead of that of β.

Before the onset of detrimental effects at high current levels (such as high level injection in the base, space-charge effects in the base-collector depletion region, and series resistance), the collector current in a bipolar transistor (including HBTs) follows the relationship

$$I_C = I_{CO}\, e^{qV_{BE}/nkT} \tag{1}$$

where q is the electron charge, V_{BE} is the base-emitter bias, k is Boltzmann's constant, T is the temperature and n is an ideality factor. For I_C, n is equal to unity except for possible small deviations in extreme conditions in HBTs due to second order effects [11,12] such as modulation of a parasitic barrier by the BE bias, which are beyond the scope of this review. In a homojunction device with a constant base doping N_A, I_{CO} can be expressed as

$$I_{CO} = \frac{A\, q\, n_i^2\, D_n}{N_A\, W_B} \tag{2}$$

where A is the area of the base-emitter junction, n_i is the intrinsic carrier concentration in the base, W_B is the neutral base width and D_n is the electron diffusion coefficient in the base. In a $Si/Si_{1-x}Ge_x/Si$ HBT with the same neutral base width and base doping as a homojunction device (BJT), the collector current (and thus I_{CO}) will be enhanced over that in a homojunction device by the following ratio (for a uniform Ge fraction in the base):

$$I_{C,HBT}/I_{C,BJT} = \frac{N_{C,SiGe}\, N_{V,SiGe}\, D_{n,SiGe}}{N_{C,Si}\, N_{V,Si}\, D_{n,Si}}\, e^{\Delta E_V/kT} \tag{3}$$

where $N_{C,i}$ and $N_{V,i}$ refer to the effective density of conduction and valence band states, respectively, in material i. Depending on the grading of the interface between the Si and the $Si_{1-x}Ge_x$, it can be argued that the valence band offset ΔE_V should be replaced by the full bandgap offset ΔE_g [13]. For strained $Si_{1-x}Ge_x$ on Si(100) substrates, however, ΔE_C is near zero for $x < 0.7$ [14], so that this distinction is not important for n-p-n HBTs. Therefore, for the rest of this Datareview, $E_{g,eff}$ and $\Delta E_{g,eff}$, which refer to the effective bandgap of the HBT and the effective reduction from that of an all-Si device, respectively, will be used instead of

ΔE_V to model the collector current enhancement. Bandgap narrowing effects due to heavy doping will also be included in $E_{g,eff}$. The term is referred to as 'effective', as opposed to the true bandgap, because its goal is to describe the minority carrier density and collector current. In the case of degenerate doping conditions in the base, band filling and Fermi-Dirac statistics effects will cause $E_{g,eff}$ to differ from the true bandgap [15]. For I_{CO} in an HBT one can then write

$$I_{CO,HBT} = \frac{A\,q\,N_{C,SiGe}\,N_{V,SiGe}\,D_{n,SiGe}}{N_A\,W_B}\,e^{-E_{g,eff}/kT} \tag{4}$$

$\Delta E_{g,eff}$ and collector current enhancement are thus a true measure of narrow bandgap base HBT effects. However, it is well established that if there is excess diffusion of boron from the heavily-doped base into the emitter and collector regions, parasitic barriers will form in the conduction band which will substantially reduce the collector current enhancement, reduce $\Delta E_{g,eff}$ and increase the base transit time [11,16-19]. Diffusion lengths as small as several nm can have substantial effects, and implantation into the overlying single crystal emitter has been found to substantially enhance the diffusion of boron in the base in later anneals [18,20].

There are two approaches to measurement of $E_{g,eff}$. The first is from the temperature dependence of the collector current, where $E_{g,eff}$ may be extracted using Eqn (4). However, it is known that the electron mobility in p-type Si (and hence D_n) is a strong function of temperature [21]. To avoid the effect of temperature in the prefactor, usually a similar temperature dependence for D_n in Si and SiGe is assumed, and $I_{CO,HBT}$ is then ratioed to that in an all-Si device (Eqn (3)), so that $\Delta E_{g,eff}$ is measured instead of $E_{g,eff}$.

The first reports of $Si/Si_{1-x}Ge_x/Si$ HBTs in 1988 contained only claims of increased current gain, not an explicit measurement of I_C enhancement [1,2]. The first explicit measure of I_C enhancement in a SiGe HBT was made by Patton et al [3], but the enhancement was far less than expected and these results have since been discounted as due to some unknown parasitic device problem [22], such as strain relaxation or parasitic barriers. $\Delta E_{g,eff}$ was then characterised by temperature dependence of I_C by several groups in more well-behaved devices [18,19,23-27]. This approach has the advantage that no information about N_A, W_B, A or D_n is required, the only assumption being that the prefactor ratio in Eqn (3) is independent of temperature. This assumption introduces a possible source of error: for example, if this prefactor ratio changes by a factor of two over a fitting range of 300 to 400 K, an error in $\Delta E_{G,eff}$ of ~75 meV will be introduced. The available data is summarised in FIGURE 1, and a best fit to $\Delta E_{g,eff}$ is found to be

$$\Delta E_{g,eff} = 0.74x \text{ meV} \tag{5}$$

When the differing base dopings and widths of the HBTs and all-Si control transistors are taken into account, the prefactor ratio of Eqn (3) is found to be ~0.4 [18], consistent with the expected difference in the effective densities of states for $x > 0.15$ [16]. No further experimental information about density of states or diffusion constant is available, however. The transistors used in the data of FIGURE 1 had dopings varying between 5×10^{17} to 7×10^{18} cm^{-3}, differing in some cases even between the Si control device and the HBTs by over a factor of ten [25], so that there is some uncertainty in the numbers due to bandgap narrowing effects due to heavy doping.

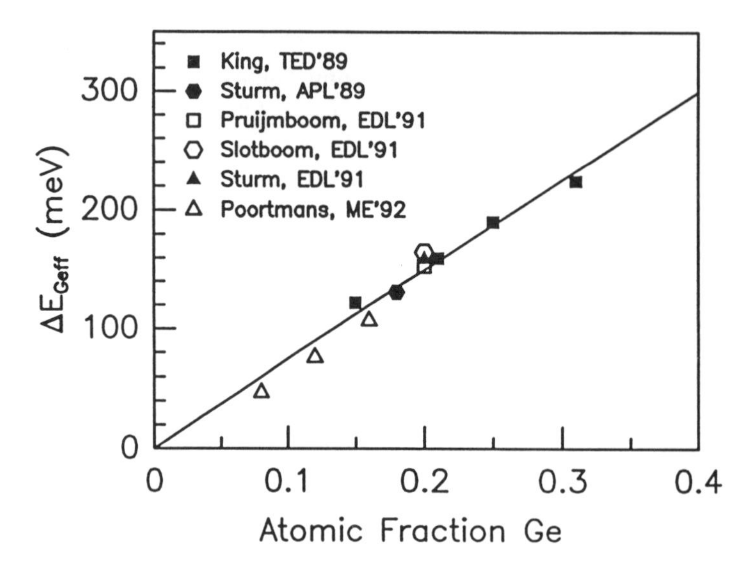

FIGURE 1 Effective bandgap reduction in Si/SiGe/Si HBTs compared to all-Si devices vs. Ge fraction as measured by the temperature dependence of I_C compared to that of an all-Si device. (The base doping is under 7 x 10^{18} cm^{-3} in all cases.) The fitted line is $\Delta E_{G,eff} = 0.74x$ meV. References for data are [18,19,24-27].

A second method for extraction of $\Delta E_{g,eff}$ is to make assumptions for $N_{C,SiGe}$, $N_{V,SiGe}$ and $D_{n,SiGe}$, and if N_A and W_B are known, $E_{g,eff}$ can be extracted directly by applying Eqn (4) to the measured I_{CO} at some temperature (e.g. room temperature). This has been done directly in the work of [28] and sufficient data is presented so that points can also be extracted from [18,20,24,25,29,30]. The model of [16] has been assumed for the ratio of N_V N_C in SiGe to that in Si, the data for electron minority carrier mobility in Si vs. doping [15] was assumed for SiGe, and an n_i^2 of 1.0 x 10^{20} cm^{-3} in Si (295 K) was assumed to extract $\Delta E_{g,eff}$. $\Delta E_{g,eff}$ is plotted vs. Ge fraction for various doping levels in FIGURE 2. General trends of a reduced

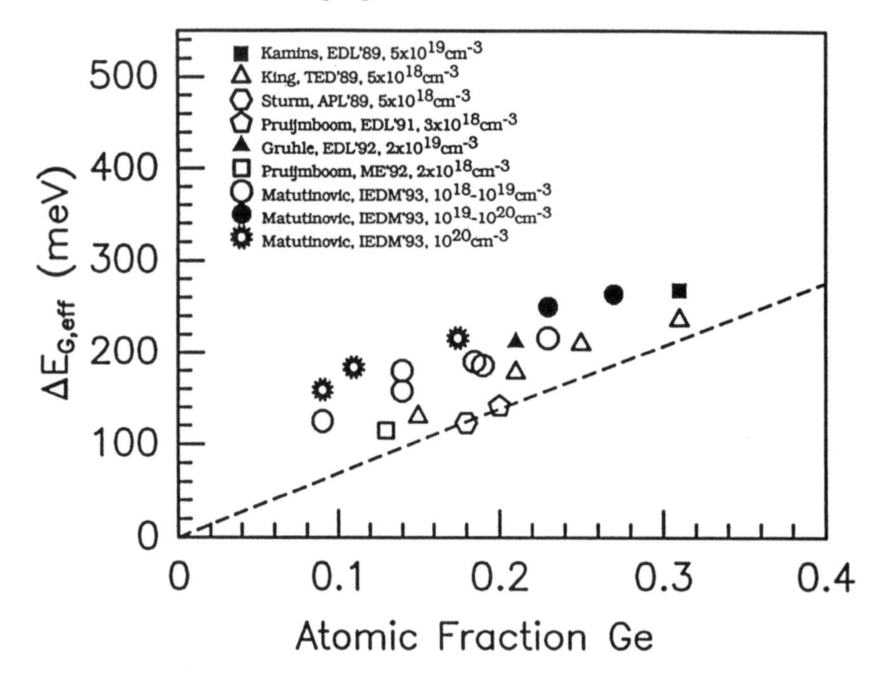

FIGURE 2 Effective bandgap reduction in HBTs vs. Ge fraction as extracted by Eqn (4) as described in the text at room temperature (295 K). The numbers next to the legends refer to the base doping. The dotted line shows the Ge dependence for similar doping from the same source of 0.69x meV. References are [18,20,24,25,28-30].

$\Delta E_{g,eff}$ at increasing Ge levels and high dopings are seen. If one defines $\Delta E_{g,eff}$ as due to the sum of both independent Ge and heavy doping components ($\Delta E_{g,Ge}$ and $\Delta E_{g,dop}$, respectively), one finds by comparing $\Delta E_{g,Ge}$ at similar doping levels from the same source that $\Delta E_{g,Ge}$ is relatively independent of doping for this method of extracting $\Delta E_{g,eff}$ and is equal to 0.69x, as indicated by the dotted line in FIGURE 2. Plotting the remaining $\Delta E_{g,eff,dop}$ vs. doping (FIGURE 3) shows that the dependence of $\Delta E_{g,eff,dop}$ is to first order independent of Ge, and a best straight line has been made, resulting in the following empirical model, valid for base dopings over 10^{18} cm^{-3}:

$$\Delta E_{g,eff}/\text{meV} = 30 + 0.69x + 23 \log_{10}(N_A \text{ cm}^3/10^{18}) \quad (6)$$

Note that a difference in $\Delta E_{g,eff}$ of ~30 meV between base doping levels of 5 x 10^{17} and 5 x 10^{18} cm^{-3} from x = 0.08 to 0.16 has been observed by temperature dependent measurements, in reasonable agreement with Eqn (6) [27]. There is still considerable variation in the points in FIGURE 3. While some is certainly due to inaccuracies in the specified device parameters (doping, Ge levels, etc.), it is interesting to note that nearly all of the points below the fitted line were from processes that involved implantation into single crystal emitters followed by annealing. This is now known to lead to enhanced base diffusion [18,20] which could lead to parasitic barriers at the BE and/or BC junctions, reducing $\Delta E_{g,eff}$. (Because of this possibility, the data of [18,24,25,29] was not included in the fit of Eqn (6).) Regarding the two methods described for extracting $\Delta E_{g,eff}$, this latter absolute fitting method should be more accurate than the method using temperature dependence for modelling room temperature collector current, but the temperature dependence method should be more accurate for scaling I_{CO} to various temperatures.

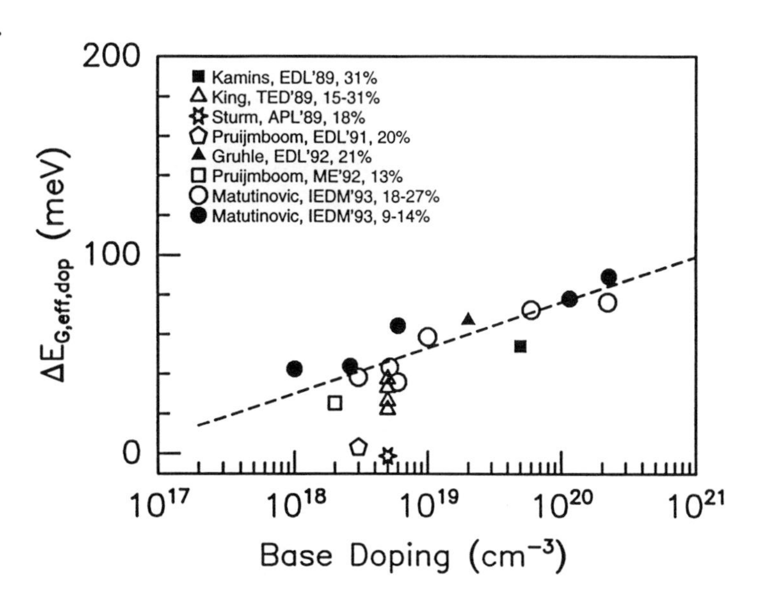

FIGURE 3 Apparent bandgap narrowing due to heavy base doping ($\Delta E_{g,eff,dop}$) after subtracting the Ge dependence from the data of FIGURE 2. References are [18,20,24,25,28-30].

To fully calculate the trade-off of current gain vs. intrinsic base resistance for various amounts of Ge, one would have to know the vertical D_n, N_V and N_C as well as the lateral hole drift (not Hall) mobility for base resistance, and heavy doping effects in $Si_{1-x}Ge_x$. Although some initial modelling and measurement of these densities of states and lateral drift mobilities has been done [16-18,28,31-34], there is not yet any systematic data on these parameters. However, a direct experimental measurement and a model of the relative collector current, e.g. I_{CO} (and hence current gain), vs. intrinsic base sheet resistance for flat-base devices (50 nm neutral base width) with various dopings and Ge levels in the base has been reported [28]. The data and accompanying model results are reproduced in FIGURE 4. The devices were designed to avoid parasitic barrier effects (spacers, no emitter implants etc.), and hence FIGURE 4 fully represents the base resistance/ current gain trade-off in flat-base HBTs. One indeed sees an increase in I_C as Ge is added to the base and as the base resistance is increased. The model is similar to Eqns (4) and (6) and includes the effects of heavy base doping on both the hole and electron mobilities and the bandgap, and good agreement with experimental data is achieved. In devices with varying amounts of Ge and doping across the base (e.g. 'grading'), one can in principle calculate the collector current enhancement [35,26]. This requires knowledge of all densities of states, diffusion coefficients, heavy doping effects, and if desired, velocity saturation [36]. Such formulae may not be applicable if parasitic barriers due to outdiffusion exist in depletion regions, however [37]. Experimental data is available in a form similar to that of FIGURE 4, but only in the case of a single linearly graded Ge profile [38].

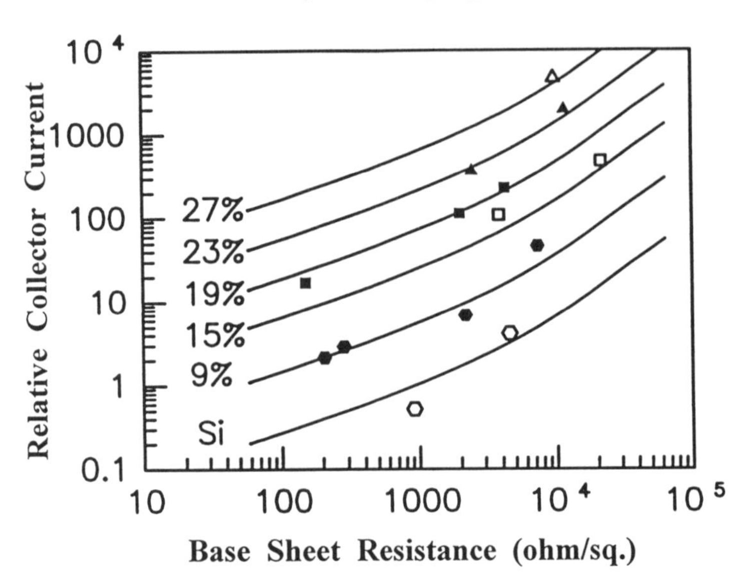

FIGURE 4 Data and model of relative collector current vs. intrinsic base sheet resistance ($R_{B,i}$) in $Si_{1-x}Ge_x$/Si HBTs with flat base and doping profiles at 295 K [28]. The neutral base thickness is 50 nm, and a relative collector current of 1 for an all-Si device of 1000 Ω/square is defined.

D AC PROPERTIES

There are several figures of merit for AC performance of HBTs. The simplest is the unity gain cut-off frequency (f_T), which is an intrinsic measure of the speed of the transport of electrons in the vertical direction through the intrinsic device. At high frequencies, various delays and charging effects have the combined effects of reducing the AC collector current and increasing the AC base current for a fixed AC V_{BE} bias modulation, and hence reducing the small signal current gain from its low frequency value. f_T is defined as the frequency at which the common-emitter short-circuit small-signal current gain decreases to unity (or is extrapolated to unity from lower frequency measurements). These various delays and charging effects can be represented by several delays internal to the device. Such delays include the emitter storage time, the base capacitance charging time, the base transit time, the collector depletion layer transit time and the collector charging time. All of these may be significant depending on the technological details of the device design. The main advantage of an HBT for a high f_T is through a reduced base transit time (τ_B) in a graded-base structure as described earlier. In the simplest model (ignoring velocity saturation), the ratio of the transit time in an HBT with a linearly graded base to that in an all-silicon device with a flat base profile of the same base width is [35,38]

$$\frac{\tau_{B,SiGe}}{\tau_{B,Si}} = \frac{2kT}{\Delta E_g}\left(1 - \frac{kT}{\Delta E_g}\left(1 - e^{-\Delta E_g/kT}\right)\right) \tag{7}$$

ΔE_g is the reduction in the bandgap from the emitter side of the neutral base to that at the collector side (which may differ from the total Ge profile due to depletion regions [12]), and the same D_n is assumed in both cases. For example, for a bandgap reduction from the emitter to collector side of the base of 75 meV, at room temperature one would expect a reduction in transit time by a factor of 0.46. However, if there are parasitic barriers due to boron outdiffusion (or valence band barriers in p-n-p SiGe HBTs [8,9,39]), base transit times are substantially increased above these amounts [18,19,11]. For more arbitrary profiles, a generalised method for calculating base transit time is given in [35]. Again, these formulae are not applicable if the highest parasitic barrier is not in a neutral region but in a depletion region [37]. Furthermore, the base transit time also increases sharply at high current levels due to space-charge effects in the base-collector depletion region. This effect can be especially severe in HBTs because of parasitic barrier formation as the base 'pushes out' into the base-collector depletion region [40,41]. Besides the reduction of the base transit time in graded-base HBTs, in some HBT structures, the emitter storage time may also be substantially reduced compared to a comparable all-silicon device due to the increased current gain of the HBT. However, because of the many other contributions to the total device delay, and because these components depend critically on the exact doping profiles in the device and the current level, f_T cannot be related to the physical parameters of the structure such as bandgaps, mobilities etc. in a simple general fashion as was done for the collector current enhancement.

Because f_T does not reflect any parasitic lateral device components, such as base resistance, it is not a good predictor of actual circuit performance, however. A better figure of merit for microwave circuit operation is therefore the unity power gain cut-off frequency (f_{max}). In a first order model, f_{max} is related to f_T as:

$$f_{max} = \left(\frac{f_T}{8\pi R_B C_C} \right)^{\frac{1}{2}} \qquad (8)$$

where R_B is the total base resistance (not just intrinsic base sheet resistance) and C_C is the collector capacitance. f_{max} depends on components which depend on lateral scaling, such as extrinsic and intrinsic base resistances, and thus depends critically on process technology factors such as minimum lithographic dimensions, self-alignment of the base and emitter contacts, etc., in addition to the vertical device profile which defines f_T. For digital applications, an ultimate figure of merit is a gate propagation delay. This depends even further on many processing issues for VLSI integration and circuit design choices. The VLSI integration issue is particularly important since the device design may have to be relaxed to withstand thermal budgets or for yield considerations.

To best summarise the AC properties, TABLE 1 has been prepared to show as much data as is available to date for both n-p-n and p-n-p devices. Of primary interest are the peak f_T, peak f_{max} and the circuit type and gate delay. Additional factors which affect these results, such as the base width, sheet resistance and Ge profile, the emitter technology and the emitter width are also included. To date, room temperature unity gain cut-off frequencies in excess of 110 GHz [7,59] have been achieved, along with a peak f_{max} of 90 GHz [61] and minimum ECL gate delay of 17 ps [60].

TABLE 1 Summary of high frequency and gate delay data for $Si_{1-x}Ge_x$ n-p-n and p-n-p HBTs at room temperature, along with some structural and fabrication information. The Ge profile is in some cases the entire profile and in some cases just for the neutral base, and a single Ge profile number indicates a flat Ge profile. The base width in some cases is the neutral base and in some cases is the entire width between the metallurgical junctions. f_T and f_{max} are peak values and propagation delay is the minimum value.

Ref Where	Date Type	Base width (nm)	Base Ge profile (E to C)	$R_{B,i}$ (kΩ/square)	$N_{A,base}$ (cm^{-3})	Emitter width (μm)	Emitter technol.	Self-align. B-E?	β_{max}	f_T (GHz)	f_{max} (GHz)	Circuit type	Prop. delay (ps)
[42] IBM	1988 p-n-p	70	0.06-0.12	<4		1.2	ex-situ doped epi	no	50	12			
[29,43] HP	1989 n-p-n	25	0.31		7×10^{18}	1	implant/ anneal	no	25	29	35		
[44] IBM	1989 n-p-n	65	0-0.11	20		1.5	in-situ doped(?) poly/anneal	no	1,000	40			
[45] IBM	1990 n-p-n	45	0-0.07	17		0.9	poly/impl./ anneal	no	135	75			
[8] IBM	1990 p-n-p	45	0?->0.15		1×10^{19}	1.5	ex-situ doped epi	no	~70	15			
[46] IBM	1990 n-p-n	60	0-0.10	8		0.6 0.4	poly/impl./ anneal	yes	100	50		ECL	28.3 24.6
[47] IBM	1990 n-p-n	<65	0-0.11 0-0.18	8.3 17		0.35 0.45	poly/impl./ anneal	yes yes	90 ?	50 63		ECL	27-28
[39] IBM	1990 p-n-p	50	0.05-0.15? 0.05-0.15?	3×10^{18} ?	8.5 >15	? ?	poly/impl./975°C ex-situ epi/850°C	no no	65 ?	30 31			
[9] IBM	1991 p-n-p	~50? ~50?	0-0.11 0-0.18	>20 20		0.8 0.8	in-situ B-doped-poly/925°C-5s	no no	85 135	55 52			
[48] Ruhr-U. Bochum	1991 n-p-n	40	0.18		1×10^{19}	5	epi/impl./ 900°C anneal	no	100	30			

TABLE 1 continued

Ref Where	Date Type	Base width (nm)	Base Ge profile (E to C)	$R_{B,i}$ (kΩ/square)	$N_{A,base}$ (cm^{-3})	Emitter width (μm)	Emitter technol.	Self-align. B-E?	β_{max}	f_T (GHz)	f_{max} (GHz)	Circuit type	Prop. delay (ps)
[49] IBM	1991 n-p-n	50	0.3-0.10	8	$\sim 10^{19}$	0.7	epi/in-situ dop.- poly/impl./anneal	yes	94	43	40	ECL NTL	24 19
[30,50] DB	1992 n-p-n	30	~0.25	~1	$2\text{-}6 \times 10^{19}$	1-3	in-situ doped epi	yes	30-550	40-46	40-53		
[51] IBM	1992 n-p-n	35	0-0.15	16		0.6	poly/P impl./ 750/860°C anneal	yes	290	73	26	NTL ECL	28 34
[52] NEC	1992 n-p-n	60	0-0.15	~4?		0.2	in-situ P-doped poly/anneal	yes	120	51	50	ECL	19
[53] IBM	1992 n-p-n	70	0-0.07	12	2×10^{18}	0.5	poly/impl./ anneal	yes	94	50	61	ECL	18.9
[54] IBM	1993 p-n-p	80?	0?-0.12	9.5		0.7	in-situ B-doped poly/935°C anneal	yes	45	31	21	ECL	44.5
[55,56] DB	1993 n-p-n	22-25	0.25?	1.2 2.0	8×10^{19} 3×10^{19}	1 1	in-situ doped epi	yes	? >50	95 91	50 65		
[57,58] IBM	1993 n-p-n	~65 ~40 <30	0-0.08 0.22 0.22	10.5 8 >20	$\sim 4 \times 10^{18}$ $\sim 2 \times 10^{19}$ $\sim 8 \times 10^{18}$	0.6 0.6 0.6	in-situ As-doped poly/800°C 15" anneal	yes yes yes	200 170 2400	44 45 64	34 48 30	 NTL NTL	 21 24
[7,59] IBM	1993 n-p-n	35	0-0.25	7	2×10^{19}	0.5	in-situ P-doped poly/800°C anneal	no	443	113			
[60] IBM	1993 n-p-n	100	0-0.14	5-7?		0.5	poly/impl./ anneal	yes .	?	48	50	ECL	17.2
[61] DB	1994 n-p-n	16-20	0.3	0.78	8×10^{19}	1?	in-situ doped epi	yes	?	59	90		

E CONCLUSION

The DC and AC performance of Si/$Si_{1-x}Ge_x$/Si n-p-n HBTs has been reviewed. The performance of the devices depends both on fundamental materials parameters such as the bandgap of the $Si_{1-x}Ge_x$ layers as well as technological issues such as doping profiles and process integration. Many of the minority and majority transport parameters in the $Si_{1-x}Ge_x$ layers remain to be characterised.

ACKNOWLEDGEMENTS

The technical expertise of Z. Matutinovic Krstelj in preparing this manuscript and the financial support of the Office of Naval Research (USA) are gratefully acknowledged.

REFERENCES

[1] T. Tatsumi, H. Hirayama, N. Aizaki [*Appl. Phys. Lett. (USA)* vol.52 (1988) p.895-7]
[2] H. Temkin, J.C. Bean, A. Antreasyan, R. Leibenguth [*Appl. Phys. Lett. (USA)* vol.52 (1988) p.1089-91]
[3] G.L. Patton, S.S. Iyer, S.L. Delage, S. Tiwari, J.M.C. Stork [*IEEE Electron Device Lett. (USA)* vol.9 (1988) p.165-7]
[4] D.-X. Xu, G.-D. Shen, M. Willander, W.-X. Ni, G.V. Hansson [*Appl. Phys. Lett. (USA)* vol.52 (1988) p.2239-41]
[5] P. Narozny, M. Hamacher, H. Dämbkes, H. Kibbel, E. Kasper [*Int. Electron Devices Meet. Tech. Dig. (USA)* (1988) p.562-5]
[6] J.F. Gibbons et al [*Int. Electron Devices Meet. Tech. Dig. (USA)* (1988) p.566-9]
[7] E.F. Crabbé, B.S. Meyerson, J.M.C. Stork, D.L. Harame [*Int. Electron Devices Meet. Tech. Dig. (USA)* (1993) p.83-6]
[8] D.L. Harame et al [*Tech. Dig. VLSI Tech. Symp. (USA)* (IEEE, New York, 1990) p.47-8]
[9] D.L. Harame et al [*Tech. Dig. VLSI Tech. Symp. (USA)* (Business Center for Acad. Soc. Japan, Tokyo, 1991) p.71-2]
[10] E.J. Prinz, J.C. Sturm [*IEEE Electron Device Lett. (USA)* vol.12 (1991) p.661-3]
[11] A. Gruhle [*IEEE Trans. Electron Devices (USA)* vol.41 (1994) p.198-203]
[12] E. Crabbé, J.D. Cressler, G.L. Patton, J.M.C. Stork, J.H. Comfort, J.Y.-C. Sun [*IEEE Electron Device Lett. (USA)* vol.14 (1993) p.193-5]
[13] H. Kroemer [*Proc. IEEE (USA)* vol.70 (1982) p.13-25]
[14] C.G. Van de Walle, R.M. Martin [*Phys. Rev. B (USA)* vol.34 (1986) p.5621-34]
[15] S.E. Swirhun, Y.-H. Kwark, R.M. Swanson [*Int. Electron Devices Meet. Tech. Dig. (USA)* (1986) p.24-7]
[16] E.J. Prinz, P.M. Garone, P.V. Schwartz, X. Xiao, J.C. Sturm [*Int. Electron Devices Meet. Tech. Dig. (USA)* (1989) p.639-42]
[17] E.J. Prinz, P.M. Garone, P.V. Schwartz, X. Xiao, J.C. Sturm [*IEEE Electron Device Lett. (USA)* vol.12 (1991) p.42-4]
[18] A. Pruijmboom et al [*IEEE Electron Device Lett. (USA)* vol.12 (1991) p.357-9]
[19] J.W. Slotboom, G. Streutker, A. Pruijmboom, D.J. Gravesteijn [*IEEE Electron Device Lett. (USA)* vol.12 (1991) p.486-8]
[20] A. Pruijmboom et al [*Microelectron. Eng. (Netherlands)* vol.19 (1992) p.427-32]
[21] S.E. Swirhun, D.E. Kane, R.M. Swanson [*Int. Electron Devices Meet. Tech. Dig. (USA)* (1988) p.298-301]

[22] S.S. Iyer, G.L. Patton, J.M.C. Stork, B.S. Meyerson, D.L. Harame [*IEEE Trans. Electron Devices (USA)* vol.36 (1989) p.2043-64]

[23] C.A. King, J.L. Hoyt, C.M. Gronet, J.F. Gibbons, M.P. Scott, J. Turner [*IEEE Electron Device Lett. (USA)* vol.10 (1989) p.52-4]

[24] J.C. Sturm, E.J. Prinz, P.M. Garone, P.V. Schwartz [*Appl. Phys. Lett. (USA)* vol.54 (1989) p.2707-9]

[25] C.A. King, J.L. Hoyt, J.F. Gibbons [*IEEE Trans. Electron Devices (USA)* vol.36 (1989) p.2093-104]

[26] J.C. Sturm, E.J. Prinz, C.W. Magee [*IEEE Electron Device Lett. (USA)* vol.12 (1991) p.303-5]

[27] J. Poortmans et al [*Microelectron. Eng. (Netherlands)* vol.29 (1992) p.443-6]

[28] Z. Matutinovic-Krstelj, E.J. Prinz, V. Vcnkataraman, J.C. Sturm [*Int. Electron Devices Meet. Tech. Dig. (USA)* (1993) p.87-90]

[29] T.I. Kamins et al [*IEEE Electron Device Lett. (USA)* vol.10 (1989) p.503-5]

[30] A. Gruhle, H. Kibbel, U. König, U. Erben, E. Kasper [*IEEE Electron Device Lett. (USA)* vol.13 (1992) p.206-8]

[31] T. Manku, A. Nathan [*J. Appl. Phys. (USA)* vol.69 (1991) p.8414-6]

[32] T. Manku, A. Nathan [*IEEE Electron Device Lett. (USA)* vol.12 (1991) p.704-6]

[33] T. Manku, A. Nathan [*Phys. Rev. B (USA)* vol.43 (1991) p.12634]

[34] J.M. McGregor, T. Manku, J.P. Noel, D.J. Roulston, A. Nathan, D.C. Houghton [*J. Electron. Mater. (USA)* vol.22 (1993) p.319-22]

[35] H. Kroemer [*Solid-State Electron. (UK)* vol.28 (1985) p.1101-3]

[36] K. Suzuki, N. Nakyama [*IEEE Trans. Electron Devices (USA)* vol.39 (1992) p.623-8]

[37] E.J. Prinz, J.C. Sturm [*Int. Electron Devices Meet. Tech. Dig. (USA)* (1991) p.853-6]

[38] G.L. Patton, D.L. Harame, J.M.C. Stork, B.S. Meyerson, G.J. Scilla, E. Ganin [*IEEE Electron Device Lett. (USA)* vol.10 (1989) p.534-6]

[39] D.L. Harame et al [*Int. Electron Devices Meet. Tech. Dig. (USA)* (1990) p.33-6]

[40] S. Tiwari [*IEEE Electron Device Lett. (USA)* vol.9 (1988) p.142-4]

[41] P.E. Cottrell, Z. Yu [*IEEE Electron Device Lett. (USA)* vol.11 (1990) p.431-3]

[42] D.L. Harame et al [*Int. Electron Devices Meet. Tech. Dig. (USA)* (1988) p.889-90]

[43] T.I. Kamins et al [*Int. Electron Devices Meet. Tech. Dig. (USA)* (1989) p.647-50]

[44] S.E. Fischer et al [*Int. Electron Devices Meet. Tech. Dig. (USA)* (1989) p.890-2]

[45] G.L. Patton et al [*IEEE Electron Device Lett. (USA)* vol.11 (1990) p.171-3]

[46] J.H. Comfort et al [*Int. Electron Devices Meet. Tech. Dig. (USA)* (1990) p.21-4]

[47] J.N. Burghartz et al [*Int. Electron Devices Meet. Tech. Dig. (USA)* (1990) p.297-300]

[48] H.-U. Schreiber, J.N. Albers [*Electron. Lett. (UK)* vol.27 (1991) p.1465-6]

[49] J.H. Comfort et al [*Int. Electron Devices Meet. Tech. Dig. (USA)* (1991) p.857-60]

[50] A. Gruhle, H. Kibbel, E. Kasper [*IEEE Trans. Electron Devices (USA)* vol.39 (1992) p.2636]

[51] E.F. Crabbé et al [*IEEE Electron Device Lett. (USA)* vol.13 (1992) p.259-61]

[52] F. Sato, T. Hashimoto, T. Tatsumi, H. Kitahata, T. Tashiro [*Int. Electron Devices Meet. Tech. Dig. (USA)* (1992) p.397-400]

[53] D.L. Harame [*Int. Electron Devices Meet. Tech. Dig. (USA)* (1992) p.19-22]

[54] D.L. Harame et al [*Tech. Dig. VLSI Tech. Symp. (USA)* (1993) p.61-2]

[55] A. Gruhle, H. Kibbel [*Electron. Lett. (UK)* vol.29 (1993) p.415-7]

[56] A. Gruhle, H. Kibbel, U. Erben, E. Kasper [*Tech. Abs. Dev. Res. Conf. (USA)* (1993)]

[57] J.N. Burghartz, T.O. Sedgwick, D.A. Grützmacher, D. Nguyen-Ngoc, K.A. Jenkins [*Tech. Proc. Bip. Circ. Tech. Mtg. (USA)* (1993) p.55-62]

[58] J.N. Burghartz [private communication]

[59] E.F. Crabbé et al [*Tech. Abs. Dev. Res. Conf. (USA)* (1993)]

[60] D.L. Harame et al [*Int. Electron Devices Meet. Tech. Dig. (USA)* (1993) p.71-4]

[61] A. Schüppen, A. Gruhle, U. Erben, H. Kibbel, U. König [*Tech. Abs. Dev. Res. Conf. (USA)* (1994)]

7.2 Strain adjustment for n-MODFETs: SiGe unstrained, Si strained (1.5%)

D.K. Nayak and Y. Shiraki

October 1993

A INTRODUCTION

Many high-performance electronic devices such as HBTs, MOSFETs (metal oxide silicon FETs) and MODFETs (modulation doped FETs) have been demonstrated in the Si/Ge material system. Recently obtained record high electron mobility in a Si/Ge modulation-doped structure has generated renewed interest in Si/Ge n-channel MODFETs. Room temperature electron mobility as high as 2830 cm^2/V s [1], and low temperature (4.2 K) electron mobility as high as 180000 cm^2/V s [2] have been achieved. This high mobility has been possible because of the breakthrough in the growth technology of high quality SiGe buffer layers [3,4]. It has been found that a high quality (defect density 10^6 cm^{-2} or less) lattice relaxed SiGe buffer layer on (100) Si can be grown by grading Ge concentration in the buffer layer. The first fabrication of a SiGe n-MODFET was made on a uniform-composition buffer layer, where Ge concentration in the buffer was kept constant. A maximum transconductance of 40 mS/mm at 300 K was found for a 1.6 μm gate device [5]. Employing new compositionally-graded SiGe buffer layer technology, a maximum transconductance as high as 340 mS/mm at 300 K has recently been reported for a 1.4 μm gate n-MODFET [6].

A schematic cross-sectional diagram of a high-performance SiGe n-MODFET is shown in FIGURE 1. This structure is similar to the one given in [6]. The basic requirement for n-channel modulation doping is to have a sufficiently large conduction band discontinuity at the strained-Si-channel/SiGe heterointerface (FIGURE 1). This is possible only when the Si-channel layer is under biaxial tensile strain, which is produced by growing the Si channel pseudomorphically on a relaxed SiGe buffer [7]. Under biaxial strain, the Si bandgap is reduced compared to its bulk value. This is given by the relation

$$E_g(x) = 1.11 - 0.4x \text{ eV}$$

where x is the Ge content in the top part of a completely relaxed SiGe buffer [8]. The conduction band discontinuity, ΔE_c, for a typical modulation-doped structure

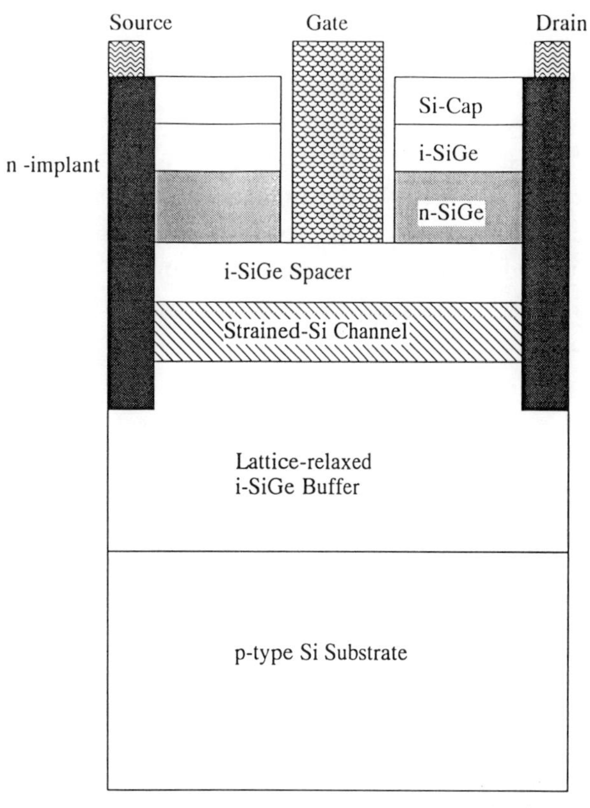

FIGURE 1 Schematic diagram of a high performance Si/Ge n-MODFET.

(strained-Si on a relaxed-$Si_{0.7}Ge_{0.3}$ buffer) is 180 meV [9]. This amount of conduction band discontinuity is sufficient to confine electrons in the strained-Si channel at 300 K. The following strain-related parameters are important for optimisation of n-MODFET performance (FIGURE 1):

(a) SiGe buffer: Ge content in the buffer, thickness, type of buffer (uniform composition versus compositionally graded) and degree of strain relaxation.

(b) Strained-Si channel: thickness, and defect density.

(c) SiGe cap layer: Ge content, total layer thickness, spacer layer thickness, and donor dopant position and concentration.

(d) Mobility of the 2D electron gas in the strained-Si channel as a function of strain.

(e) Concentration of the 2D electron gas in the strained-Si channel as a function of strain.

In this Datareview, electronic and material properties of the above parameters are presented. In Section A, the n-MODFET structure has been described. In Section B, we present theoretical and experimental data on electron mobility of modulation-doped Si/Ge structures as a function of strain. Strain dependent performances of n-MODFETs are given in Section C. A short conclusion is provided in Section D.

B IN-PLANE ELECTRON MOBILITY OF Si/Ge MODULATION-DOPED STRUCTURES

B1 Theoretical Data

When a thin Si epilayer is grown commensurately on a relaxed SiGe buffer, the sixfold degenerate valleys of the conduction band of Si are split into two groups. The twofold degenerate valleys that exhibit longitudinal mass axis perpendicular to the growth plane are lowered in energy, whereas the fourfold degenerate valleys that exhibit longitudinal mass axis parallel to the growth plane are raised in energy. The splitting energy between these two groups is

$$\Delta E = 0.6x \text{ eV}$$

where x is the Ge content in the top part of the relaxed SiGe buffer [8]. Consequently, at a low field only the twofold degenerate valleys, lying lower in energy, are occupied. Electrons in these two lowered valleys have in-plane effective mass equal to the transverse mass ($m_t = 0.19m_0$) of bulk Si. Due to this small in-plane effective mass coupled with reduced intervalley scattering resulting from band splitting, in-plane electron mobility is significantly improved with strain.

TABLE 1 gives theoretical electron mobility data for strained Si [8] as well as for modulation-doped strained Si structures [9-11]. For 1% biaxial in-plane strain in Si, room temperature mobility between 3000 and 4000 cm^2/V s can be expected. This value is more than two times larger than the bulk mobility of intrinsic Si (1500 cm^2/V s). For 1% strain, electron mobility can reach 20000 - 23000 $cm^2/V\,s$ at 77 K and 180000 $cm^2/V\,s$ at 1.5 K (TABLE 1). However, the saturation electron velocity of strained Si is found to be approximately the same as that of bulk Si at 300 K and 77 K [8,11].

TABLE 1 Theoretical calculations of electron mobility in Si/Ge modulation-doped structures at different temperatures: * denotes values extracted from graphs given in the reference.

Ref	Strain in Si-channel (%)	Ge conc. in SiGe buffer (%)	Temperature (K)	Electron density (10^{12} cm^{-2})	Low-field mobility ($cm^2/V\,s$)	Saturation velocity (10^7 cm/s)	Remarks
[10]	1	25	300	1	2000 2300		60 Å-well 120 Å-well
[9]	1.2	30	1.5	0.8	180000*		Background dop. 10^{14} cm^{-3}, spacer-100 Å
				0.7	130000*		Background dop. 10^{15} cm^{-3}, spacer-100 Å
[11]	1.2	30	300	2	3000*	Same as bulk Si	Background dop. 10^{14} cm^{-3}
			80		20000*	Same as bulk Si	
[8]	≥0.66	≥16.6	300 77		4000 23000	1 1.3	

For a low level of strain in Si and for low electric fields, electron mobility increases with increasing strain (TABLE 2). No further improvement in mobility is observed at 300 K for strain levels exceeding 0.7 - 0.8% [8,11]. At 77 K, improvement in mobility with strain is seen only up to a strain level of 0.2% [11].

TABLE 2 Dependence of low-field electron mobility (theory) on strain level in Si at 300 K and 77 K: * denotes values extracted from graphs given in the reference.

Ref	Strain in Si (%)	Ge concentration in the buffer (%)	Temperature (K)	Low-field mobility enhancement factor*
[11]	0.1	2.5	300	1.14
	0.2	5		1.27
	0.4	10		1.5
	0.6	15		1.65
	0.8	20		1.73
	1	25		1.74
	0.1	2.5	77	1.28
	0.2	5		1.36
	0.4	10		1.36
[8]	0.66	16.6	300	2.67
	1.33	33.3		2.67
	0.66	16.6	77	1.35
	1.33	33.3		1.35

B2 Experimental Data

Experimental data on electron mobility for Si/Ge modulation-doped structures are presented in TABLE 3. These data can be broadly divided into two groups: one with the uniform-composition buffer, where Ge content is kept constant, and the other with the compositionally-graded buffer, where Ge content is varied.

TABLE 3 Experimental electron mobility in Si/Ge modulation-doped structures at different temperatures: * denotes values extracted from the graphs in the reference; ** denotes electron density per well in multiple quantum well structures; *** strain level may be less than 1% due to the use of this type of uniform-composition buffer.

Ref	Strain in Si-channel (%)	Temp. (K)	Spacer thickness (Å)	Electron density (10^{12} cm^{-2})	Low-field mobility (cm^2/V s)	Type of buffer layer used	Defect density (cm^{-2})
[7]	1***	300 2		4**	600* 2300*	Uniform-comp. 0.2 µm $Si_{0.75}Ge_{0.25}$	-
[12]	1	300 77	100	2.4**	1280 3000*	Uniform-comp. 0.2 µm $Si_{0.68}Ge_{0.32}$	-
[13]	1	1.5	100	1.1**	14,000	Uniform-comp. $Si_{0.7}Ge_{0.3}$	10^9 - 10^{10}
[14]	1.2	300 77 1.5	100	1.2 0.67 0.65	1260 4400 13100	Uniform-comp. 0.3 - 0.5 µm $Si_{0.7}Ge_{0.3}$	2 x 10^9
[14]	1.2	300 77 1.5	100	1.1 0.84 0.69	1780 16600 173000	Comp.-graded 2 - 3 µm $Si_{0.7}Ge_{0.3}$	<5 x 10^7
[15]	1.2	300 4.2	100	1.2 0.78	1600 125000	Comp.-graded 4 µm $Si_{0.7}Ge_{0.3}$	4 x 10^6
[2]	1.2	300 4.2	200	0.6	2100 180000	Comp.-graded 4 µm $Si_{0.7}Ge_{0.3}$	106
[16]	1.2	300	100	0.2	2,640	Stepwise Comp.-graded $Si_{0.7}Ge_{0.3}$	10^6 - 10^8
[1]	1.2	300 77	40 - 150	1	2830 18000	Comp.-graded 0.6 - 1.5 µm $Si_{0.7}Ge_{0.3}$	-

In the case of the uniform-composition buffer [7,12,13], strain relief is a function of buffer layer thickness. In order to achieve a strain level of 1% in Si, which corresponds to a completely relaxed $Si_{0.75}Ge_{0.25}$ buffer, a partially relaxed 0.2 µm-$Si_{0.68}Ge_{0.32}$ uniform-composition buffer is required (TABLE 3). For an effective strain level of 1% in Si on a uniform-composition buffer, record high electron mobilities of 1280 cm^2/V s at 300 K [12] and 14000 cm^2/V s at 1.5 K [13] have been reported. In this type of buffer, mobility is limited by the presence of a large number of defects (10^9 - 10^{10} cm^{-2}) in the buffer layer.

TABLE 4 Dependence of n-MODFET performance on strain in Si and quality of SiGe buffer layer at different temperatures: * strain level may be less than 1% due to the use of this type of uniform-composition buffer.

Ref	Strain in Si channel (%)	Type of SiGe buffer used	Gate length (μm)	Temp. (K)	Electron density (10^{12} cm^{-2})	Low-field mobility (cm^2/V s)	Transconductance (mS/mm)	Remarks
[5]	1*	Uniform-comp. $Si_{0.75}Ge_{0.25}$ 0.2 μm	1.6	300	0.1 - 0.36	1,550	40 (extrinsic) 70 (intrinsic)	
[17,18]	1.3	Uniform-comp. $Si_{0.68}Ge_{0.32}$ 0.3 μm + $Si_{0.5}Ge_{0.5}$/Si superlattice	1.4	300	1 - 2	1,090	80 (extrinsic) 88 (intrinsic) 155 (extrinsic)	deep-channel upper-channel
[6]	1.2	Comp.-graded $Si_{0.7}Ge_{0.3}$ 1.5 μm	1.4	300 77 300 300 77 77			60 - 72 (extrinsic) 100 - 133 (extrinsic) 340 (extrinsic) 380 (intrinsic) 670 (extrinsic) 800 (intrinsic)	depletion-mode enhancement-mode
[19]	1.2	Stepwise Comp.-graded $Si_{0.7}Ge_{0.3}$ (defect density 10^4 cm^{-2})	0.25	300 77	2.5 1.5	1,500 9,500	330 (extrinsic) 600 (extrinsic)	
[20]	1.2	Stepwise Comp.-graded $Si_{0.7}Ge_{0.3}$	0.5	300 77	1.5	2,600	390 (extrinsic) 520 (extrinsic)	

C STRAIN-DEPENDENT PERFORMANCE OF n-MODFETs

Employing modulation-doped structures described in Section B2, n-MODFETs have been fabricated. TABLE 4 summarises important strain-related properties of n-MODFETs. Using grouping similar to that used in Section B2 for modulation-doped structures, n-MODFETs can be broadly divided into two groups: one with the uniform-composition buffer and the other with the compositionally-graded buffer.

In the case of the uniform-composition buffer, room temperature transconductance of 40 mS/mm (1.6 μm gate) at 1% strain [5], and 80 mS/mm (1.4 μm gate) at 1.3% strain [18] has been found (TABLE 4). Transconductance in these devices is limited by defect centres in the buffer layer [5,17,18] and access resistance to the 2D electron gas [5].

Employing a compositionally-graded buffer layer instead, n-MODFET performance has been dramatically improved (TABLE 4). For 1.2% strain in Si, room temperature transconductance of 340 mS/mm for a 1.4 μm gate device [6], 330 mS/mm for a 0.25 μm gate device [19], and 390 mS/mm for a 0.5 μm gate device [20] has been obtained. 2D electron gas density of (1 - 3) x 10^{12} cm^{-2} has been measured in these devices. At 77 K and for 1.2% strain, transconductances of 670 mS/mm for a 1.4 μm gate device [6], 600 mS/mm for a 0.25 μm gate device [19], and 520 mS/mm for a 0.5 μm device [20] have been measured.

D CONCLUSION

Using compositionally-graded SiGe buffer layers, high quality strained Si layers can now be grown pseudomorphically. For a moderate strain level of 1% in Si, room-temperature mobility close to 3000 cm^2/V s and low-temperature mobility close to 180000 cm^2/V s have been experimentally demonstrated in Si/Ge modulation-doped structures. These values are close to theoretically predicted values, implying that high-quality strained Si with very low background doping can be produced. n-MODFETs made on these structures show high transconductances of 340 mS/mm at 300 K and 670 mS/mm at 77 K for a 1.4 μm gate device. Thus, it can be concluded that the required strain adjustment for Si/Ge n-MODFETs is now possible. However, technological problems such as gate leakage, device isolation, self-alignment process and device reliability are to be solved in order that the n-MODFET becomes a commercial product.

REFERENCES

[1] K. Ismail, S.F. Nelson, J.O. Chu, B.S. Meyerson [*Appl. Phys. Lett. (USA)* vol.63 (1993) p.660-2]

[2] Y.H. Xie, E.A. Fitzgerald, D. Monroe, P.J. Silverman, G.P. Watson [*J. Appl. Phys. (USA)* vol.73 (1993) p.8364-70]

[3] F.K. LeGoues, B.S. Meyerson, J.F. Morar [*Phys. Rev. Lett. (USA)* vol.66 (1991) p.2903-6]

[4] E.A. Fitzgerald et al [*Appl. Phys. Lett. (USA)* vol.59 (1991) p.811-3]

[5] H. Daembkes, H.-J. Herzog, H. Jorke, H. Kibbel, E. Kasper [*IEEE Trans. Electron Devices (USA)* vol.33 (1986) p.633-8]

[6] U. Konig, A.J. Boers, F. Schäffler, E. Kasper [*Electron. Lett. (UK)* vol.28 (1992) p.160-2]

[7] G. Abstreiter, H. Brugger, T. Wolf, H. Jorke, H.-J. Herzog [*Phys. Rev. Lett. (USA)* vol.54 (1985) p.2441-4]

[8] H. Miyata, T. Yamada, D.K. Ferry [*Appl. Phys. Lett. (USA)* vol.62 (1993) p.2661-3]

[9] F. Stern, S.E. Laux [*Appl. Phys. Lett. (USA)* vol.61 (1992) p.1110-2]

[10] P.K. Basu, S.K. Paul [*J. Appl. Phys. (USA)* vol.71 (1992) p.3617-9]

[11] Th. Vogelsang, K.R. Hofmann [*Appl. Phys. Lett. (USA)* vol.63 (1993) p.186-8]

[12] H.-J. Herzog, H. Jorke, F. Schäffler [*Thin Solid Films (Switzerland)* vol.184 (1990) p.237-45]

[13] G. Schuberth, F. Schäffler, M. Besson, G. Abstreiter, E. Gornik [*Appl. Phys. Lett. (USA)* vol.59 (1991) p.3318-20]

[14] F. Schäffler, D. Tobben, H.-J. Herzog, G. Abstreiter, B. Hollander [*Semicond. Sci. Technol. (UK)* vol.7 (1992) p.260-6]

[15] Y.J. Mii et al [*Appl. Phys. Lett. (USA)* vol.59 (1991) p.1611-3]

[16] S.F. Nelson, K. Ismail, J.O. Chu, B.S. Meyerson [*Appl. Phys. Lett. (USA)* vol.63 (1993) p.367-9]

[17] U. Konig, F. Schäffler [*Electron. Lett. (UK)* vol.27 (1991) p.1405-7]

[18] U. Konig, A.J. Boers, F. Schäffler [*IEEE Electron Device Lett. (USA)* vol.14 (1993) p.97-9]

[19] K. Ismail, B.S. Meyerson, S. Rishton, J. Chu, S. Nelson, J. Nocera [*IEEE Electron Device Lett. (USA)* vol.13 (1992) p.229-31]

[20] K. Ismail, S. Rishton, J.O. Chu, K. Chan, B.S. Meyerson [*IEEE Electron Device Lett. (USA)* vol.14 (1993) p.348-50]

7.3 SiGe p-MOSFETs: SiGe strained (2.0%)

D.K. Nayak

January 1994

A INTRODUCTION

The performance of CMOS VLSI circuits is being improved steadily by aggressively scaling the device dimensions to the submicron regime. However, the scaling is becoming increasingly more difficult due to many technological and fundamental limitations. In order to circumvent these difficulties, new device structures and materials based on Si technology have been proposed. The high-performance heterostructure SiGe p-MOSFET, which exhibits higher channel mobility than its bulk Si counterpart, was first proposed by Nayak et al [1,2]. The channel mobility in this device is improved by employing a strained $Si_{0.8}Ge_{0.2}$ layer as the conducting channel. With further improvement in device structure, record high channel mobilities of 240 and 1500 cm^2/V s at 300 and 77 K, respectively, have been achieved for $Si_{0.5}Ge_{0.5}$ p-MOSFETs [3]. For a short-channel (0.25 μm) $Si_{0.8}Ge_{0.2}$ device, a saturation transconductance of 167 mS/mm has been reported [4].

FIGURE 1(a) shows a schematic diagram of a SiGe p-MOSFET [1]. The structure is similar to a conventional Si p-MOSFET except that a pseudomorphic SiGe channel is used for channel conduction. FIGURE 1(b) gives the band diagram of this transistor when the device is 'ON' [1]. The biaxial compressive stress in SiGe lifts the degeneracy between heavy- and light-hole bands, and the spin-orbit band is lowered in energy [5]. This results in a reduced bandgap [5] and high in-plane hole mobility for the SiGe strained layer [6-9]. The reduction in bandgap of strained SiGe compared to bulk Si appears mostly as the valence band discontinuity at the Si/SiGe heterointerface

$$\Delta E_c = 0.02 \text{ eV}$$

$$\Delta E_v = 0.74x \text{ eV}$$

where x is the Ge content in the SiGe layer [5]. The improvement in channel mobility of the SiGe device is achieved by confining holes in the buried SiGe channel (FIGURE 1(b)). The enhancement in channel mobility is due to two main factors [1,2]:

(i) Strain-induced high in-plane hole mobility of the SiGe channel.

(ii) Reduced SiO_2/Si surface scattering as the SiGe channel is separated from gate SiO_2 by the Si cap layer.

Factors affecting the strain level in the SiGe channel are:

(1) Ge concentration, x, in the channel.

(2) Thickness of the strained SiGe channel.

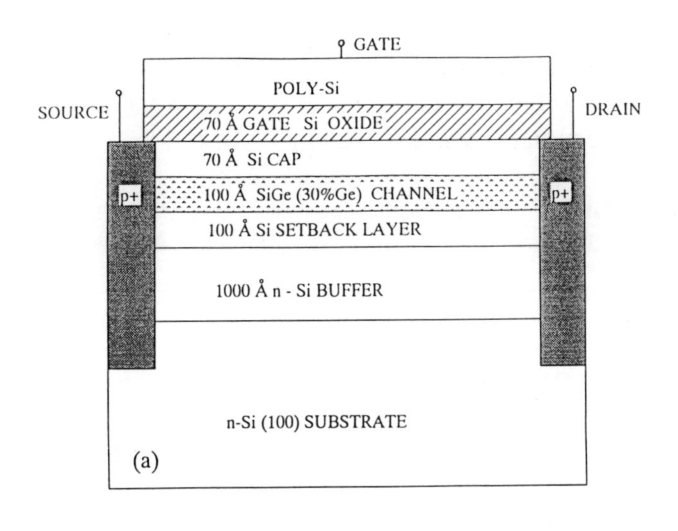

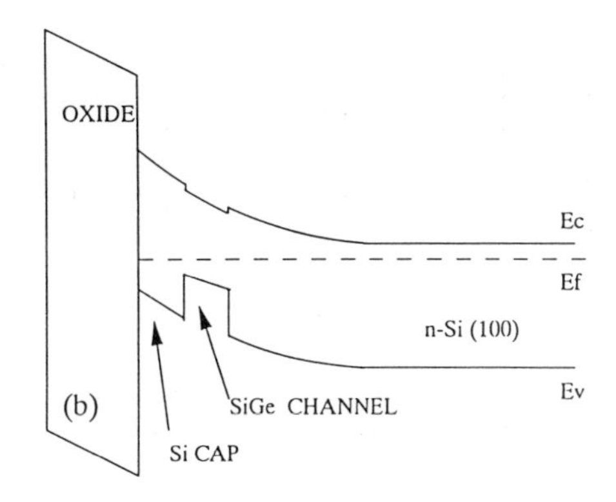

FIGURE 1 (a) Schematic device cross-section of a strained SiGe p-MOSFET, and (b) band diagram of this device when the device is 'ON'.

A high Ge content is desirable to improve the device performance, because the in-plane low-field hole mobility and high-field carrier saturation velocity increase with strain. Furthermore, for fixed Si cap layer and SiGe channel thickness, hole confinement in the SiGe channel improves with increasing strain [2,10,11]. However, the thermodynamically stable critical thickness of strained SiGe on Si(100) decreases rapidly to practically unacceptable values with strain: 80, 40 and 25 Å for $Si_{0.7}Ge_{0.3}$, $Si_{0.5}Ge_{0.5}$ and $Si_{0.3}Ge_{0.7}$, respectively [12]. For a very thin SiGe channel (<40 Å), Si/SiGe interface scattering may limit channel mobility [13].

In this Datareview, electronic and material properties of strained SiGe layers used in SiGe MOSFETs are reviewed. In Section A, the SiGe p-MOSFET structure has been presented. Theoretical data on in-plane hole mobility of SiGe with strain are given in Section B. Section C gives the experimental results of channel mobility as a function of strain. A brief conclusion is provided in Section D.

B IN-PLANE HOLE MOBILITY OF STRAINED SiGe LAYERS

Due to the lifting of band degeneracy in the valence band and change of directional effective mass with strain, the in-plane hole mobility of strained SiGe on Si(100) has been found to be higher than that of bulk Si (TABLE 1). For a 1.6% strain, an enhancement factor of 2.5 - 4.5 can be expected. The low-field mobility increases monotonically with strain. The theoretical values given in TABLE 1 are calculated assuming that the alloy interaction potential, U, of

SiGe is 0.2 - 0.27 eV. With increasing U, however, the mobility value has been found to decrease [6]. Recent measurement results indicate that the value of U can be as high as 0.6 eV, which will lower the mobility values given in TABLE 1.

TABLE 1 Theoretical calculation of in-plane hole mobility of strained SiGe layer on Si(100) substrate.

Ref	Strain in SiGe (%)	Hole mobility enhancement factor ($\mu_{strained\text{-}SiGe}/\mu_{bulk\text{-}Si}$)	Remarks
[6,7]	0.4	1.5	No doping
	0.8	2.0	U = 0.2 eV
	1.2	3.3	
	1.6	4.7	
[8]	0.4	1.2	No doping
	0.8	1.4	U = 0.27 eV
	1.2	1.8	
	1.6	2.4	
	2.0	3.6	
	2.4	5.1	
[9]	0.4	1.5	Low-doping
	0.8	2.3	(10^{13} cm^{-3})
	1.2	3.4	U = 0.2 eV
	1.6	4.6	

In addition to the enhancement in low-field hole mobility, saturation hole velocity also increases with strain. It is shown that for a 1.6% strain in SiGe, the high-field characteristics of SiGe approach those of bulk Ge [7].

C EXPERIMENTAL CHANNEL MOBILITY OF SiGe p-MOSFETs

The fabrication of SiGe MOSFETs has been reported by many groups [1-4,14-17]. As summarised in TABLE 2, the channel mobility increases with increasing strain in SiGe. However, when the SiGe channel thickness considerably exceeds the thermodynamically stable critical thickness, degradation of mobility due to strain relaxation has been observed. Severe degradation of channel mobility has been reported for a 100 Å channel with 1.6% strain [14] and a 70 Å channel with 2.8% strain [3].

For a 1 - 1.2% strain level, the channel mobilities of SiGe MOSFETs have been found to be 180 - 220 and 750 - 1000 cm^2/V s at 300 and 80 K, respectively. At a 2% strain level, the channel mobility reached 240 cm^2/V s at 300 K and 1500 cm^2/V s at 77 K [3]. Using a SIMOX substrate instead of a Si substrate, the hole confinement in the SiGe channel can be enhanced. SiGe SIMOX MOSFETs on SIMOX show 90% higher channel mobility than SiGe MOSFETs on bulk Si [17].

TABLE 2 Experimental low-field channel mobility of SiGe p-MOSFETs. All the values are taken from the figures given in the references. Due to the buried channel nature of the SiGe device, the definition used for effective gate capacitances varies from author to author. The mobility values given in this table are taken directly from the graphs available in the references. * The percentage increase only provides an approximate comparison of the performance of different devices. ** The mobility corresponds to a conventional Si p-MOSFET on SIMOX substrate.

Ref	Strain in SiGe channel (%)	SiGe channel thickness (Å)	Temp. (T)	Bulk-Si p-MOS channel mobility (cm^2/V s)	Strained-SiGe p-MOS channel mobility (cm^2/V s)	Improvement* in mobility (%)	Remarks
[1,2]	0.8	150	300	122	155	27	
[14,15]	0.8	100	300	120	166	38	
	1.2	100			186	55	
	1.6	100					Relaxed
	0.8	100	90	372	573	54	
	1.2	100			760	104	
[16]	1.0	200	300	-	220	-	Ge graded (15 - 25%)
	1.0	200	82	-	980	-	Mod-doping
[4]	0.8	125-300	300	95	150	57	
	0.8	125-300	82	250	400	60	
[17]	1.2	100	300	97**	184	90	on SIMOX
[3]	2.0	70	300	141	240	70	
	2.8	70			160	13	Relaxed
	2.0	70	77	600	1500	150	
	2.8	70			550	-8	Relaxed

D CONCLUSION

It has been shown theoretically and experimentally that the channel mobility of SiGe MOSFETs increases with increasing strain in SiGe. Using strain as high as 2%, channel mobilities of 240 and 1500 cm^2/V s at 300 and 77 K, respectively, have been achieved experimentally. However, when the SiGe channel thickness considerably exceeds the thermodynamically stable critical thickness, strain relaxation has been observed in SiGe MOSFETs. Therefore, the highest strain level in this device will be limited to about 2%. Other technological problems, such as SiGe p-MOS and n-MOS integration, device isolation, low thermal budget VLSI processing, implant damage and annealing of strained SiGe, and alloying of SiGe with Al and other metals, are to be tackled so that SiGe CMOS becomes a commercially attractive product.

REFERENCES

[1] D.K. Nayak, J.C.S. Woo, J.S. Park, K.-L. Wang, K.P. MacWilliams [*IEEE Electron Device Lett. (USA)* vol.12 (1991) p.154-6]

[2] D.K. Nayak [Physics and Technology of GeSi Quantum-Well PMOSFETs (PhD Dissertation, Department of Electrical Engineering, University of California, Los Angeles, USA, August 1992)]
[3] K. Goto et al [*Jpn. J. Appl. Phys. (Japan)* vol.32 (1993) p.438-41]
[4] V.P. Kesan et al [*Int. Electron Devices Meet. Tech. Dig. (USA)* (1991) p.25-8]
[5] R. People [*IEEE J. Quantum Electron. (USA)* vol.22 (1986) p.1696-710]
[6] J.M. Hinckley, V. Sankaran, J. Singh [*Appl. Phys. Lett. (USA)* vol.55 (1989) p.2008-10]
[7] J.M. Hinckley, J. Singh [*Phys. Rev. B (USA)* vol.41 (1990) p.2912-26]
[8] T. Manku, A. Nathan [*IEEE Electron Device Lett. (USA)* vol.12 (1991) p.704-6]
[9] S.K. Chun, K.-L. Wang [*IEEE Trans. Electron Devices (USA)* vol.39 (1992) p.2153-64]
[10] S.S. Iyer et al [*IEEE Electron Device Lett. (USA)* vol.12 (1991) p.246-8]
[11] P.M. Garone, V. Venkataraman, J.C. Sturm [*IEEE Electron Device Lett. (USA)* vol.12 (1991) p.230-2]
[12] R. People, S.A. Jackson [*Semicond. Semimet. (USA)* vol.32 (1990) p.119-74]
[13] A. Gold [*Phys. Rev. B (USA)* vol.35 (1987) p.723-33]
[14] P.M. Garone, V. Venkataraman, J.C. Sturm [*IEEE Electron Device Lett. (USA)* vol.13 (1992) p.56-8]
[15] P.M. Garone, V. Venkataraman, J.C. Sturm [*Int. Electron Devices Meet. Tech. Dig. (USA)* (1991) p.29-32]
[16] S. Verdonckt-Vandebroek et al [*IEEE Electron Device Lett. (USA)* vol.12 (1991) p.447-9]
[17] D.K. Nayak, J.C.S. Woo, G.K. Yabiku, K.P. MacWilliams, J.S. Park, K.-L. Wang [*IEEE Electron Device Lett. (USA)* vol.14 (1993) p.520-2]

7.4 Strain symmetrisation for ultrathin SiGe superlattices

T.P. Pearsall

September 1993

A INTRODUCTION

Epitaxial growth of Ge-Si heterostructures on Si is impeded only by the large lattice mismatch between these two materials. Strain symmetrisation is an important technique that enables the strain introduced by growing an epitaxial layer of Ge to be compensated by growing a subsequent layer of Si. This method has made it possible to produce device-quality Si-Ge layers many thousands of angstroms in thickness.

B PRINCIPLES OF STRAIN SYMMETRISATION

Although Ge and Si share many properties, their fundamental cubic lattice parameters differ by more than 4%. Epitaxial growth of Ge on Si introduces large amounts of strain. The energy associated with this strain exceeds the threshold for creation of threading misfit dislocations after deposition of only five monolayers. This is just slightly more than one unit cell of Ge in thickness. Once this critical thickness is reached, it is necessary to stop growth entirely or to deposit Si on Ge in order to maintain single crystal structure. Deposition of Si reduces the strain energy per unit volume and makes it possible to redeposit additional Ge layers. This alternation between Ge and Si creates a superlattice, whose period is often quite close to the fundamental unit cell dimension of four monolayers that is characteristic of both Si and Ge. However, this superlattice also has a critical thickness limit that depends mainly on the average Ge content. For example, the critical thickness limit of a short period superlattice consisting of an alternation of three monolayers of Si and three monolayers of Ge deposited on a Si substrate is the same as that of a $Ge_{0.5}Si_{0.5}$ alloy, about 50 Å.

The principle of strain symmetrisation permits the synthesis of such superlattices with a thickness far greater than the limit of 50 Å for growth directly on silicon substrates. In this method of Ge-Si strained-layer growth, the Si substrate is deliberately decoupled from the epitaxial layer by interposing a buffer layer of Ge-Si alloy whose thickness is over the critical thickness limit. The lattice mismatch strain is relieved by the formation of misfit dislocations in the buffer layer. Further increases in the buffer layer thickness result in a gradual annealing and subsequent reduction of the misfit dislocation density. Several reviews of the principles and applications of Ge-Si strained-layer epitaxy have appeared recently [1-5]. A detailed discussion of symmetrically strained Ge-Si growth is given in [1].

The critical thickness limit for growth of some strained-layer materials is shown in FIGURE 1(a). In the lower right-hand corner of this figure, it can be seen that the thickness limit for commensurate growth of Ge directly on Si is about 10 Å [1,5]. However, by the method of symmetrically-strained growth the thickness limit, lying in the centre of the diagram, is much larger. Complete relaxation of the strain imposed by the substrate (Si) and annealing of the defects associated with that relaxation are the keys to this approach. The inclusion of an Sb surfactant layer at the interface of the substrate and the first Ge-Si buffer layer causes the threading defects, introduced during the initial phase of strain relief, to self-annihilate upon continued growth, leaving behind a dislocation network only at the Ge-Si interface. Use of an Sb surfactant is now standard procedure in symmetrically-strained epitaxy.

Presuming a buffer layer composition of $Ge_{0.5}Si_{0.5}$, it is then possible to grow a $Ge_{0.5}Si_{0.5}$ alloy with no net strain. It is equally possible to deposit an alternating sequence of three monolayers of Si and three monolayers of Ge with very little net strain energy, provided each sequence of Si monolayer deposition is compensated with a deposition of a similar number of monolayers of Ge. In this method, the critical layer thickness for contiguous Ge deposition on Si is no longer 10 Å, but closer to 50 Å. This means that a wider variety of strained superlattice compositions and structures can be obtained using this method. A high resolution lattice image of a Ge-Si (5:5) superlattice [6] demonstrating these features is shown in FIGURE 1(b). The superlattice thickness shown here is about 130 Å or twice the critical thickness if this superlattice had been grown commensurate to (001) Si. The total sample thickness is about ten times larger than shown in the figure.

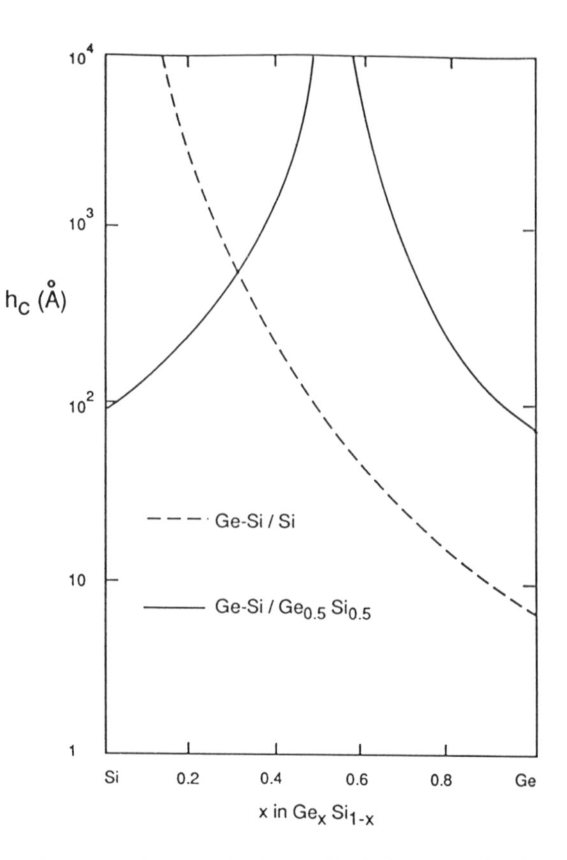

FIGURE 1 Critical thickness limit for growth of Ge-Si alloys and superlattices on Ge-Si substrates of various compositions. (a) The critical thickness limit depends only on the average composition, so that the limit for a Ge-Si(3:3) superlattice is the same as that for a $Ge_{0.5}Si_{0.5}$ alloy. Growth beyond this limit is possible, but the strain energy is partially relieved by the formation of misfit dislocations.

FIGURE 1 Critical thickness limit for growth of Ge-Si alloys and superlattices on Ge-Si substrates of various compositions. (b) High resolution lattice image of a Ge-Si(5:5) superlattice. The electron beam is parallel to the interfaces in the <110> direction. (Reproduced from Jäger et al [6], by permission of the authors.)

The symmetrically-strained growth method means starting epitaxial growth on a buffer with an elevated dislocation concentration (10^6 cm^{-2} is typical of the present state of the art). Subsequently, Ge-Si alloys can be grown epitaxially with reduced levels of strain compared to those produced by commensurate growth. This strain may be positive or negative, whereas for commensurate growth on Si, only compressive strain (negative strain) can be obtained. Ge-Si superlattices consisting of elemental layers of Ge and Si can be grown, and there is no limit in principle to the thickness of these short-period superlattices as long as the positive strain energy introduced by the Ge layers is compensated by the negative strain energy of the Si layers.

Following the epitaxial growth of the Ge-Si material, the strain-free capping layer must be a Ge-Si alloy of the same composition as that of the buffer. Growth of a high-quality SiO_2 layer requires a Si overlayer on the surface of device structures. The symmetrically-strained growth method places limits of 50 Å on the thickness of this critical Si overlayer. This is too thin to be useful. On the other hand, the commensurate epitaxy method can be used to produce the required Si overlayer without thickness limits.

The advantages of the symmetrically-strained approach are therefore:

(1) minimised strain energy in the epitaxial layer of interest;
(2) no limit on the thickness of the strained superlattice structure;
(3) thickness of individual Ge and Si layers can be 50 Å instead of 10 Å;
(4) both compressive strain and tensile strain can be realised.

C ELECTRONIC STRUCTURE OF SHORT-PERIOD SUPERLATTICES

The band structure of Ge and Si is based on the cubic diamond symmetry of the lattice and the chemistry of group IV elements of the periodic table. The short-period superlattice structure changes the symmetry from cubic to tetragonal without changing the chemistry. Alonso et al have made a systematic investigation of the symmetry that can be created in Si-Ge superlattices grown along one crystal direction [7,8]. There are six groups that are easily classified by the number of atomic monolayers present in each basic period (see TABLE 1).

The unit cell symmetry determines directly important optical and electronic properties of the superlattice structure [9]. For example, superlattices lacking a centre of inversion symmetry have substantial non-linear optics coefficients and can be used for second-harmonic generation or as optical modulators. Other superlattices for which the unit cell is composed of ten monolayers of Ge and Si possess a direct fundamental bandgap. Note that there are five possible superlattice structures that are in principle direct bandgap materials with non-linear optical capability.

An example of such a superlattice that has been studied both in experiment and in theory is Ge-Si (5:5) [10-12]. The electronic band structure of this superlattice is shown in FIGURE 2(a). The band structure of Si is shown for comparison in FIGURE 2(b). The electronic structure of the superlattice is derived from that of Si through the folding of the Brillouin zone by the superlattice symmetry. It can be seen that the superlattice possesses a global minimum in the bandgap energy at Γ, the zone centre. The transition matrix element for the bandgap is

TABLE 1 Symmetry groups for Ge-Si(l:n) short-period superlattices, adapted from Alonso et al [7,8].

Space group		Point group	1, n	Example
T^2	(F43m)	(43m) no inversion symmetry	1 = 1 = n zincblende	Ge-Si (1:1)
D^5	(Pmma)	(mm) inversion symmetry	l even, m even 1 + n = 4 K	Ge-Si (2:2) Ge-Si (4:4)
D^{28}	(lmma)	(mm) inversion symmetry	l even, m even 1 + n = 4 K + 2	Ge-Si (2:4)
D^5	(P42m)	(42m) no inversion symmetry	l odd, m odd 1 + n = 4 K	Ge-Si (3:5)
D^9	(142m)	(42m) no inversion symmetry	l odd, m odd 1 + n = 4 K + 2	Ge-Si (3:3) Ge-Si (5:5)
D^{19}	(141/amd)	(42m) inversion symmetry	1 + n = odd	Ge-Si (3:4)

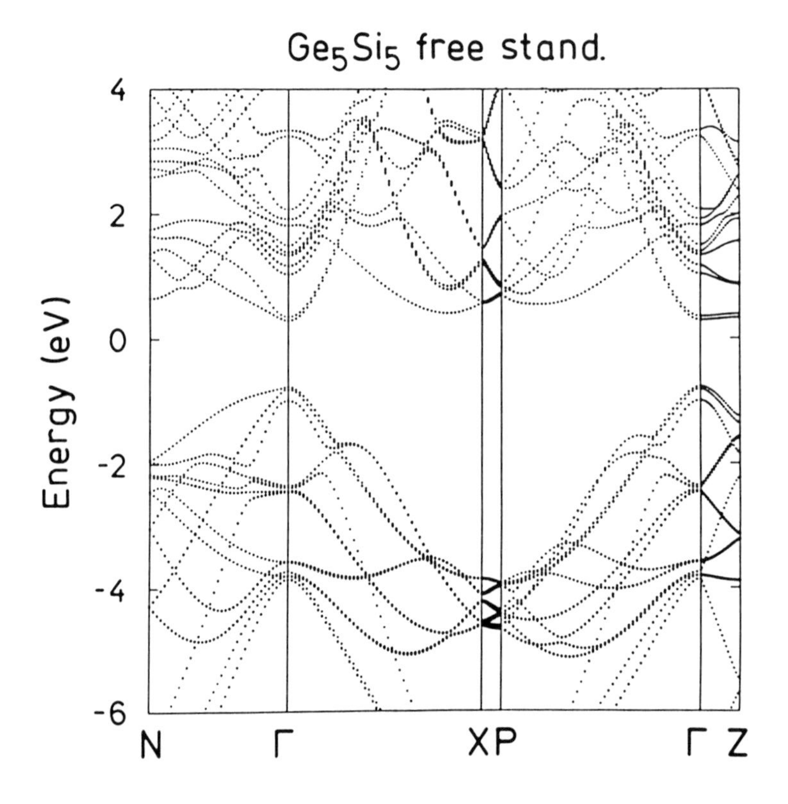

FIGURE 2 (a) Electronic band structure of the Ge-Si(5:5) superlattice, grown on a $Ge_{0.5}Si_{0.5}$ relaxed alloy buffer layer. Because of the symmetric strain imposed by the substrate, there is no limit to the thickness of this superlattice. This superlattice has a global energy minimum for the conduction band at the zone centre. The transition matrix element for the lowest energy transition is finite. The magnitude is estimated by theory to be a factor of ten less than that of GaAs. This superlattice behaves like a direct bandgap semiconductor. (From Schmid et al [10], reproduced by permission of the authors.)

intermediate between that of Si and that for zincblende semiconductors such as GaAs. The first-order non-linear optic coefficient, $\chi^{(2)}$, for this superlattice has been calculated and is shown in FIGURE 3 [11].

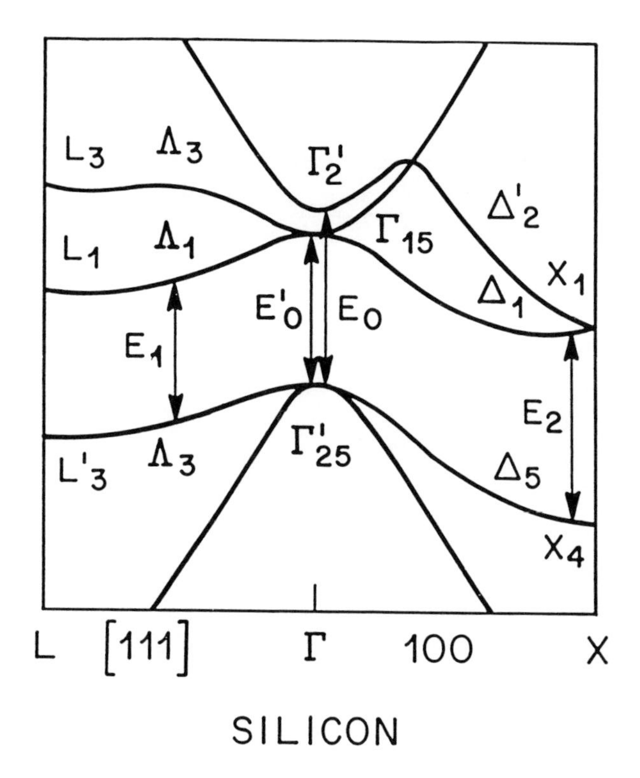

FIGURE 2 (b) Electronic band structure of Si. By comparison with (a), it can be seen that the band structure of Si is simpler, indicating a higher degree of symmetry. The global energy minima of Si lie along the <100> directions, so that Si is an indirect bandgap material.

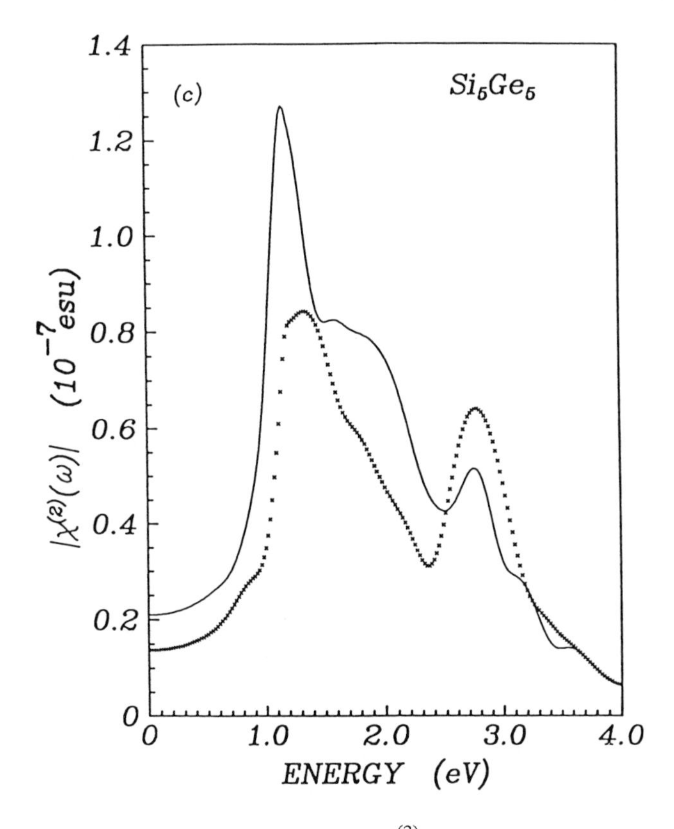

FIGURE 3 The first-order nonlinear optic coefficient, $\chi^{(2)}$, for the Ge-Si(5:5) superlattice. For a centro-symmetric material such as Si or Ge-Si alloys, this is strictly zero. In this case, it is within one order of magnitude of that for GaAs. Nonlinear optical activity and direct bandgap behaviour are independent properties. The 5:5 superlattice possesses both. (From Ghahramani et al [11], reproduced by permission of the authors.)

The lifting of cubic symmetry in these superlattices is a consequence of the unit cell period being greater than four atomic monolayers. For these tetragonal structures the valence band degeneracy is removed affecting electronic transport properties as well as optical properties. The light- and heavy-hole energy surfaces in Si and Ge are strongly warped. In the presence of a uniaxial element of symmetry (introduced by either strain or quantum confinement or short-period superlattice structure) the two surfaces are separated in energy, and these surfaces become more spherical in shape [13,14]. These changes affect the mobility through a modification of carrier effective mass. There have been no measurements of effective mass in symmetrically strained superlattices. Conventional techniques make it difficult to separate the effective mass of carriers in the substrate and buffer layers from that in the superlattice. Recent measurements of effective mass taken in strained Ge-Si alloys [15] show that measurements in superlattice structures may be feasible.

A similar effect occurs in the conduction band for epitaxial growth on the (001) surface. The breaking of cubic symmetry by strain or superlattice structure separates the <001> energy band minimum from the minima in the <010> and <100> directions. If the Ge-Si is under net compressive strain, these two latter minima lie at lower energy. As a result, the effective mass tensor is composed only of transverse electron mass components for transport in the plane of the superlattice. The resulting effective mass for conduction in the plane is reduced, raising the conductivity of the superlattice material. However, at the same time, the conductivity of the superlattice material is lowered by a reduced scattering time because of alloy disorder.

Using the strained symmetry approach during growth makes it possible to apply uniaxial strain to silicon alone, thus breaking the cubic symmetry of bulk Si. However, this strain is tensile and not compressive. The mobility of elemental Si can thus be raised for transport perpendicular to the plane of growth without having to pay a penalty of alloy disorder scattering. This configuration is useful for fabrication of the emitter region of Ge-Si heterojunction bipolar transistors.

D DEVICE APPLICATIONS FOR SHORT-PERIOD SUPERLATTICE STRUCTURES

The precise effect of the short-period superlattice is to modify the fundamental symmetry of the crystalline unit cell. With each distinguishable symmetry class, there is a different combination of electronic wavefunctions that form the electronic states on the valence and conduction bands. The short-period superlattice method therefore opens the route for the synthesis of materials with a custom-tailored electronic band structure. The selection of a new basis set of electronic states is called 'wavefunction engineering' [16]. The symmetrically-strained epitaxy permits this application of wavefunction engineering to be extended over several thousand angstroms or even microns. This makes it possible to produce epitaxial layers that can be used in real devices.

Wavefunction engineering affects directly three important properties of electrons and holes near the fundamental gap: the density of states, the interband transition matrix element, and the effective mass. Some of these properties, notably the effective mass, can also be changed by applying strain alone either to bulk silicon or to a Ge-Si alloy. However the optical properties such as the transition matrix element are significantly changed only by modification of the wavefunction itself. For this reason, short-period superlattices are particularly interesting for their promise as optoelectronic semiconductors based on silicon. Light-emitting diodes, lasers, and electro-optic modulators are examples of the more promising applications of short-period superlattices. However, it must be emphasised that no realistic or practical device using these materials has been demonstrated as of this writing.

With this background in mind, it is helpful to outline what properties Ge-Si short-period superlattice devices ought to have in order to be useful.

(1) Compatibility with Si VLSI processing methods:
Basically this means that both exposed sides of the epitaxial wafer are pure Si so that growth of a high-quality oxide is possible.

(2) Clock rate superior to 10 GHz:
It is frequently suggested that Ge-Si light-emitting diodes could be used in chip-to-chip optical interconnects, or to distribute the clock in integrated circuits. LEDs are slow, even when using direct bandgap materials. In Ge-Si direct bandgap superlattices, the minority carrier lifetime is expected to be longer than that of direct bandgap semiconductors. Uses for Ge-Si LEDs in this domain seem quite limited. However, lasers are intrinsically faster than LEDs, and if such devices can be realised, they will open the way to a new generation of integrated circuits.

(3) Low-loss ($\alpha < 1\ cm^{-1}$) optical waveguides:
Current conceptions of guided-wave optics are based on the idea of single mode transmission (laser required!). Switching and modulation are achieved by electric-field induced changes in the index of refraction. This can be achieved by incorporating a forward-biased p-n junction in the circuit (slow and power-hungry) or by using Ge-Si materials with functional optical nonlinearities (yet to be demonstrated).

We conclude by summarising recent promising results that support the concept of short-period Ge-Si superlattices as optoelectronic materials.

(a) Confirmation of the nanostructure of Ge-Si short-period superlattices [6,17,18]:
It is possible to confirm the size and periodicity of the new unit cell created by the short-period superlattice. However, the coherence length of the periodicity remains short (~150 Å).

(b) New electronic band structure induced by superlattice symmetry [19,20]:
Two groups have shown independently that new optical transitions with substantial matrix elements are induced at the fundamental bandgap of Ge-Si superlattice structures. However, it has not yet been possible to prove that these transitions are direct.

(c) Enhanced optical absorption at the bandgap of short-period superlattice structures [17,20,21]:
Two independent groups have measured optical absorption coefficients in Ge-Si short-period superlattices that are superior to those of Si, Ge or Ge-Si alloys. The absorption coefficient increases linearly with energy above the bandgap in agreement with a theoretical analysis. However, the absorption coefficient appears to be one order of magnitude less than that of bulk direct gap materials.

(d) Band to band photoluminescence and electroluminescence [22-24]:
Several research groups have shown strong, excitonic optical recombination at the bandgap energy of Ge-Si short-period superlattice materials. Some results are shown in FIGURE 4. However, luminescence remains weak at room temperature, and no evidence of laser action under optical or electrical pumping has been seen.

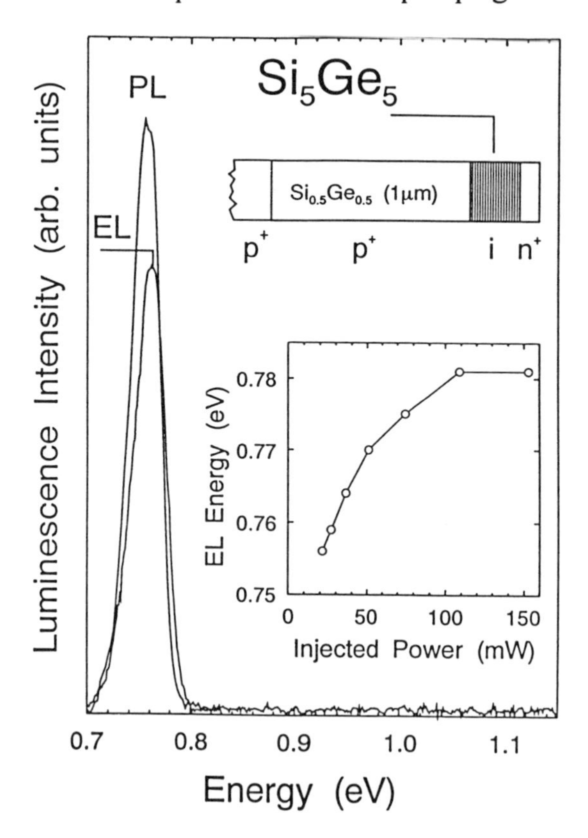

FIGURE 4 Band edge photoluminescence and electroluminescence from a Ge-Si(5:5) short-period symmetrically-strained superlattice. Measurements are taken at low temperature near 10 K. Electroluminescence has been obtained at room temperature [24]. Progress in obtaining strong luminescence at room temperature appears to be linked to improvements in defect density. (From Menczigar et al [22], reproduced by permission of the authors.)

E CONCLUSION

Growth of extended superlattices composed of Ge and Si is enabled by the method of strain symmetrisation. This method is equally useful for the epitaxial growth of extended regions of Si-Ge alloys. This method is a key technology in the evolution of Ge-Si heterostructures from laboratory curiosities to commercial electronic and optoelectronic products.

REFERENCES

[1] E. Kasper, F. Schäffler [*Semicond. Semimet. (USA)* vol.33 (1991) p.223-304]
[2] S.C. Jain, J.R. Willis, R. Bullough [*Adv. Phys. (UK)* vol.39 (1990) p.127-90]
[3] S.C. Jain, W. Hayes [*Semicond. Sci. Technol. (UK)* vol.6 (1991) p.547-76]
[4] T.P. Pearsall [*CRC Crit. Rev. Solid State Mater. Sci. (USA)* vol.15 (1989) p.551-600]
[5] T.P. Pearsall [*Prog. Quantum Electron. (UK)* vol.18 (1994) p.97-152]
[6] W. Jäger et al [*Thin Solid Films (Switzerland)* vol.222 (1992) p.221-6]

[7] M.I. Alonso, M. Cardona, G. Kanellis [*Solid State Commun. (USA)* vol.69 (1989) p.479-83]
[8] M.I. Alonso, M. Cardona, G. Kanellis [*Solid State Commun. (USA)* vol.70 (1989) p.i-ii (after p.784)]
[9] M. Jaros [*Semicond. Semimet. (USA)* vol.32 (1990) p.175]
[10] U. Schmid, N.E. Christensen, M. Alouani, M. Cardona [*Phys. Rev. B (USA)* vol.43 (1991) p.14597-614]
[11] E. Ghahramani, D.J. Moss, J.E. Sipe [*Phys. Rev. B (USA)* vol.43 (1991) p.8990-9002]
[12] U. Schmid, N.E. Christensen, M. Cardona [*Phys. Rev. B (USA)* vol.41 (1990) p.5919-30]
[13] T. Manku, A. Nathan [*Phys. Rev. B (USA)* vol.43 (1991) p.12634-7]
[14] T. Manku, A. Nathan [*IEEE Electron Device Lett. (USA)* vol.12 (1991) p.704-6]
[15] J.-P. Cheng, V.P. Kesan, D.A. Grutzmacher, T.O. Sedgwick, J.A. Ott [*Appl. Phys. Lett. (USA)* vol.62 (1993) p.1522-4]
[16] T.P. Pearsall [*Mater. Res. Soc. Symp. Proc. (USA)* vol.160 (1990) p.623-30]
[17] T.P. Pearsall, C.C.M. Bitz, L.B. Sorensen, H. Presting, E. Kasper [*Thin Solid Films (Switzerland)* vol.222 (1992) p.254-8]
[18] E. Koppensteiner et al [*Appl. Phys. Lett. (USA)* vol.62 (1993) p.1-3]
[19] K. Asami, K. Miki, K. Sakamoto, T. Sakamoto, S. Gonda [*Jpn. J. Appl. Phys. (Japan)* vol.29 (1990) p.L381-4]
[20] T.P. Pearsall, J.M. Vandenberg, R.H. Hull, J.C. Bonar [*Phys. Rev. Lett. (USA)* vol.63 (1989) p.2104-7]
[21] J. Olajos, J. Engvall, H.G. Grimmeiss, H. Kibbel, E. Kasper, H. Presting [*Thin Solid Films (Switzerland)* vol.222 (1992) p.243-5]
[22] U. Menczigar et al [*Thin Solid Films (Switzerland)* vol.222 (1992) p.227-33]
[23] U. Menczigar et al [*Phys. Rev. B (USA)* vol.47 (1993) p.4099-102]
[24] J. Engvall, J. Olajos, H. Grimmeiss, H. Presting, H. Kibbel, E. Kasper [*Appl. Phys. Lett. (USA)* vol.63 (1993) p.491-3]